U.S.NRC

United States Nuclear Regulatory Commission

Protecting People and the Environment

NUREG-2116

Safety Evaluation Report for the International Isotopes Fluorine Products, Inc. Fluorine Extraction Process and Depleted Uranium Deconversion Plant in Lea County, New Mexico

Docket No. 40-9086

International Isotopes Fluorine Products, Inc.

Office of Nuclear Material Safety and Safeguards

AVAILABILITY OF REFERENCE MATERIALS
IN NRC PUBLICATIONS

NUREG-2116

United States Nuclear Regulatory Commission

Protecting People and the Environment

Safety Evaluation Report for the International Isotopes Fluorine Products, Inc. Fluorine Extraction Process and Depleted Uranium Deconversion Plant in Lea County, New Mexico

Docket No. 40-9086

International Isotopes Fluorine Products, Inc.

Manuscript Completed: May 2012
Date Published: May 2012

Office of Nuclear Material Safety and Safeguards

ABSTRACT

This report documents the U.S. Nuclear Regulatory Commission's (NRC's) staff review and safety and safeguards evaluation of International Isotopes Fluorine Products, Inc.'s (IIFP), application for a license to construct a Fluorine Extraction Process & Depleted Uranium Deconversion Plant (FEP/DUP) and possess and use source materials. INIS proposes that the FEP/DUP be located in Lea County, New Mexico, approximately 24 kilometers (15 miles) west of the city of Hobbs. The facility will possess natural and depleted uranium, and significant quantities of hazardous chemicals derived from licensed material, e.g., hydrogen fluoride.

The objective of this review is to evaluate the facility's potential adverse impacts on worker and public health and safety, under both normal operations and accident conditions. The review also considers the management organization, administrative programs, and financial qualifications provided to ensure safe design and operation of the facility.

The NRC staff concludes, in this safety evaluation report, that the applicant's descriptions, specifications, and analyses provide an adequate basis for safety of facility operations and that operation of the facility does not pose an undue risk to the worker and public health and safety.

Potential environmental impacts associated with the proposed facility and its reasonable alternatives are addressed in a separate document, "Environmental Impact Statement for the Proposed Fluorine Extraction Process and Depleted Uranium Deconversion Plant in Lea County, New Mexico," NUREG-2116. The draft Environmental Impact Statement was published in the *Federal Register*, Vol. 77, dated Friday, January 13, 2012, see pages 2096-2098 and the final EIS is scheduled to be published in the fall of 2012.

CONTENTS

TABLES

EXECUTIVE SUMMARY

On December 30, 2009, International Isotopes Fluorine Products, Inc. (IIFP), a subsidiary of International Isotopes, Inc. (INIS), submitted to the U.S. Nuclear Regulatory Commission (NRC) an application requesting a license, under Title 10 of the *Code of Federal Regulations* (10 CFR) Part 40, "Domestic Licensing of Source Material." Subsequently, beginning February 9, 2011, and concluding October 17, 2011, IIFP also provided 26 separate responses to NRC's requests for additional information (RAI). This Safety Evaluation Report (SER) primarily references the information which was incorporated into the Revision B of the IIFP application, submitted in April 2012. When applicable, the SER directly references the appropriate RAIs directly. If approved, the license would authorize construction and operation of a fluoride extraction and depleted uranium deconversion facility to possess and use byproduct and source material. The applicant proposed to locate the facility in Lea County, NM, approximately 24 kilometers (15 miles) west of Hobbs, NM. The proposed facility would be capable of deconverting approximately 3.3 million kilograms per year (7.3 million pounds per year) of depleted uranium hexafluoride into fluoride products and uranium oxide for disposal. The facility will possess significant quantities of depleted uranium hexafluoride, silicon tetrafluoride, boron trifluoride, and anhydrous hydrogen fluoride.

The license application (LA) also indicates that IIFP anticipates expanding the capacity of the facility after 3–4 years of operating experience. The Environmental Impact Statement addresses this future expansion as a reasonably foreseeable action. However, the LA does not request or evaluate the expansion, which would require a separate licensing action and environmental review.

IIFP submitted an application for a source material license regulated in accordance with 10 CFR Part 40, Domestic Licensing of Source Material, among other regulations. The NRC staff conducted its safety review of the LA, which describes the proposed operations, safety commitments, and other required programs (e.g., decommissioning, environmental protection, management measures). The applicant submitted an integrated safety analysis (ISA), which the NRC staff reviewed in accordance with 10 CFR Part 70, "Domestic Licensing of Special Nuclear Material," as directed by the Commission in SRM-SECY-07-0146, "Staff Requirements– Regulatory Options for Licensing New Uranium Conversion and Depleted Uranium Deconversion Facilities," dated October 10, 2007. The NRC staff notified the applicant that the 10 CFR Part 70 ISA requirements would eventually be superseded by similar requirements in 10 CFR Part 40, which is undergoing rulemaking to incorporate ISA requirements. The NRC staff also reviewed the applicant's Quality Assurance Program Description, Emergency Plan (EP), and Security Plan.

The NRC staff used several guidance documents to evaluate the LA and to complete the SER for the IIFP facility. The 10 CFR Part 40 deconversion application does not have a dedicated Standard Review Plan, but many of the regulatory requirements for 10 CFR Part 40 fuel cycle facilities are similar to 10 CFR Part 70 fuel cycle facilities. The NRC staff used primarily NUREG-1520, Revision 0, "A Standard Review Plan for the Review of a License Application for a Fuel Cycle Facility," to review the application. The updated revision of the NUREG was not available until after IIFP had submitted its application in 2009.

NUREG-1520 is the Standard Review Plan for 10 CFR Part 70 reviews, and the hazards at the proposed IIFP facility are similar to 10 CFR Part 70 fuel cycle facilities. These hazards include processing of depleted uranium hexafluoride, use of hazardous chemicals (e.g., hydrogen fluoride), and the associated safety systems. Certain portions of the NUREG do not apply to 10 CFR Part 40 facilities, such as criticality, because of the absence of special nuclear material. In addition, the NRC staff used Regulatory Guide (RG) 5.59, Revision 1, "Standard Format and Content for a Licensee Physical Security Plan for the Protection of Special Nuclear Material of Moderate or Low Strategic Significance," issued February 1983, to review the security-related portions of the application.

In conducting its safety review, the NRC staff assessed, among other things, whether the applicant's proposed equipment, facilities, and procedures would adequately protect public health and safety and the environment. The NRC staff evaluated the applicant's facility designs, procedures, and stated commitments. The Commission will only issue a license if it makes a determination that the proposed facility complies with the regulatory requirements, as demonstrated in the application.

Once a license is granted, construction of the facility may begin. In accordance with the ISA annual update requirements, the applicant (then licensee) will submit to the NRC annual updates to the ISA Summary along with a brief summary of the nonsafety-related changes made throughout the year. The NRC reviews these submissions as well as any license amendment requests that may be submitted.

The applicant also submitted an Environmental Report, which the NRC staff used to prepare NUREG-2113, "Draft Environmental Impact Statement for the Proposed Fluorine Extraction Process and Depleted Uranium Deconversion Plant" issued February 27, 2012 (NRC, 2012).

The LA contains IIFP's commitments which demonstrate compliance with the NRC's regulatory requirements, and the entire application is tied down in Safety Condition S-1 of the license. A summary of NRC's review and findings in each of the review areas follows.

General Information

The applicant adequately described the facility and processes, providing the NRC staff with an overall understanding of the relationships of the facility features and their functions. The application properly explained and outlined the financial qualifications. The description of the site included important information about regional hydrology, geology, meteorology, the nearby population, and potential effects of natural phenomena at the facility.

Organization and Administration

The applicant adequately described the responsibilities and associated resources for the design, construction, and operation of the facility and its plans for managing the project. The plans and commitments described in the application provide reasonable assurance that an acceptable organization; appropriate administrative policies; and sufficient resources have been established or committed for the design, construction, and safe operation of the facility.

Integrated Safety Analysis and Summary

The applicant provided sufficient information about the site, facility processes, hazards, and types of accident sequences. The information provided addressed each credible event, the potential radiological and chemical consequences of the event, and the likelihood of the event. No mitigated event consequence exceeds the performance requirements of 10 CFR 70.61, "Performance Requirements." The applicant also provided adequate information about items relied on for safety (IROFS).

Radiation Protection

The applicant provided sufficient information to evaluate the Radiation Protection Program. The application adequately describes the (1) qualification requirements, (2) written radiation protection procedures, (3) radiation work permit program, (4) program for ensuring that worker and public doses are as low as reasonably achievable, and (5) necessary training for all personnel who have access to radiologically restricted areas. The radiation survey and monitoring program is adequate to protect workers and members of the public who may be potentially exposed to radiation.

Chemical Process Safety

The applicant adequately described and assessed accident consequences that could result from the handling, storage, or processing of licensed materials and that could have potentially significant chemical consequences and effects. The applicant performed hazard analyses that identified and evaluated chemical process hazards and potential accidents and established safety controls that meet the regulatory requirements.

Fire Safety

The applicant committed to reasonable engineered and administrative controls to adequately minimize the risk of fires and explosions. The IROFS and defense-in-depth protection discussed in the applicant's ISA Summary, along with safety basis assumptions and the planned programmatic commitments in the LA, meet safety requirements and provide reasonable assurance that the facility is protected against fire hazards.

Emergency Management

The applicant provided an adequate EP for the facility that meets the regulatory requirements. The applicant committed to maintaining and executing an EP for responding to the radiological and chemical hazards resulting from potential release of radioactive or chemically hazardous materials incident to the processing of licensed material. Approved, written procedures implement the requirements of the EP.

Environmental Protection

The applicant committed to adequate environmental protection measures, including (1) environmental and effluent monitoring and (2) effluent controls to maintain public doses as low as reasonably achievable as part of the Radiation Protection Program. The applicant's proposed controls are adequate to protect the environment and the health and safety of the public, and they comply with the regulatory requirements.

Decommissioning

The applicant provided a conceptual Decommissioning Plan for the facility that adequately addresses (1) contamination control, (2) control of worker exposures and waste volumes, (3) waste disposal, (4) the final radiation survey, (5) control of special nuclear material, (6) control of classified matter, and (7) recordkeeping for decommissioning.

The applicant provided a Decommissioning Funding Plan (DFP) for the facility that adequately demonstrates that adequate funding will be available for decommissioning and that decommissioning will not pose a threat to public health and safety or the environment. The NRC staff will add a license condition, if a license is issued, to ensure that the applicant provides final copies of the proposed financial assurance instruments to the NRC for review at least 6 months before the planned date for possessing licensed material, as well as final executed copies of the reviewed financial assurance instruments before the receipt of licensed material. The DFP is acceptable because it provides sufficient funding to ensure that decommissioning and decontamination of the facility can be accomplished, even if the licensee is unable to meet its financial obligations.

Management Measures

The applicant provided sufficient information about management measures that will be applied to the project. The information describes (1) the overall configuration management program and policy, (2) the maintenance program, (3) training, and (4) the process for the development, approval, and implementation of procedures. The applicant explained the audits and assessments program, as well as incident investigations and the records management system. The applicant committed to establishing and documenting surveillances, tests, and inspections to provide reasonable assurance of satisfactory performance of the IROFS. The proposed management measures are acceptable and meet the regulatory requirements in 10 CFR 70.62(d).

Materials Control and Accountability

The applicant provided adequate information describing the fundamental nuclear material control plan for the project. The plan describes the programs to be used to control and account for licensed material in the facility. The program meets the applicable regulatory requirements in 10 CFR Part 74, "Material Control and Accounting of Special Nuclear Material."

Physical Protection

The applicant provided adequate information on the policies, methods, and procedures to be implemented to protect the facility. This information is acceptable and meets the requirements in 10 CFR Part 73, "Physical Protection of Plants and Materials."

Reference

(NRC, 2012) U.S. Nuclear Regulatory Commission, "Draft Environmental Impact Statement for the Proposed Fluorine Extraction Process and Depleted Uranium Deconversion Plant," NUREG-2113, 2012.

ABBREVIATIONS

ADAMS	Agencywide Documents Access and Management System
AEGL	acute exposure guideline level
AHF	anhydrous hydrogen fluoride
ALARA	as low as reasonably achievable
ALOHA	Area Locations of Hazardous Atmospheres (computer code)
ANSI	American National Standards Institute
ASCE	American Society of Civil Engineers
ASME	American Society of Mechanical Engineers
ASTM	American Society for Testing and Materials
BDC	baseline design criteria
BF_3	boron trifluoride
B_2O_3	diboron trioxide
Bq	becquerels
C	Celsius
CAA	controlled access area
CaF_2	calcium fluoride
CAP	corrective action program
CEDE	committed effective dose equivalent
CFR	*Code of Federal Regulations*
Ci	curie
cm	centimeter
CM	configuration management
CEO	Chief Executive Officer
COO	Chief Operations Officer
DAC	derived air concentration
DB	design and build
DCE	Decommissioning Cost Estimate
DFP	Decommissioning Funding Plan
DOE	U.S. Department of Energy
DOT	U.S. Department of Transportation
DR	damage ratio
DUF_4	depleted uranium tetrafluoride
DUF_6	depleted uranium hexafluoride
DUO	depleted uranium oxide
EAL	environmental assessment lead
EMT	emergency medical technician
EOC	emergency operations center
EP	Emergency Plan
EPA	U.S. Environmental Protection Agency
EPIP	emergency plan implementation procedure
EPP	environmental protection process
ER	environmental report

ERO	emergency response organization
ERPG	emergency response planning guideline
ESH	environmental safety and health
ESO	environmental and safety officer
F	Fahrenheit
FAA	functional allocation analysis
FEMA	Federal Emergency Management Agency
FEP	fluorine extraction process
FEP/DUP	fluorine extraction process and depleted uranium deconversion plant
FHA	fire hazards analysis
FNMCP	fundamental nuclear material control plan
FSRC	Facility Safety Review Committee
ft	feet
ft/s	feet per second
g	acceleration of gravity
gal	gallon
gpm	gallons per minute
H_3BO_3	boric acid
HBF_4	fluoroboric acid
HDBK	handbook
HED	human engineering discrepancies
HF	hydrogen fluoride
HFE	human factors engineering
HPS	Health Physics Society
HS&E	health, safety, and environment
HVAC	heating, ventilation, and air conditioning
ICRP	International Commission on Radiological Protection
IEEE	Institute for Electrical and Electronics Engineering
IIFP	International Isotopes Fluorine Products, Inc.
in.	inch
LLW	low-level radioactive waste
INIS	International Isotopes, Inc.
lpm	liters per minute
IROFS	item relied on for safety
ISA	integrated safety analysis
I&C	instrumentation and control
KF	potassium fluoride
kg	kilogram
km	kilometer
km/h	kilometer per hour
KOH	potassium hydroxide
kPa	kilopascal
L	liter

LA	license application
lb	pound
m	meter
m_b	body wave magnitude
M_L	local magnitude
m^3	cubic meter
MC&A	material control and accounting
μCi	microcurie
MDC	minimum detectable concentration
MEI	maximally exposed individual
mi	mile
MIL	military
mg	milligram
mm	millimeter
MM	Modified Mercalli
MOA	military operation area
MOU	memorandum of understanding
mph	miles per hour
mrem	millirem
mrem/yr	millirem per year
m/s	meter per second
mSv	millisievert
mSv/yr	millisiever per year
MSDS	material safety data sheet
M&TE	measuring and test equipment
NCS	Nuclear Criticality Cafety
NELAC	National Environmental Laboratory Accreditation Conference
NFPA	National Fire Protection Association
NIST	National Institute of Standards and Technology
NMCBC	New Mexico Commercial Building Code
NMED	State of New Mexico Environment Department
NPDES	National Pollutant Discharge Elimination System
NOAA	National Oceanic and Atmospheric Administration
NRC	U.S. Nuclear Regulatory Commission
NVLAP	National Voluntary Laboratory Accreditation Program
OSHA	U.S. Occupational Safety and Health Administration
OER	operational experience review
P&E	purge and evacuation
PFD	process flow diagram
PFPE	perfouropolyether
PHA	process hazards analysis
PM	plant manager
PM	preventive maintenance
psi	pounds per square inch
psig	pounds per square inch gauge

PSB	payment surety bond
PSP	physical security plan
QA	quality assurance
QAPD	quality assurance program description
QL	quality level
QMS	quality management system
RAI	request for additional information
RCRA	Resource Conservation and Recovery Act
rem	roentgen equivalent man
REMP	Radiological Environmental Monitoring Program
RQAD	Regulatory Affairs and Quality Assurance Director
RPM	radiation protection manager
RWP	radiation work permit
SEI	Structural Engineering Institute
SER	safety evaluation report
SiF_4	silicon tetrafluoride
SiO_2	silicon dioxide
SNM	special nuclear material
SNM-LSS	special nuclear material—low strategic significance
SSC	structure, system, and components
STA	standby trust agreement
STD	standard
Sv	sievert
TA	task analysis
TEDE	total effective dose equivalent
TSV	task support verification
UF_4	uranium tetrafluoride
UF_6	uranium hexafluoride
UO_2	uranium dioxide
UO_2F_2	uranyl fluoride
U_3O_8	triuranium octoxide
USGS	U.S. Geological Survey
V&V	verification and validation
WIP	work in process

1.0   GENERAL INFORMATION

1.1   Facility and Process Description

The purpose of the U.S. Nuclear Regulatory Commission's (NRC's) review of the International Isotopes Fluorine Products, Inc.'s (IIFP or the applicant) facility and process description is to evaluate whether the license application (LA) (IIFP, 2012a) includes an overview of the facility layout and a summary description of the proposed processes. The Integrated Safety Analysis (ISA) Summary (IIFP, 2012b) contains a more detailed description of the facility and processes.

Some information related to the ISA Summary for the site description has been marked by the applicant as "Security-Related Information," pursuant to Title 10 of the *Code of Federal Regulations* (10 CFR) 2.390, and is not included in the public version of this report.

1.1.1   Regulatory Requirements

The regulations in 10 CFR 40.32, entitled, "General Requirements for Issuance of Specific Licenses," require each application for a license to include information on the proposed activity and the equipment and facilities that the applicant will use to protect health and minimize danger to life and property. In addition, the regulations in 10 CFR 70.65, "Additional Content of Applications," require each application to include a general description of the facility, with emphasis on those areas that could affect safety, including identification of the controlled area boundaries. The LA, including license commitments, is incorporated directly into the license by reference as Safety Condition S-1.

1.1.2   Regulatory Guidance and Acceptance Criteria

Chapter 1 of NUREG-1520, Revision 0, "Standard Review Plan for the Review of a License Application for a Fuel Cycle Facility," issued March 2002 (NRC, 2002), contains the guidance applicable to NRC's review of the facility and process description section of the LA (IIFP, 2012a). This chapter of NUREG-1520 is applicable to the IIFP application in its entirety. In particular, NUREG-1520, Revision 0, Section 1.1.4.3, contains the acceptance criteria applicable to this portion of the review.

1.1.3   Staff Review and Analysis

The first section of Chapter 1 of the LA (IIFP, 2012a) provides a basic overview of the proposed IIFP facility with sufficient information to understand the process. The applicant plans to build the plant 24 kilometers (15 miles) west of Hobbs, NM, in Lea County. The facility will provide deconversion services to the uranium enrichment industry throughout the United States (U.S.) to convert depleted uranium hexafluoride (DUF_6) into uranium oxide for disposal. In addition, the fluoride extracted in the deconversion of DUF_6 will be used to produce specialty fluoride gas products, such as silicon tetrafluoride (SiF_4), boron trifluoride (BF_3), and anhydrous hydrogen fluoride (AHF). The plant is designed to process 3.3 million kilograms (kg) per year of DUF_6, which will result in production of up to 0.68 million kg of SiF_4, 0.23 million kg of BF_3, and 0.45 million kg of AHF annually. These products will be sold commercially for various industrial applications.

The basic process involves receipt of the DUF_6 via tractor trailers in 14-ton cylinders from any of the U.S. commercial enrichment facilities. The DUF_6 is gasified in an autoclave. The DUF_6 gas then reacts with hydrogen gas to produce depleted uranium tetrafluoride (DUF_4) and AHF. The DUF_4 is transferred into a calciner with silicon dioxide or boron trioxide. As the calciner heats and mixes the materials, the mixture reacts to produce depleted uranium oxidize (DUO) powder and SiF_4 or BF_3 gas, depending on the process line. The DUO powder is collected in a disposal container for transportation to a licensed low-level waste disposal facility. The fluoride gases are purified, concentrated, and collected into storage cylinders for resale. These are stored temporarily, packaged for transportation, and sold commercially.

The proposed facility is to be built on 40 acres of a 640-acre plot in a rural area of Lea County, NM. Figure 1-4 of the LA (IIFP, 2012a) provides an aerial photograph of the entire property, with the outline of the plant superimposed in the top left quadrant of the grid at its approximate location on the site. The terrain is primarily flat and underdeveloped, but overlays oil and gas geologic formations which are developed extensively in the area. Land use within 5 miles of the facility consists primarily of cattle grazing and oil and gas production. The nearest industrial facility is a gas-fired power plant on the west boundary of the site, and the nearest resident lives 1.6 miles northwest of the facility. Chapter 1 of the ISA Summary (IIFP, 2012b) contains an expanded description of the facility.

The IIFP facility, as described in the LA (IIFP, 2012a) and ISA Summary (IIFP, 2012b), consists of multiple buildings; each one of which performs a specific function. A list of the primary buildings and their specific functions is provided below in the order in which the material flows through the plant.

- The DUF_6 storage pad is used to store 14-ton cylinders full of DUF_6, which arrive at the IIFP facility via tractor trailers. These cylinders are inspected; assayed; and, if accepted, offloaded to the storage pad. Shipments of silicon dioxide (SiO_2), diboron trioxide (B_2O_3), and DUF_4 are also received via tractor trailers and are surveyed and stored at separate locations.

- The DUF_6 autoclave building receives cylinders from the storage pad. They are placed via crane into steam autoclaves which vaporize the DUF_6. The gaseous material is then piped into the process building. Empty uranium hexafluoride (UF_6) cylinders are moved to a cold box for removing cylinder tails.

- The DUF_6 gas is piped to the DUF_4 process building for conversion into DUF_4 and AHF. Hydrogen gas is injected into a reaction vessel and interacts exothermically with the DUF_6 to produce DUF_4 powder and AHF gas. The DUF_4 is collected in storage hoppers for transfer to the fluoride extraction process. The AHF gas is condensed and transferred to a storage tank.

- The AHF staging containment building is used to store liquid AHF before shipment. The AHF is stored in 8,000-pound (lb) tanks, which are surrounded by dikes capable of holding the entire contents of a ruptured tank. The building is totally enclosed and contains leak detection and water spray systems to mitigate the airborne release of spills.

- A DUF_4 container staging building is located just outside the DUF_4 process building. The building is used to store and organize additional inventory of DUF_4 before deconversion. The building will also be used to receive external shipments of DUF_4 from offsite sources.

- The fluoride extraction process (FEP) building is used to further deconvert the DUF$_4$. The SiO$_2$ or B$_2$O$_3$ is combined with the DUF$_4$ and then heated and mixed in a calciner. The products consist of SiF$_4$ or BF$_3$ gas, respectively, and DUO powder. The DUO is collected and packaged in a container approved by the U.S. Department of Transportation (DOT) for shipment to a licensed low-level radioactive waste disposal site. The SiF$_4$ or BF$_3$ gas is condensed in a cold trap and transferred to a gas storage cylinder.

- A number of storage, shipping, and environmental protection activities are conducted in ancillary buildings onsite. The LA, Section, 1.1.2.1 (IIFP, 2012a) describes these activities. Secondary facilities for the IIFP include data processing facilities, emergency response facilities, electrical distribution systems, security fencing and portals, a pump house, an air generation plant, a cooling tower, a boiler system, a training facility, a maintenance facility, storage facilities, and waste accountability facilities.

The first chapter of the LA (IIFP, 2012a) and ISA Summary (IIFP, 2012b) provide additional descriptions, drawings, and process details for each of the areas described above. The IIFP facility is primarily a chemical facility which processes depleted uranium, among other operations. The chemical reactions involve the conversion of DUF$_6$ to DUO. This process can be summarized as DUF$_6$ reacting with hydrogen gas to produce DUF$_4$ and AHF. The DUF$_4$ is then chemically reacted with SiO$_2$ or B$_2$O$_3$ to produce gaseous SiF$_4$ or BF$_3$, respectively, and DUO powder. These processes have a radiological component because of the natural radioactivity of uranium and daughter products. The facility does not process special nuclear material.

The various fluoride and DUO streams result in residual offgases of hydrogen fluoride, SiF$_4$, BF$_3$, and other trace gas contaminants. The facility uses particulate filters and liquid scrubbers to minimize the release of effluents to the environment. These offgases are processed through a three-stage potassium hydroxide (KOH) scrubbing system. The system is designed to remove 99.9 percent of fluoride-bearing components before being vented to the plant stack. The effluents released from the stack are continuously monitored. Once the KOH scrubbing liquid has absorbed significant quantities of fluorides, the KOH is reacted with a lime slurry to regenerate the solution. The fluorides in the solution produce calcium fluoride which is collected, dried, and packaged for sale or disposal. The regenerated KOH is reused in the scrubbing system.

The proposed facility produces both solid and liquid wastes. Solid waste consists of industrial, mixed (radioactive and chemical), and hazardous waste. These wastes are collected, volume reduced if feasible, and shipped to a licensed disposal facility. Liquid wastes containing uranium are transferred to the decontamination building to recover the uranium, if possible. Other process liquids, including sanitary sewage, are treated, sampled and used for landscaping—or are evaporated. The primary byproduct from the facility is DUO, which is shipped offsite to a licensed disposal facility. The LA, Section, 1.1.7 (IIFP, 2012a) provides a summary list of the raw materials, byproducts, wastes, and finished products involved in the process.

The NRC staff reviewed the general description of the proposed facility and process using Section 1.1 of NUREG-1520, Revision 0 (NRC, 2002).

IIFP adequately described the general facility description according to Section 1.1 of the Standard Review Plan. IIFP adequately described (1) the facility and its processes so that the NRC staff has an overall understanding of the relationships of the facility features and (2) the function of each feature. IIFP cross-referenced its general description with the more detailed descriptions elsewhere in the application. The NRC staff also reviewed comparable sections in both the ISA Summary and the Emergency Management Plan and, based on that review, the NRC staff determined that the information is consistent with the information provided in the LA. Therefore, the NRC staff finds that IIFP has complied with the requirements of 10 CFR 40.31, "Application for a specific license," and 10 CFR 70.65(b)(1-2), as applicable to this section.

1.2   Institutional Information

The purpose of this review is to establish whether the LA includes adequate information identifying the applicant, the applicant's characteristics, and the proposed activity.

1.2.1   Regulatory Requirements

The regulations in 10 CFR 40.32(b) require the applicant to provide sufficient information to demonstrate that it is technically and financially qualified to operate the facility safely. Consistent with this requirement, the following institutional information is necessary: (1) the identity of the applicant; (2) the identity of the licensed material—including name, chemical and physical form—and maximum amount that will be possessed; and (3) the purpose for which the licensed material will be used.

1.2.2   Regulatory Guidance and Acceptance Criteria

Section 1.2.4 of NUREG-1520 (NRC, 2002), and the regulations in 10 CFR 40.32 contain the acceptance criteria applicable to NRC's review of the institutional information section of the application.

1.2.3   Staff Review and Analysis

1.2.3.1   Corporate Identity

In Section 1.2.1 of the LA (IIFP, 2012a), the applicant provided information on the corporate organization. IIFP is a wholly owned subsidiary of International Isotopes Inc. (INIS). INIS was originally organized as a Texas corporation in 1995, but has since moved operations to Idaho Falls, ID. In 2001, the company obtained a license under 10 CFR Part 30, "Rules of General Applicability to Domestic Licensing of Byproduct Material," to manufacture high quality radiochemicals for medical, industrial, and research applications (Docket 030-35486 and License 11-27680-01). In 2005, INIS received a license under 10 CFR Part 40, "Domestic Licensing of Source Material," to produce fluoride compounds from DUF_4 (Docket 040-09058 and License SUB-1587). In December 2009, INIS submitted the application for the proposed IIFP facility in New Mexico. Its wholly owned subsidiaries are International Isotopes Idaho, Inc.; IIFP; and International Isotopes Transportation Services, Inc., all of which are Idaho corporations.

The IIFP's principal office is located at 4137 Commerce Circle, Idaho Falls, ID 83401. The facility is planned to be built in Lea County, NM, 24 kilometers (km) [15 miles (mi)] west of Hobbs, NM, and approximately 17 mi (27 km) west of the Texas state border. The

LA, Section, 1.2.1 (IIFP, 2012a) provide additional information on the IIFP corporate identity. This description meets the guidance in Section 1.2.4.3(1) of NUREG-1520 (NRC, 2002), and is, therefore, acceptable.

1.2.3.2   Financial Qualifications

Section 1.2.2 of the LA (IIFP, 2012a) states that the estimated project cost to build and operate the facility is $75–90 million dollars in 2009 dollars. IIFP plans to obtain these funds from a number of capital investors. In addition, a surety bond is planned for financial assurance for decommissioning. The LA, Chapter 10 (IIFP, 2012a), provides additional information on the Decommissioning Funding Plan (DFP), which is reviewed separately in Chapter 10 of this safety evaluation report (SER). The information provided by the applicant meets the guidance in Section 1.2.4.3(2) of NUREG-1520 (NRC, 2002), and is, therefore, acceptable.

1.2.3.3   Type, Quantity, and Form of Licensed Material

Source material makes up the vast majority of licensed material at the IIFP. Table 1-4 of the LA (IIFP, 2012a) lists the types, quantities, and forms of the licensed material. The maximum uranium onsite at any one time is limited to 1,650,000 lbs (750,000 kg) of depleted uranium in any chemical form. The source material will be in the chemical form of UF_6, UF_4, uranium oxides, and other trace uranium compounds. All of the process material will consist of DUF_6. IIFP only uses commercial enrichment plant suppliers and does not accept cylinders that contain technetium or transuranics, contaminants from facilities related to the U.S. Department of Energy (DOE).

During shipment from the enrichment facilities and storage onsite, the DUF_6 is in a solid crystalline form. In the autoclaves, a percentage of the DUF_6 may become liquefied during gasification, but it is in the gaseous state during processing. The DUF_4, both received from other facilities and produced onsite is a powder during shipment, storage, and processing. The waste products, depleted uranium dioxide (DUO_2), and depleted triuranium octoxide (DU_3O_8), are powders which are collected in containers for offsite disposal. Precautions, including ventilation and personal protective equipment, are in place to protect against small quantities of depleted uranyl fluoride (DUO_2F_2) formed by the reaction of DUF_6 with moisture in the air. Waste liquids may contain this chemical before uranium recovery.

The applicant identified the maximum quantity and chemical and physical forms of the licensed material present at the facility. The information provided by the applicant is consistent with Section 1.2.4.3(3) of NUREG-1520 (NRC, 2002), and the regulatory requirements and is, therefore, acceptable.

1.2.3.4   Authorized Uses

The application is for the issuance of a license under 10 CFR Part 40. As indicated above, IIFP does not store or process special nuclear material.

If granted, the proposed source material license would authorize the deconversion of DUF_6 into oxides. This process would involve the recovery of the fluoride products for commercial resale and the disposal of the oxides as waste for licensed disposal. The deconversion services would be marketed to enrichment facilities as a means for them to dispose of depleted uranium.

The applicant has requested a 40-year license term. The maximum extent allowed for license terms are not prescribed by regulation, but are determined by Commission policy. Previously, the Commission approved, in the NRC staff requirements memorandum to SECY-06-0186, "Increasing Licensing Terms for Certain Fuel Cycle Facilities," the issuance of up to a 40-year license for facilities which implement an ISA, consistent with the requirements in 10 CFR Part 70, "Domestic Licensing of Special Nuclear Materials," Subpart H, "Additional Requirements for Certain Licensees Authorized To Possess a Critical Mass of Special Nuclear Material." The ISA Summary serves as one of the primary safety bases for approval of the application, and it must be updated annually. The management measures and change processes required for the ISA mitigate the aging phenomena, thereby allowing longer license terms of up to 40 years. The DFP is also required to be updated every 3 years or as stipulated in the license. Additional information on the decommissioning review is available in Chapter 10. IIFP is implementing the 10 CFR Part 70, Subpart H, requirements until similar requirements are included in 10 CFR Part 40, Subpart H, which is being developed via rulemaking at the time of this review. The NRC staff has concluded that the ISA requirements for IIFP are consistent with the Commission's approval for up to a 40-year license term.

As stated above, the applicant provided a summary—nontechnical, narrative description—for each activity or process in which it proposed to acquire, deliver, receive, possess, produce, use, process, transfer, or store licensed material. The authorized uses of licensed material proposed for the facility are described and are consistent with the Atomic Energy Act of 1954, et seq. The description is also consistent with more detailed process descriptions submitted as part of the ISA Summary (IIFP, 2012b), as reviewed in Chapter 3 of this SER. The information provided by the applicant meets the guidance in Section 1.2.4.3(4) of NUREG-1520 (NRC, 2002), and is, therefore, acceptable.

1.2.3.5   Special Exemptions or Special Authorizations

1.2.3.5.1   Use of ICRP 68 and 72 DAC and ALI Values

In Chapter 1, Section 1.5, of the LA (IIFP, 2012a), IIFP requested an exemption from the derived air concentration (DAC) and annual limit on intake (ALI) values provided in 10 CFR Part 20, Appendix B. In their place, the exemption requested authorization to use the International Committee on Radiological Protection (ICRP) Publication 68, entitled "Dose Coefficients for Intake of Radionuclides by Workers" (ICRP, 1994c) for determining dose to workers and ICRP Publication 72, entitled "Age-dependent Doses to the Members of the Public from Intake of Radionuclides Part 5, Compilation of Ingestion and Inhalation Coefficients" (ICRP, 1996) for determining dose to the public due to effluents.

The purpose of this request is to allow for the use of updated radiation dose conversion factors for DAC and ALI values calculated using the internal dosimetry models described in ICRP Publications 68 and 72. The current NRC regulations governing airborne radioactivity are derived from the inhalation dose conversion factors (DCF) in ICRP Publications 26 (ICRP, 1977) and 30 (ICRP, 1982), as disseminated by the U.S. Environmental Protection Agency in the Federal Guidance Report 11. Subsequent work by the ICRP refined the dosimetry models used to derive ICRP Publications 26 and 30. ICRP Publications 68 and 72 contain improved dose conversion factors based on updated scientific data, but the derived documents for 10 CFR 20 are still based on the outdated ICRP 26 and 30 data. This exemption request proposes to replace the use of ICRP 26 and 30 data with the updated ICRP 68 and 72 data as applicable for internal dose calculations, determining the allowed limits for intakes, and calculating derived air concentration and effluent concentration limits.

The Commission has determined that requests to use ICRP 68 and 72 must be considered on a case-specific basis as exemptions. On April 21, 1999 (SRM-SECY-99-077), the Commission approved the NRC staff's granting of exemptions for the use of ICRP 68 and 72 on a case-by-case basis based on the precedent set by the Commission's decision in the OSRAM, Inc.'s, exemption request (SECY-99-077). In addition, the majority of fuel fabrication facilities and several source material facilities have been granted this exemption.

In a rulemaking to 10 CFR 20, finalized May 21, 1991 (56 FR 23391), the NRC revised the regulation of radioactive materials so that it would be based on dose instead of intake. A consequence of the dose-based rule is that compliance would not necessarily be constrained by use of a specific set of parameters to calculate dose. Title 10 CFR Part 20, in fact, allows certain adjustments to be made to the model parameters if specific information is available, such as adjustments when the particle size of the airborne radioactive material is known, rather than using a default particle size. However, because of the way Part 20 was written, NRC licensees are not permitted to use the revised and updated internal dosimetry models, unless permission is specifically granted in the license.

In the same year that the Commission approved the final 10 CFR Part 20 rule based on ICRP 26 and 30, ICRP published a major revision of its radiation protection recommendations in ICRP 60. In the several years following the revision of ICRP 60, ICRP published a series of reports in which it described the components of an extensive updated and revised internal dosimetry model. These reports include ICRP Publications 60 (ICRP, 1991), 68 (ICRP, 1994a), 66 (ICRP, 1994b), 71, (1995a), 72 (ICRP, 1996) and 78 (ICRP, 1997).

Although the dose per unit intake calculated using the new models (ICRP 68 and 72) does not differ by more than a factor of about two from the values in 10 CFR Part 20 for most radionuclides, the differences are substantial for some, particularly for the isotopes of thorium (Th), uranium (U), and some of the transuranic nuclides. Dose calculations using the DAC and ALI values from ICRP 26 and 30 (i.e., 10 CFR Part 20, Appendix B) are several times higher than real life; therefore, 10 CFR Part 20 requires significantly more protective measures for Th and U than would be warranted based on the updated models in ICRP 68 and 72. This is IIFP's primary concern, and they have requested to be allowed to use DAC and ALI values based on the coefficients from ICRP 68 and 72. Through NRC staff reviews of IIFP's Radiation Safety Protection program, the NRC staff concludes that the program is sufficiently sophisticated by training and expertise to utilize the ICRP models in a manner equivalent to those listed in 10 CFR 20, Subpart C and D, i.e., IIFP's program is capable of properly calculating the total effective dose equivalent (TEDE) for its occupational workers and the public using ICRPs 68 and 72.

IIFP's use of ICRP 68 and 72 in place of existing values in Appendix B to Part 20 gives its workers and the public equivalent radiological protection. Thus, the exemption is authorized by law and will not result in undue hazard to life or property.

The NRC staff has developed an Environmental Impact Statement (EIS) (NRC, 2011) separate from this SER which includes this exemption request. The exemption is described in the Environmental Report (ER) (IIFP, 2011a), which states that ICRP 68 and 72 are used for calculating dose estimates.

The NRC staff finds that IIFP's request for an exemption is acceptable under 10 CFR 20.2301 and 10 CFR 40.14, which state that the Commission may—upon application by a licensee or upon its own initiative—grant an exemption from the requirements of the regulations if the NRC determines the exemption is authorized by law and would not result in undue hazard to life or property. This determination is based on the NRC staff's safety evaluation, as described above, and the EIS (NRC, 2011) conducted for this application.

Consistent with the 10 CFR 2301 and 10 CFR 40.14, IIFP is granted an exemption from DAC and ALI values in 10 CFR 20.1003 and 10 CFR Part 20, Appendix B. This exemption authorizes IIFP to use the ALIs and DACs based upon dose coefficients published in the ICRP Publication 68, entitled "Dose Coefficients for Intake of Radionuclides by Workers," (ICRP, 1994a) for determining dose to workers. The exemption also authorizes IIFP to use ICRP Publication 72, entitled "Age Dependent Doses to the Members of the Public from Intake of Radionuclides Part 5, Compilation of Ingestion and Inhalation Coefficients," (ICRP, 1996) for determining dose to the public due to effluents.

This exemption does not change, in any way, the NRC dose limits which the applicant must maintain. Rather, the exemption changes the methodology IIFP uses to calculate the doses, and assures the calculations are based on updated and internationally recognized standards published in the ICRP 68 and 72, as described above.

1.2.3.5.2 Memorandum of Understanding between IIFP and New Mexico Environmental Department

On October 22, 2009, IIFP finalized a memorandum of agreement (MOA) with the State of New Mexico Environment Department (NMED) to limit the maximum quantities of depleted uranium onsite (IIFP, 2009). The limits established in the MOA (4,850,169 lbs [2,200,000 kg] of uranium) are significantly higher than the possession limits requested in the LA (1,653,500 lbs [750,000 kg] of uranium of any form) (IIFP, 2012a). The uranium possession limit requested in the LA represents approximately 60 full cylinders of DUF_6. The NRC staff finds this possession limit to be reasonable for the types of operations presented in the LA (IIFP, 2012a), which include DUF_6 storage before processing, uranium within the process, and DUO awaiting dispatch. The IIFP/NMED agreement is not binding on the NRC and represents an independent agreement between IIFP and NMED.

1.2.3.5.3 Financial Assurance Exemption

In Section 1.6 of the LA (IIFP, 2012a), the applicant requested an exemption from the requirements of 10 CFR 40.36(d) to provide an alternate timeframe for implementing the financial instruments for decommissioning.

The NRC requirements for financial assurance are described in 10 CFR 40.36(d) and include a requirement that a licensee certify that financial assurance has been provided in the amount of the cost estimate for decommissioning at the time of licensing. IIFP's cost estimate for decommissioning is based on a 40-year operating life and includes the estimated decommissioning costs of the facility and site, and the expected costs associated with the disposition of the depleted uranium onsite. The proposed exemption would allow IIFP to submit its financial instrument for decommissioning to the NRC for review 6 months before the startup of operations and fully fund the instrument at least 21 days before operations. Because of the time required for construction, the alternate schedule proposed in the exemption request for finalizing the financial assurance funding instrument differs significantly from the schedule required by 10 CFR 40.36(d) (i.e., full funding at the time the license is issued).

IIFP has requested the exemption on the basis that no significant quantity of licensed material will be onsite before the financial assurance funding instrument is approved by the NRC or funded by IIFP (IIFP, 2012a). Therefore, the cost of maintaining the financial assurance funding instrument for an extended period before operations, when licensed material is not onsite, would represent a significant financial burden to the applicant. Submittal of the financial assurance funding instrument 6 months before operations would allow sufficient time for the NRC to verify the adequacy of the instrument. Fully funding the instrument 21 days before operations would ensure that the funds for decommissioning are in place before operations. Thus, the requested exemption would allow IIFP to satisfy the applicable decommissioning funding assurance requirements without imposing an unnecessary financial burden.

Under the provisions of 10 CFR 40.14, "Specific Exemptions," the Commission may, upon application by any interested person or upon its own initiative, grant exemptions from the requirements of 10 CFR Parts 40, respectively, when the exemptions are authorized by law, will not endanger life or property or the common defense and security, and are otherwise in the interest of the public.

The NRC staff evaluated the exemption request. The proposed exemption is authorized by law because the Atomic Energy Act of 1954, as amended, contains no provisions prohibiting an applicant from providing decommissioning funding commensurate with the quantity of licensed material onsite. The NRC staff also determined that, because the funding will be reviewed by the NRC and will be provided before receipt of licensed material, the approach will not endanger life or property or the common defense and security. Reducing the time and cost associated with maintaining the financial assurance before operations has no negative impact on decommissioning. Therefore, the NRC staff has determined that the proposed approach will be in the public's interest by reducing unnecessary regulatory costs.

The exemption granted from 10 CFR 40.36(d) financial assurance requirements meets the criteria for categorical exclusion under 10 CFR 51.22(c)(25), "Criterion for Categorical Exclusion; Identification of Licensing and Regulatory Actions Eligible for Categorical Exclusion or Otherwise Not Requiring Environmental Review." The granting of this exemption from regulatory requirements meets the following criteria for categorical exclusion described in 10 CFR 51.22(c)(25):

(i)   "There is no significant hazards consideration;
(ii)   There is no significant change in the types or significant increase in the amounts of any effluents that may be released offsite;
(iii)   There is no significant increase in individual or cumulative public or occupational radiation exposure;
(iv)   There is no significant construction impact;
(v)   There is no significant increase in the potential for or consequences from radiological accidents; and
(vi)   The requirements from which an exemption is sought involve:
        ....
        (H) Surety, insurance or indemnity requirements; ..."

Therefore, no Environmental Assessment or EIS is required for this action.

The NRC staff grants the requested exemption. The license will include a license condition that will address IIFP's commitments to submit financial assurance documents for review and approval. Chapter 10 of this SER discusses this license condition further.

1.2.3.5.4   License Conditions for Construction and Operational Readiness

The NRC will implement the following license conditions into the IIFP license to allow NRC staff to conduct routine inspections during the construction of the IIFP facility and evaluate the Operational Readiness Review (ORR). The inspectors will verify that IIFP is constructing the facility consistent with internal written procedures, commitments in the application, and relevant regulations. The second license condition will require IIFP to conduct an ORR. IIFP will be required to summarize the ORR in a report or series of reports and provide the report(s) to the NRC for review at least 60 days prior to operations. The applicant reviewed each of the following license conditions and agrees to their requirements.

Construction License Condition
During construction of the IIFP facility, the licensee shall afford to the NRC at all reasonable times, opportunity to inspect the premises and facilities and to verify that the facility has been constructed in accordance with the requirements of the license and onsite implementing procedures.

Operational Readiness Review License Condition
In addition, prior to commencing operations of the deconversion facility, IIFP shall conduct an operational readiness and management measures verification review and document the results in a report(s) to the NRC. The review must verify and document that IROFS and their management measures have been implemented to ensure compliance with the performance requirements of the integrated safety analysis. IIFP shall provide the NRC with a copy of the ORR report(s). The report(s) may address the entire facility as a whole or be submitted in parts. Each part must address major subsections of the facility with clearly defined boundaries, e.g., the uranium tetrafluoride building. The report(s) must be provided to the NRC no later than 60 days in advance of the start of commercial operations. (i.e., excluding testing) for that section of the facility. IIFP shall provide NRC opportunity to verify, through review and inspection, the report's(s') findings.

1.2.3.6   Security of Classified Information

All processes, materials, and information at the IIFP are unclassified. Therefore, 10 CFR Part 95 is not applicable as stated in the acceptance criteria in NUREG-1520, Section 1.2.4.3 (NRC, 2002). Security-related information will be controlled consistent with internal written procedures.

The NRC staff reviewed the institutional information for the proposed IIFP according to Section 1.2 of NUREG-1520 (NRC, 2002). On the basis of the review, the NRC staff has determined that the applicant has adequately described and documented the corporate structure and financial information, and is in compliance with those parts of 10 CFR 40.31 and 70.65 related to other institutional information. In addition, the applicant has adequately described the types, forms, quantities, and proposed authorized uses of licensed materials to be permitted at this facility:

Table 1-1 Types, Forms, Quantities, and Proposed Authorized Uses Of Licensed Materials

Material	Form	Quantity	Authorized Use(s)
Uranium (depleted) and daughter products	Physical: solid, liquid, and gas Chemical: UF6, UF4, UO2F2, oxides, and other compounds	750,000 Kilograms as uranium	IIFP will be authorized to acquire, deliver, receive, possess, produce, use, transfer, and/or store source material consistent with their license.

The applicant's proposed activities are consistent with the Atomic Energy Act of 1954, as amended. The applicant has provided all institutional information necessary to understand the ownership, financial qualifications, location, planned activities, and nuclear materials to be handled in connection with the requested license.

1.3   Site Description

The purpose of the NRC's review of the applicant's site description is to evaluate whether the application adequately describes the geographic, demographic, meteorological, hydrologic, geologic, and seismologic characteristics of the site and the surrounding area. The site description summarizes the information that the applicant used in preparing the LA, ER, Emergency Plan, and ISA Summary.

1.3.1 Regulatory Requirements

The regulations in 10 CFR 40.32, and 10 CFR 70.65(b)(1) require each application to include a general description of the site, with emphasis on those factors that could affect safety (i.e., nearby facilities, meteorology, and seismology). In addition, 10 CFR 70.61(f) requires each licensee to establish a controlled area, as defined in 10 CFR 20.1003, "Definitions," and to retain authority to exclude or remove personnel and property from the area.

1.3.2 Regulatory Guidance and Acceptance Criteria

The acceptance criteria applicable to NRC's review of the site description section of the LA and ISA Summary (IIFP, 2012a, b) are contained in 10 CFR 40.32, 10 CFR 70.65(b)(1), and Section 1.3.4.3 of NUREG-1520, Revision 0 (NRC, 2002). Chapter 1 of NUREG-1520, Revision 0 (NRC, 2002), applies to the IIFP facility in its entirety.

1.3.3 Staff Review and Analysis
1.3.3.1
1.3.3.2   Site Geography

The site location of the proposed facility is described in Section 1.7.1.1 of the LA (IIFP, 2012a) and Section 1.1.1 of the ISA Summary (IIFP, 2012b). These sections indicate that the proposed site is in Lea County, New Mexico, approximately 19 km (12 mi) west of Hobbs, New Mexico and 26 km (16 mi) west of the Texas border. The IIFP facility lies approximately 1.6 km (1.0 mi) from U.S. Highway 62/180 and along the east side of New Mexico Highway 483.

The proposed site has an average elevation of approximately 1,158 m (3,800 ft) higher than the Gulf of Mexico and a relatively flat topology. The IIFP site is currently an open, vacant land. The only agricultural activity in the site vicinity is domestic livestock ranching.

1.3.3.2.1   Nearby Highways and Railroads

The applicant discussed the highways located near the proposed IIFP facilities in the ISA Summary, Section 1.2.4.1 (IIFP, 2012b). According to the applicant, the IIFP site is located north of U.S. Highway 62/180 and east of New Mexico Highway 483. New Mexico Highway 483 is the nearest transportation route to the IIFP safety structure, more than 0.94 km (0.58 mi) away. The nearest railroad is more than 8 km (5 mi) away from the IIFP facility.

The applicant stated that the IIFP structures will be designed to withstand an overpressure of 6.9 kilopascals (kPa) (1 pound per square inch [psi]). With this design basis, the applicant estimated that the safe separation distance for an explosion not to affect an IIFP safety significant structure is approximately 0.4 km (0.24 mi), assuming a postulated explosion of 4,536 kg (10,000 lbs) of propane. The applicant also calculated the safe separation distance of 0.5 km (0.3 mi) from an explosion at a transportation route using the maximum probable hazardous solid cargo for a single highway truck suggested in Regulatory Guide 1.91, Revision 1, "Evaluations of Explosions Postulated to Occur on Transportation Routes Near Nuclear Power Plants," issued February 1978 (NRC, 1978), assuming this solid is subject to a vapor effective explosion (IIFP, 2012a). In addition, the applicant estimated the safe separation distance of 0.94 km (0.58 mi) from a rail transportation accident using the maximum explosive cargo in a single railroad box car and assuming a vapor effective explosion for the cargo. As discussed earlier, the nearest highway to any IIFP safety-significant structure is more than 0.94 km (0.58 mi) away, and the nearest railroad is more than 8 km (5 mi) away from the IIFP facility. Because these distances exceed the corresponding safe separation distance the applicant estimated, IIFP concluded that potential explosion hazards from nearby transportation routes will not pose safety concerns to the IIFP site.

The NRC staff reviewed the information the applicant provided regarding the potential hazards caused by explosions at nearby highways and railroads. The NRC staff finds that the applicant used the approach recommended in Regulatory Guide 1.91, Revision 1 (NRC, 1978), to estimate the safe separation distance from the effects of explosions of hazardous materials that may be carried on nearby transportation routes. Regulatory Guide 1.91, Revision 1 (NRC, 1978), provides an acceptable approach and was used for licensing nuclear-related facilities in the past. Therefore, the NRC staff finds that the applicant's assessment of explosion effects from nearby highways and railroads using the methodology in Regulatory Guide 1.91. Revision 1 (NRC, 1978), is acceptable. In addition, the NRC staff accepts the applicant's conclusion that nearby transportation route explosion accidents do not require further consideration because the distance between the nearest highways and railroads and the IIFP facility site is greater than the corresponding safe separation distance. Based on this review, the NRC staff finds that the explosion accidents from nearby transportation routes are not a safety concern to the IIFP facility. Therefore, the applicant meets the regulatory requirements in 10 CFR 70.65(b)(1).

1.3.3.2.2   Nearby Industrial Facilities

The applicant discussed the potential hazards associated with the nearby industrial facilities to the proposed IIFP facilities in the ISA Summary, Section 1.2.4 (IIFP, 2012b). Three gas-fueled, electric-generating power plants and a gas-processing facility are located near the proposed

IIFP site: (1) Xcel Energy Cunningham Station has four units 1.6 km (1.0 mi) west of the site, (2) Xcel Energy Maddox Station is 3.5 km (2.2 mi) east-southeast of the site, (3) Colorado Energy Hobbs Generating Station is 3.1 km (1.9 mi) east-northeast of the site, and (4) DCP Midstream Linam Ranch Plant is 5.8 km (3.16 mi) southeast of the site (IIFP, 2012b).

The Xcel Energy Cunningham Station consists of two steam-electric generating units and two natural gas combustion turbine units. The Xcel Energy Maddox Station consists of a natural gas-fired, steam-electric generating unit and two natural gas combustion turbine units. The Colorado Energy Hobbs Generating Station uses a combined-cycle technology, including both a gas combustion turbine and a steam-driven turbine. DCP Midstream Linam Ranch Plant is a natural gas processing facility.

The applicant pointed out that the potential hazards to the proposed IIFP facility from these industrial facilities are facility explosions caused by the natural gas at the facilities themselves and an aircraft crash into the facilities, facility fires, and power outages (IIFP, 2012b). The applicant considered the fire and power outage events resulting from the nearby industrial facilities in its ISA, and suitable measures and controls are identified to alleviate the potential impact (IIFP, 2012b). The review of the fire protection measures are addressed further in Chapter 7, "Fire Safety," and the review of the electrical systems are addressed in Appendix D, "Electrical Systems" of this SER.

The applicant stated that the explosions caused by an aircraft crash into a nearby industrial facility would be highly unlikely events. The applicant made the assessment based on the analysis of a potential aircraft crash into the IIFP facility. Even if an aircraft were to crash into a nearby facility, the induced explosion would not be a close enough distance to damage the IIFP facility. Based on its analysis of a natural gas pipeline explosion at the Cunningham Power Station (closest to the IIFP facility compared to other nearby facilities), the applicant determined that the natural gas explosion would not affect the operation's safety of the IIFP facility because of the distance between these facilities and the proposed IIFP facility. Consequently, the applicant concluded that no industrial facilities are close enough to the proposed facility site to pose a safety concern regarding explosions to the IIFP facility operations.

The NRC staff reviewed the information regarding the nearby industrial facilities and finds that the applicant provided sufficient information to support evaluation of potential hazards associated with the nearby industrial facilities. The NRC staff finds that the applicant's conclusion that explosion-related hazards do not pose a safety concern to the proposed facility is acceptable because these facilities are located at a substantial distance from the IIFP facility, avoiding any appreciable damage, according to Regulatory Guide 1.91, Revision 1 (NRC, 1978). The NRC staff also finds that the applicant adequately considered the nearby facility-related fire and power outage accident events in its ISA. Based on this review, the NRC staff concludes that the applicant has met the regulatory requirements in 10 CFR 70.65(b)(1).

1.3.3.2.3   Nearby Gas Pipelines

The applicant discussed the nearby gas pipeline hazards in the ISA Summary, Section 1.2.4.2 (IIFP, 2012b). The applicant stated that, based on the easement records of Lea County, one petroleum gas pipeline and 12 natural gas pipelines are located within 1.6 km (1 mi) of the proposed IIFP site. The pipeline diameter and gas pressure data are known for all but two pipelines. Among these pipelines, the largest pipeline diameter is 30.5 cm (12 in), and the largest gas pressure is 10 megapascals (MPa) (1,500 psi). IIFP verified that these values are conservative and bound the diameter and pressures of the unknown piblines (IIFP, 2012b).

The applicant determined the design basis for external explosion-induced overpressure caused by dynamic effects to be 6.9 kPa (1 psi) for the IIFP process building. The applicant's approach to assessing the potential hazards of a gas pipeline explosion was to determine the annual probability of a nearby gas pipeline explosion (following rupture of the pipelines and subsequent detonation) that could generate an overpressure greater than the design-basis, blast-induced overpressure for the IIFP process building using the approach suggested in Regulatory Guide 1.91, Revision 1 (NRC, 1978). The approach in Regulatory Guide 1.91, Revision 1 (NRC, 1978), requires the following information: (1) explosion rate of gas pipelines in question in explosions per mile, (2) number of gas pipelines, and (3) exposure distance in miles. In conducting the analysis, the applicant estimated the explosion rates per year per pipeline mile for natural gas pipelines and petroleum gas pipelines using the gas pipeline safety data from the DOT Web site ("Data & Statistics" at http://phmsa.dot.gov/pipeline/library/data-stats).

The applicant estimated the annual probabilities for explosions of the natural and petroleum gas pipelines to be less than 3×10^{-6}. Therefore, the applicant concluded that gas pipeline explosion accidents are highly unlikely near the proposed IIFP facility.

The NRC staff reviewed the discussions the applicant provided on gas pipeline hazards near the proposed IIFP facility and finds that the applicant used reliable DOT data to estimate the gas pipeline explosion rates. The NRC staff finds that the applicant's assumptions used to determine the average release rate, produce a conservative result. The NRC staff also finds that use of the Area Locations of Hazardous Atmospheres (ALOHA) computer code to calculate blast radius of a pipeline explosion is acceptable because the ALOHA computer code, which the U.S. Environmental Protection Agency endorses, is commonly used by the industry. In addition, the NRC staff finds that the applicant estimated the exposure distance of a gas pipeline consistent with that defined in Regulatory Guide 1.91, Revision 1 (NRC, 1978). Furthermore, the NRC staff finds that the applicant's approach of estimating annual weighted exposure distance for each pipeline using the annual frequency of the atmospheric conditions in the area is acceptable because the annual frequency of the atmospheric conditions was determined based on local specific data (IIFP, 2012b). Therefore, the NRC staff concludes that the applicant acceptably determined the annual probability of a gas pipeline explosion that could result in an overpressure greater than the design basis for a process building and that nearby gas pipelines are not a safety concern to the proposed IIFP facility. Based on this review, the NRC staff concludes that the applicant has met the regulatory requirements in 10 CFR 70.65(b)(1).

1.3.3.2.4   Air Transportation

The applicant discussed aircraft crash hazards to the proposed IIFP facility in the ISA Summary, Section 1.2.4.3 (IIFP, 2012b). Seventeen airports are located within 161 km (100 mi), but more than 8 km (5 mi) away from, the IIFP site. Among these airports, Lea County Regional Airport, Hobbs Industrial Airpark, and Lea County Zip Franklin Memorial Airport are within 32 km (20 mi) of the IIFP site. The Lea County Regional Airport is located approximately 13 km (8 mi) east-southeast, the Hobbs Industrial Airpark is 13.7 km (8.5 mi) east-northeast, and the Lea County Zip Franklin Memorial Airport is 27 km (17 mi) north-northwest of the proposed IIFP site.

There are four military routes within a 48 km (30 mi) radius of the proposed IIFP site, with the closest approach approximately 27 km (17 mi) southwest of the IIFP site (IIFP, 2012b). The number of military operations at the Lea County Regional Airport is 561 annually. In addition, a special-use airspace for two military operations areas (MOAs) are located north of the IIFP site. The closest edge of the MOA is approximately 9.3 km (5.8 mi) from the IIFP facility.

In assessing the probability of aircraft crash hazards, the applicant (IIFP, 2012b) used the first proximity acceptance criterion provided in NUREG-0800, "Standard Review Plan for the Review of Safety Analysis Reports for Nuclear Power Plants: LWR Edition," Section 3.5.1.6, "Aircraft Hazards" (NRC, 2010). NUREG-0800 provides three proximity acceptance criteria. If all three criteria are met, the probability of aircraft accidents with potential radiological consequences is considered to be less than about 1×10^{-7} per year. The first proximity criterion states that, for the aircraft crash hazards related to the airport operations to be insignificant to the site, the annual number of operations for the airports with a distance (1) between 8 and 16 km (5 and 10 mi) from the site of interest should be less than $500 \times D^2$ and (2) more than 16 km (10 mi) from the site should be less than $1,000 \times D^2$ (NRC, 2010). Note that D is the distance between the airport of interest and the IIFP site. In assessing the probability of hazards associated with airport operations near the IIFP facilities, the applicant used the first proximity acceptance criterion. By comparing the annual number of operations with the first proximity acceptance criterion, the applicant determined that the annual number of operations for all of the 17 airports located within 161 km (100 mi) of the IIFP site is less than the maximum aircraft activities stipulated in the first proximity acceptance criterion (IIFP, 2012b). Therefore, the applicant concluded that, consistent with NUREG-0800 (NRC, 2010), the probability of aircraft accidents with potential radiological consequences is considered to be less than 1×10^{-7} per year.

So that military-related aircraft activities do not affect a site of interest, the second proximity acceptance criterion states that the facility of interest should be at least 8 km (5 mi) from the nearest edge of military training routes, including low-level training routes. Facilities outside this range may still be affected if there are greater than 1,000 flights per year or where activities (such as practice bombing) may create an unusual stress situation (NRC, 2010). As the applicant indicated (IIFP, 2012b), the nearest military route is approximately 27 km (17 mi) southwest, and the closest edge of the MOA is approximately 9.3 km (5.8 mi) from the IIFP site. Therefore, the applicant determined that the activities associated with the military training route and the MOA are not a safety concern to the IIFP site. In addition, the applicant indicated that the number of military operations (561 annually) at the Lea County Regional Airport, which is 13 km (8 mi) east-southeast of the IIFP site, is less than 1,000 flights per year as specified in the second proximity acceptance criterion. Consequently, the applicant concluded that military operations are not a safety concern to the IIFP facility.

The third proximity acceptance criterion in NUREG-0800, Section 3.5.1.6 (NRC, 2010), states that the facility of interest should be at least 3 km (2 mi) beyond the nearest edge of a Federal airway, holding pattern, or approach pattern. The applicant indicated that four high-level airways en route are within 56 km (35 mi) of the proposed IIFP facility, with the closest airway (Q20) passing at 16.7 km (10.4 mi) from the IIFP facility (IIFP, 2012b). The applicant further indicated that three low-level airways en route pass through the nearby air space, with the closest airway (V68) passing at 5.1 km (3.2 mi) of the IIFP facility. Based on the information regarding the airway distances from the IIFP facility, the applicant determined that all of the Federal airways meet the third proximity acceptance criterion. For further confirmation, the applicant conducted additional analysis to calculate the annual probability of an aircraft on the V68 airway crashing onto the proposed IIFP facility area using the methods from both NUREG-0800, Section 3.5.1.6 (NRC, 2010), and DOE-STD-3014-2006, "Accident Analysis for Aircraft Crash into Hazardous Facilities" (DOE, 2006). The annual aircraft crash probability is 2.7×10^{-8} using the NUREG-0800 method and 3.3×10^{-7} using the DOE method. Based on these probabilities, the applicant determined that an aircraft crash from Federal airways onto the IIFP facility is highly unlikely. Therefore, the applicant concluded that the nearby Federal airways are not a safety concern to the IIFP facility.

For the holding and approach patterns, the applicant evaluated air traffic at the Lea County Regional Airport, Hobbs Industrial Airpark, and Lea County Zip Franklin Memorial Airport, which are within 32 km (20 mi) of the IIFP site (IIFP, 2012b). The applicant stated that Lea County Regional Airport has seven instrument flight rule and two visual flight rule procedures for holding and approach patterns, and all holding and approach patterns are more than 3 km (2 mi) from the IIFP site. The closest landing and takeoff would be 12.1 km (7.5 mi) from the site. The applicant indicated that the Hobbs Industrial Airpark has no air carrier, general aviation, or military operations. The only operations at the Airpark are from 32 airpark-based aircraft. This Airpark does not have instrument flight rule procedures or specific holding patterns. The applicant estimated that an aircraft could come within 6 km (4 mi) of the IIFP site during a visual landing approach. For the Lea County Zip Franklin Memorial Airport, the applicant evaluated the airport's instrument flight rule and visual flight rule procedures and determined that holding and approach patterns would be more than 30 km (18 mi) from the IIFP site. Based on the evaluations discussed in this paragraph, the applicant concluded that the holding and approach patterns for the nearby airports are more than 3 km (2 mi) away from the IIFP site and satisfy the third proximity acceptance criterion in NUREG-800 (NRC, 2010),. Therefore, the applicant concluded that aircraft crash hazards do not pose any safety concern to the IIFP facility.

The NRC staff reviewed the information the applicant presented on aircraft crash hazards to the IIFP facility. The NRC staff finds that the applicant's use of the proximity acceptance criteria specified in NUREG-0800, Section 3.5.1.6 (NRC, 2010), to assess the aircraft crash hazards to the IIFP facility is acceptable because these criteria are an NRC-accepted approach to determining potential aircraft crash hazards to nuclear power plants in the United States. The NRC staff verified the applicant's evaluation of airport operations on the three airports nearest to the IIFP site—Lea County Regional Airport, Hobbs Industrial Airpark, and Lea County Zip Franklin Memorial Airport—using the airport operations information from the Federal Aviation Administration's, "Air Traffic Activity System (ATADS): Airport Operations," at http://aspm.faa.gov/opsnet/sys/Airport.asp and from Airnav.com's, "Airport Information," at http://www.airnav.com/airports/.

The NRC staff concludes that the applicant's acceptably demonstrated that the operations of the airports near the IIFP facility site do not present a safety concern to the facility by showing that the annual number of operations at the nearby airports is less than the maximum operation number, as specified in the first proximity acceptance criterion in NUREG-0800 (NRC, 2010). The NRC staff also finds that the applicant provided sufficient information to demonstrate that the nearest edge of military training routes, including low-level training routes, meets the second proximity acceptance criterion. The aircraft holding and approach patterns at various airports are of sufficient distance from the IIFP site to satisfy the third proximity criterion. Based on this review, the NRC staff concludes that the aircraft crash hazards are not a safety concern to the proposed IIFP facility. The NRC staff finds that the applicant has met the regulatory requirements in 10 CFR 70.65(b)(1).

1.3.3.2.5   Nearby Bodies of Water

The applicant indicated that there are two playas on the site, but no significant bodies of water or rivers are within 80 km (50 mi) of the site (IIFP, 2012a, b). The applicant stated that the nearest river to the proposed site is Pecos River approximately 80 km (50 mi) south to southeast from and at an elevation approximately 213 m (700 ft) below the proposed site. The nearest dams are Brantley Dam, approximately 98 km (61 mi), and Avalon Dam, approximately 106 km (66 mi) from the proposed site. Both dams are at an elevation more than 168 m (550 ft) below the site. Because of this information, the NRC staff determined—based on the large distances to the nearest bodies of water—as described in the LA, that such bodies of water do not present a credible disruptive event for the proposed IIFP facility. The NRC staff finds the applicant's conclusions acceptable that there are no intermediate– or high–consequence events related to nearby bodies of water. The NRC staff finds that the applicant has met the regulatory requirements in 10 CFR 70.65(b)(1).

1.3.3.3   Demographics

There is no residential population or recreational areas within the 5-mile radius of the site. The nearest residential center and the closest town to the site is Hobbs, located in Lea County, 19 km (12 mi) east of the site on U.S. Highway 62/180. Hobbs is the largest city in Lea County with a population of 28,657 in the 2000 census. The tri-county area surrounding the site includes Lea County, New Mexico; Gaines County, Texas; and Andrews County, Texas. Lea County has the largest population with 67,727 in the 2010 census, and the three counties had a combined population of approximately 100,000 in 2010 compared to 83,000 in the 2000 census. Individuals nearest the facility include IIFP employees onsite, a local resident 2.74 km (1.7 mi) northwest of the site, employees who work at the four industrial facilities within 5 miles of IIFP, and occasional ranchers or oil field workers.

The industrial facilities located within 5 miles of the IIFP consist of three power plants and one natural gas recovery facility. The power stations include Cunningham Station on the facilities west boundary, Colorado Energy Station 3.1 km (1.9 mi) northeast, and Xcel Energy Maddox Station 3.5 km (2.2 mi) east. DCP Midstream Linam Ranch Plant is a natural gas recovery facility which operates 5.8 km (3.6 mi) southeast of the plant. IIFP projects a shift workforce of approximately 140 employees spread over three 8-hour shifts.

The nearest local school facilities, daycare, nursing homes and hospitals are located in Hobbs, NM. The educational institutions include three colleges, a high school and an alternative high school, three middle schools, and twelve elementary schools—as well as two private schools. The Lea Regional Medical Center is the nearest hospital at 31.4 km (19.5 mi) from the site and has a capacity of 221-licensed beds. There are no school facilities or hospitals located within 8 km (5 mi) of the facility.

The applicant provided a summary of demographic information based on the most recent census data that showed the population distribution as a function of distance from the proposed facility. The applicant's descriptions are consistent with the more detailed information in the ISA Summary (IIFP, 2012b), the ER (IIFP, 2011a), and the EP (IIFP, 2012c). The NRC staff finds that the applicant has met the regulatory requirements in 10 CFR 70.65(b)(1).

1.3.3.4 Meteorology

The LA (IIFP, 2012a) and Section 1.3 of the ISA Summary (IIFP, 2012b) provide a meteorological description of the site and its surrounding area. The information includes several tables on prevailing winds and precipitation. The descriptions also address severe weather, thunderstorms, lightning, wind, and snow. The section provides the bounding meteorological conditions for the local area of the proposed facility, and the ISA Summary contains the evaluation of the potential high and intermediate accident sequences.

1.3.3.4.1 Tornado Hazard and Tornado-Generated Missiles

The applicant discussed tornadoes at the proposed IIFP site in Section 1.7.3.3 of the LA (IIFP, 2012a); Section 1.3.2.3 of the ISA Summary (IIFP, 2012b; and Section 3.6.1.6 of the ER (IIFP, 2011a). Ninety-two tornadoes were recorded in Lea County, NM, where the IIFP facility is located, during 1950–2010 (IIFP, 2012b). Among these recorded tornadoes, eight were F2 (Fujita Scale) tornadoes and one was an F3 tornado (IIFP, 2012a).

Using the methodology in NUREG/CR-4461, Revision 2, "Tornado Climatology of the Contiguous United States," issued February 2007 (NRC, 2006), the applicant estimated that the tornado strike probabilities for the 1-, 2-, and 4-degree boxes that contain the IIFP site are greater than 10^{-5} per year (highly unlikely) (IIFP, 2012b); therefore, a potential tornado hazard to the IIFP facility needs to be considered. The applicant indicated that the design-basis wind speed of 217 kilometers per hour (km/h) (135 miles per hour [mph]) is proposed for buildings and structures containing licensed material or chemicals or processes that may affect licensed material is greater than the estimated tornado wind speed corresponding to an annual probability of 10^{-5} per year. Therefore, the potential hazard from tornado to the IIFP facility does not have to be considered further. Section 1.3.3.3.2 of this SER evaluates and discusses the applicant's design-basis wind speed.

Using Figure 8-1 of NUREG/CR-4461, Revision 2 (NRC, 2006), the NRC staff verified the applicant's conclusion and determined that the estimated tornado wind speed is less than the design-basis wind speed of 217 km/h (135 mph). NUREG/CR-4461, Revision 2 (NRC, 2006), provides recommended tornado design-basis wind speed maps for the continental United States. Therefore, the NRC staff concludes that the applicant used appropriate meteorological data and methodology to assess the probability of a tornado strike. The NRC staff further concludes that the applicant appropriately determined that no special consideration of tornado hazard is necessary in the design basis for the IIFP facility. Based on this review, the NRC staff concludes that the applicant meets the requirements in 10 CFR 70.65(b)(1).

1.3.3.4.2 High Winds

The applicant discussed extreme winds at the proposed IIFP site in Section 1.7.3.3 of the LA (IIFP, 2012a) and Section 1.3.2.3 of the ISA Summary (IIFP, 2012b). The applicant estimated the extreme wind hazard curve for the IIFP site (IIFP, 2011b) using the gust wind speed data established in Table 3-2 of DOE-1020-2002, "Natural Phenomena Hazards Design and Evaluation Criteria for Department of Energy Facilities," issued in 2002 (DOE, 2002), for both Sandia National Laboratories and Los Alamos National Laboratory in New Mexico. Based on this hazard curve, the applicant selected a straight gust wind speed of 217 km/h (135 mph) as the design-basis wind speed for its IIFP process building. This gust wind speed corresponds to an annual probability of 10^{-4}.

The NRC staff reviewed the extreme wind information the applicant provided and finds that the applicant adequately characterized the extreme wind hazard at the IIFP site. The NRC staff finds the applicant's use of the gust wind speed data established in Table 3-2 of DOE-1020-2002 (DOE, 2002) for DOE's facilities to be acceptable because this standard is widely used for wind design of the structures, systems, and components of the DOE facilities to safely withstand the high wind hazard. The NRC staff also finds that the applicant's use of the straight gust wind speed of 217 km/h (135 mph) for an annual probability of 10^{-4} as a high-wind design basis is greater than that recommended in Revision 1 of the Coats and Murray (Coats and Murray, 1985) report entitled, "Natural Phenomena Hazards Modeling Project: Extreme Wind/Tornado Hazard Models for Department of Energy Sites." This provides additional conservatism in the wind protection design. Based on this review, the NRC staff finds that the applicant meets the requirements in 10 CFR 70.65(b)(1).

1.3.3.4.3   Extreme Precipitation

The applicant discussed extreme precipitation at the proposed IIFP site in Section 1.7.3.3 of the LA and Section 1.3.2.2 and 1.3.2.6 of the ISA Summary (IIFP, 2012a, b) and Section 3.6.1.3 of the ER (IIFP, 2011a). The IIFP site is located in a semiarid climate with an annual average rainfall of 31.37–40.46 centimeters (cm) (12.35–15.93 inches [in]) (IIFP, 2012b). Frequent rain showers and intense thunderstorms occur during summer and account for more than half the annual precipitation (IIFP, 2012b). The estimated average rainfall for a 1-hour event in Hobbs, NM, is 8.51–8.66 cm (3.35–3.41 in); the estimated average rain fall for a 24-hour event is 16.43–17.96 cm (6.47–7.07 in) (IIFP, 2011a).

Using the National Oceanic and Atmospheric Administration's (NOAA) precipitation data (Bonnin, et al., 2011), the applicant calculated the 1-hour, 24-hour, and 48-hour all-season precipitations for a return period of 100,000 years using a linear least-square regression technique. The estimated 1-hour, 24-hour, and 48-hour all-season precipitations are 18.3 cm (7.2 in), 36.6 cm (14.4 in), and 43.2 cm (17.0 in), respectively.

The NRC staff reviewed the applicant's information regarding extreme precipitation. The NRC staff finds that the applicant's estimation of extreme precipitations for a return period of 100,000 years is acceptable because the site-specific precipitation data the applicant used are from a reliable Federal Government agency, NOAA, and were peer reviewed. In addition, the applicant used the linear least square regression technique to estimate extreme precipitations with an annual probability of 10^{-5}, which is an acceptable method to estimate behavior that is outside the data range. Based on this review, the NRC staff finds that the applicant provided sufficient information on extreme precipitation to support conduct of the ISA and meets the requirements in 10 CFR 70.65(b)(1)

1.3.3.4.4   Flood

The applicant discussed potential flooding at the proposed IIFP site in Section 1.7.3.3 of the LA, Section 1.3.2.6 of the ISA Summary, and Section 3.4.11 of the ER (IIFP, 2012a,b and 2011a). According to the applicant, the IIFP site grade is above the 100-year flood elevation

(IIFP, 2011a). The applicant indicated that the Federal Emergency Management Agency (FEMA) classifies the IIFP site to be in FEMA Flood Zone D (IIFP, 2012b). By FEMA definition, for the areas located in Flood Zone D, area flooding is possible with undetermined flood hazard (FEMA, 2011).

No significant bodies of water or rivers are within 80 km (50 mi) of the site. Therefore, the applicant indicated that any flooding at the site would be caused by short-term precipitations resulting in flash flooding. Sixty-eight flood events were recorded in Lea County, NM, between 1950–2010. Twenty-nine of the 68 events were reported in Hobbs, NM (in FEMA Flood Zone D) (IIFP, 2012a). Hobbs, NM, is approximately 18.3 km (11.4 mi) east of the IIFP site and is at an elevation of 38–52 meters (m) [125–170 feet (ft)] lower than the IIFP site.

In general, parts of the area west and north of the IIFP site are at a slightly higher elevation than the IIFP site, and the topography gradually slopes down to the east and south directions for several miles. Therefore, the applicant stated that the IIFP site could receive drainage water from the northwest direction with a slope of approximately 0.21 percent and would discharge to the northeast, southeast, and southwest directions with a slope ranging from 0.35 to 0.46 percent (IIFP, 2012b).

The applicant conducted a preliminary flood screening analysis for the IIFP site using the guidelines in DOE-1023-95, "Natural Phenomena Hazards Assessment Criteria" (DOE, 1995) and concluded that flooding is not a design-basis event (IIFP, 2012b). In the screening analysis, the applicant assessed the flooding potential from the Pecos River and failure of the Brantley and Avalon Dams. The Pecos River is approximately 80 km (50 mi) away from and at an elevation of 213 m (700 ft) below the IIFP site. Brantley Dam is approximately 98 km (61 mi) away from and 168 m (550 ft) below the IIFP site. Avalon Dam is approximately 106 km (66 mi) away from and 192 m (630 ft) below the IIFP site. Because of the substantial distances of these features and the large differences in their elevation from the IIFP site, the applicant concluded that the Pecos River, Brantley Dam, and Avalon Dam are not flood hazards. The only plausible flooding hazard to the IIFP site is from storm water runoff during rain events. The applicant further excluded hurricanes and tsunamis from being potential flood hazards to the IIFP site because the site is approximately 805 km (500 mi) away and 1,158 m (3,800 ft) higher than the Gulf of Mexico.

To estimate the potential effects of rainfall-induced storm water runoff, the applicant estimated the 1-hour, 24-hour, and 48-hour all-season precipitation corresponding to a 100,000-year return period by extrapolating the NOAA precipitation data (Bonnin, et al, 2011) using a linear least-square procedure. Considering the 1-hour, 24-hour, and 48-hour all-season precipitation for a 100,000-year return period, the applicant estimated the maximum flood (standing water) level to be 12.2 cm (4.8 in) (IIFP, 2012a). This estimate was determined using the general site contour information with a gentle slope of 2.1 percent northwest of the site and assuming that the probable maximum area with rainfall water that might affect the process building is 65,154 square kilometers (km^2) (1.61 acres). The applicant indicated that the building and structures containing items relied on for safety (IROFS) and the plant roadway levels will be constructed at least 15.2 cm (6 in) above grade so that the potential flood water (at an estimated level of 12.2 cm [4.8 in]) will not prevent the IROFS from performing the intended safety functions.

The NRC staff reviewed the information on potential flood hazard at the IIFP site and determined that the applicant acceptably screened the potential sources for flooding hazards because the applicant conducted the assessment using reasonable distance and elevation information for the features analyzed and the IIFP site. The NRC staff also determined that the applicant's precipitation estimates for rainfall with an annual probability of 10^{-5} are acceptable because the applicant used the rainfall data from a reliable source (NOAA) and extrapolated the precipitation data via a commonly used method. In addition, the NRC staff determined that the applicant's maximum flood level is reasonable because the applicant used site-specific topographic information. Consequently, the NRC staff finds that the applicant's flood hazard assessment considering only that the site precipitation is reasonable, and flooding is not a hazard to the IIFP site. Based on this review the NRC staff finds that the applicant meets the regulatory requirements in 10 CFR 70.65(b)(1)

1.3.3.4.5   Snow

The applicant discussed snow in Hobbs, NM, in Section 1.7.3.3 of the LA (IIFP, 2012a), Section 1.3.2.7 of the ISA Summary (IIFP, 2012b), and Section 3.6.1.3 of the ER (IIFP, 2011a). The mean annual snowfall is 13.0 cm (5.1 in) recorded at the Hobbs weather station. The maximum recorded snow accumulation for Hobbs, NM, is 31.0 cm (12.2 in), and a 100-year, 2-day snowfall is 30.7 cm (12.1 in) (IIFP, 2011a).

The applicant stated that it will consider the snow loads in the design of the IIFP facility. The applicant proposed to use the sum of the snow load associated with a 100-year snowfall and the load from the estimated 48-hour all-season precipitation as the design-basis extreme environmental ground snow load. This snow load has an annual probability of 10^{-5} for the IIFP site. The applicant stated that it will use the guidance specified in American Society of Civil Engineers (ASCE) 7-05, "Minimum Design Loads for Buildings and Other Structures," to calculate roof snow load using the design-basis ground snow load for the IIFP building structures (IIFP, 2011a).

The NRC staff reviewed the applicant's snow hazard discussion for the IIFP facility and determined that the snow information is adequate for the applicant to assess potential snow hazard in its ISA. Based on this review, the NRC staff finds that the applicant meets the requirements in 10 CFR 70.65(b)(1).

1.3.3.4.6   Lightning

The applicant discussed lightning phenomena in Section 1.7.3.3 of the LA (IIFP, 2012a); Section 1.3.2.5 of the ISA Summary (IIFP, 2012b); and Section 3.6.1.6 of the ER (IIFP, 2011a). The applicant indicated that, from 1950–2004, two lightning events were reported to have caused loss of life, injury, and property damage in Lea County, NM. One occurred in Hobbs with minor property damage, and the other occurred in Lovington with two deaths.

The applicant evaluated the potential for a lightning strike to be credible at the proposed site. The applicant accounted for the lightning hazard by providing lightning protection to the IIFP facility, as indicated in Chapter 7 of the ISA Summary (IIFP, 2012b). In Table 7-1 of the ISA Summary, the applicant listed the National Fire Protection Association (NFPA) 780, "Standard for the Installation of Lightning Protection Systems," for lightning protection (NFPA, 2004).

Based on the inclusion of NFPA 780 for lightning protection of the IIFP structures, the NRC staff finds that the applicant adequately considered lightning strike hazard through lightning protection design. Based on this review, the NRC staff finds that the applicant meets the requirements in 10 CFR 70.65(b)(1).

1.3.3.4.7   Meteorology Conclusions

As discussed in Sections 1.3.3.3.1 through 1.3.3.3.6 of this SER, the applicant provided appropriate meteorological data, including a summary of design-basis values for accident analysis of maximum snow loads and probable maximum precipitation, as presented in the ISA Summary (IIFP, 2012b). The applicant also provided appropriate design-basis information for lightning, high winds, tornadoes, hurricanes, and extreme precipitation. The applicant's descriptions are consistent with the more detailed information in the ISA Summary (IIFP, 2012b) and the ER (2011a). Based on this review, the NRC staff finds that the applicant meets the requirements in 10 CFR 70.65(b)(1).

1.3.3.5 Hydrology

Site surface water and groundwater hydrology is discussed in Sections 1.7.3.4 of the LA (IIFP, 2012a); Section 1.4 of the ISA Summary (IIFP, 2012b); and 3.4.11 of the ER (IIFP, 2011a).

The applicant stated that there is a small stream running through the property. This stream is predominantly dry during the year (IIFP, 2012b). There are also two playas on the site, but no significant bodies of water or rivers are within 80 km (50 mi) of the site (IIFP, 2012a, b). The applicant stated that surface water is lost through evaporation, resulting in high salinity conditions in the waters in and soils associated with the playas. Because IIFP is located in a semi-arid region, evaporation and transpiration dominate resulting minimal surface water occurrence or groundwater recharge.

The IIFP site sits on the Ogallala (High Plains) Aquifer underground reservoir system. This aquifer provides water supply to the region, including the proposed IIFP site. The water table of this aquifer generally varies from 15 to 90 m (50 to 200 ft) below ground. Specifically, for the IIFP site, the Ogallala Aquifer is at a depth of 9 to 47 m (30 to 150 ft). The applicant stated that, although the Ogallala Aquifer contains hard water, the quality of the water is good. The applicant further indicated that withdrawals from the aquifer exceed recharge to it. Therefore, the Ogallala Aquifer is being depleted.

The applicant stated the IIFP site is above the elevation of the 100-year and the 500-year flood plains. The IIFP site storm system will be designed to accommodate a 100-year return period precipitation event.

The NRC staff reviewed the information provided in the LA (IIFP, 2012a), ISA Summary (IIFP, 2012b), and the ER (IIFP, 2011a). Based on the review of the information concerning site hydrology, the information provided by the applicant meets the regulatory acceptance criteria in Sections 1.3.4.3(4) and (5) of NUREG-1520 (NRC, 2002) and is, therefore, acceptable because: (1) the information is accurate and is from reliable sources; and (2) the applicant provides design-basis information for hydrological conditions applicable to the site. In addition, the NRC staff reviewed the applicant's hydrological data in the ISA Summary (IIFP, 2012b) and ER (IIFP, 2011a) and finds that they provide sufficient information to assess site flooding hazards and to assess the ground and surface water impacts. Based on this review, the NRC staff finds that the applicant meets the requirements in 10 CFR 70.65(b)(1).

1.3.3.6 Site Geology

The applicant summarized the geology of the IIFP site in Section 1.7.4.1 of the LA (IIFP, 2012a); Section 1.5.1 of the ISA Summary, (IIFP, 2012b); and Section 3.3 of the ER (IIFP, 2011a).

1.3.3.6.1   Physiography

As the applicant documented in the aforementioned references, the IIFP site is located within the Pecos Valley section of the Great Plains Physiographic Province of North America. The site is near the southwestern margin of the Llano Estacado; a flat, semiarid table and that covers 97,000 km^2 (37,500 square miles [mi^2]) of western Texas and eastern New Mexico. The tableland rises in elevation from 900 m (3,000 ft) in the southeast to 1,500 m (5,000 ft) in the northwest. Erosion of the tableland forms a prominent topographic escarpment along its western and northwestern edge called the Mescalero Ridge, which lies just to the west of the IIFP site. The Llano Estacado is composed of porous late Tertiary age fluvial gravels and eolian sands of the Ogallala Formation, which are covered by late Pleistocene and Holocene dark-brown to reddish brown sands, sandy loams, and clay loams (Trimble, 1980). The Ogallala Caprock, an erosion-resistant caliche unit, holds up the tableland. Based on this review, the NRC staff finds that the applicant meets the requirements in 10 CFR 70.65(b)(1).

1.3.3.6.2   Basin Geology

The proposed IIFP site is located within the Central Basin Platform of the Permian Basin, between the Midland and Delaware sub-basins. The Permian Basin is a downward flexure of a large thickness of originally flat-lying bedded, sedimentary rock. Downward flexure and overall development of the Permian Basin is a relic of geological activity that is estimated to have taken place between 65 and 350 million years ago. The top of the Permian section is approximately 434 m (1,425 ft) below ground surface. The Permian Basin strata are overlain by sedimentary rocks of the Triassic age Dockum Group. The upper formation of the Dockum Group is the Chinle Formation, locally overlain by the Tertiary Ogallala, Gatuña, or Antlers Formations or by Quaternary alluvium. Based on this review, the NRC staff finds that the applicant meets the requirements in 10 CFR 70.65(b)(1).

1.3.3.6.3   Soils

As discussed in the ER (IIFP, 2011a), soils for this part of Lea County generally comprise shallow to deep gravelly and loamy horizons to deep sandy soils formed from eolian and fluvial processes in the recent geologic past. Because of the arid environment, zones of soft and hard caliche are common. Loamy and sandy soils have low shrinking and swelling potential, which limits the impact that soil swelling and shrinkage may have on building foundations. The NRC staff determined that the applicant has made a conservative assumption based on this description that the soil profiles at the site will include several meters of loose eolian sands above dense to very dense and partially cemented sands and silts.

The applicant commits to implement a site-specific geotechnical and geophysical investigation and analysis plan (soil characterization plan) as described in Section 3.2.4.3 of the LA (IIFP, 2012a) to obtain accurate values for the soil liquefaction, settlement, bearing capacity and static and dynamic soil properties based on soil measurements at the IIFP site. The NRC staff reviewed the soil characterization plan and the commitments to implement the plan consistent with the appropriate ASTM standards listed in Section 3.1.5.3 of the LA (IIFP, 2012a). The NRC determined the commitments to implement the site characterization plan are adequate to obtain accurate data for the detailed design. The NRC staff concludes that commitments to construct the facility consistent with the data obtained using the soil characterization plan committed to by IIFP in the LA (IIFP, 2012a) meet the baseline design criteria for natural phenomena hazards described in 10 CFR 70.64(a)(2) and the content requirement in 70.65(b)(1), and are, therefore, acceptable.

The applicant's geotechnical and geophysical investigation and analysis plan, as described in the LA, Section 1.7.4 (IIFP, 2012a), requires the applicant to acquire additional detailed information on the site soil class, seismic site response, liquefaction potential, soil settlement potential, and allowable bearing capacity for the IIFP site (see ISA Summary, Section 1.5.4 [IIFP, 2012b]). The field activities include dilatometer soundings, seismic cone penetration tests, crosshole seismic tests, standard penetration tests, auger borings, and soil sampling. The applicant commits to implement the geotechnical and geophysical investigation analysis plan in accordance with the appropriate American Society for Testing and Materials (ASTM) standards listed in the LA, Section 3.1.5 (IIFP, 2012a). The NRC staff reviewed the plan and determined its implementation will provide accurate data. The applicant's commitments to refine the soil calculations using site-specific data further supports the NRC staff's finding that IIFP meets the baseline design criteria in 10 CFR 70.64 and the content requirement in 70.65(b)(1).

The NRC staff reviewed the site geology description for the IIFP site in accordance with Section 1.3 of NUREG-1520, Revision 0 (NRC, 2002), and finds that the applicant adequately described and summarized the geological information needed to support evaluations of the potential natural hazards that could impact the safe operation of the proposed facility. The information on which the applicant relied is consistent with the NRC staff's understanding of the site's geologic conditions based on information provided by the applicant, as well as information available from the geologic literature. The review verified that the site description is consistent with the information used as a basis for the LA, ER, and ISA Summary. Based on this review, the NRC staff finds that the applicant has met the requirements in 10 CFR 70.65(b)(1).

1.3.3.6.4   Seismic Hazard

The applicant described seismic hazards of the IIFP site in Section 1.7.4.2 of the LA (IIFP, 2012a); Section 1.5.2 of the ISA Summary (IIFP, 2012b); and Section 3.3.1 of the ER (IIFP, 2011a).

The NRC staff reviewed the following areas concerning the seismic hazard, which are applicable to the safety analysis and design of the proposed IIFP facility:

- seismic source characterization

- seismic hazard calculation

- site-specific hazard spectra

- surface faulting

1.3.3.6.5   Seismic Source Characterization

Geological and Tectonic Settings

As noted in the ER (IIFP, 2011a), the area around Hobbs, NM, is seismically quiet and structurally stable. The Laramide Orogeny (late Cretaceous to Early Tertiary time) uplifted the region to its present elevation, and no substantial tectonic activity has occurred since this early Tertiary deformation. The Permian Basin has subsided slightly since the Laramide Orogeny. However, this subsidence is believed to have resulted from dissolution of the Permian evaporite deposits by groundwater or possibly compaction from the extraction of oil and gas deposits. No active faults have been identified at the site, and the only evidence of past active faulting is related to the development of the Permian Basin and the Laramide Orogeny. The nearest evidence of Quaternary faulting (fault displacements within approximately the last 2 million years) is 161 km (100 mi) west of the site in the Basin and Range tectonic province.

Historic Seismicity

Earthquakes in the region of the proposed IIFP site include isolated and small clusters of low-to-moderate magnitude events toward the Rio Grande Valley of New Mexico and in Texas, southeast of the proposed site. According to the applicant, no earthquakes in the site region are known to be correlated to specific faults, and many of the small magnitude events are thought to be related to water injections and withdrawal used for secondary recovery operations in oil fields in the Central Basin Platform area.

The largest earthquake known to occur within 322 km (200 mi) of the site was the August 16, 1931, earthquake near Valentine, TX. This earthquake had an estimated body wave magnitude (m_b) of 5.6 to 6.4 and produced a maximum Modified Mercalli (MM) intensity of VIII (Dumas, 1980). This earthquake occurred approximately 260 km (160 mi) from the proposed IIFP site location. Within 80 km (50 mi) of the site, the largest earthquake was a local magnitude (M_L) 4.7 event in 1992, which occurred approximately 50 km (30 mi) southeast of Hobbs, NM. No injuries and only minor damage were reported from this event (*The New York Times*, 1992).

The ER (IIFP, 2011a) provides a catalog of the largest earthquakes to occur in New Mexico since 1869. The catalog was compiled by Yarger (2009) at the New Mexico Institute of Mining and Technology and includes 30 events with estimated magnitudes between 4.5 and 5.5 (all magnitudes in the catalog are reported as duration magnitudes, which closely approximate M_L). These earthquakes have maximum estimated MM intensity values of VI to VIII. MM values of VIII can result in considerable damage to poorly constructed structures, but only slight damage to specially designed structures.

The NRC staff reviewed the site geology description for the IIFP site in accordance with Section 1.3 of NUREG-1520, Revision 0 (NRC, 2002), and finds that the applicant adequately described and summarized the seismic source information needed to support the seismic hazards evaluation. The information relied upon by the applicant is consistent with the NRC staff's understanding of the potential seismic sources in eastern New Mexico and western Texas, based on information from the geologic literature. The review verified that the description of the site is consistent with the information used as a basis for the LA, environmental review, and ISA Summary. Based on this review, the NRC staff finds that the applicant meets the requirements in 10 CFR 70.65(b)(1).

1.3.3.6.6   Seismic Hazard Calculation

Table 1-8 of the LA (IIFP, 2012a) and Table 1-3 of the ISA Summary (IIFP, 2012b) provide the seismic hazard criteria for the proposed IIFP facility. The criteria specify peak ground accelerations for 2,500-year return periods based on the U.S. Geologic Survey (USGS) 2002, National Seismic Hazard Maps (Frankel, et al., 2002). Although not specified in the LA or ISA Summary, the applicant indicated in the RAI response to RAI#SS-5-2, Revision C, page 33 (IIFP, 2011b), that the seismic design of all structures will be based on the ground motions from the USGS National Seismic Hazard Map for the 2-percent probability of exceedence in 50 years, which approximately equates to a 2,500-year return period. The applicant also noted that the specific design methodology used to ensure a sufficient design margin necessary to achieve the beyond-design-basis performance objectives will depend on the consequence classification. Buildings that contain material that have the potential for high radiological consequences to the public will be designed as PC-3 structures, as described in DOE-1020-2002 (DOE, 2002). According to the DOE standard, PC-3 structures designed to the 2,500-year return period earthquake using nuclear grade design codes and standards will retain their safety functions, even if subjected to earthquakes loads from the 10,000-year return period earthquake.

For the proposed IIFP site, the USGS hazard maps show that the 2,500-year return period peak ground acceleration is 0.12 times the acceleration of gravity (g). The 2,500-year return period spectral accelerations for 1.0 second and 0.2 second corner frequencies are 0.041 g and 0.216 g, respectively. The USGS earthquake hazard maps were developed assuming relatively firm bedrock conditions (Site Class B) with average shear-wave velocity (Vs30) in the first 30 m (100 ft) of subsoil of 760 m per second (m/s) (2,500 ft per second (ft/s)). Vs30 serves as a measure of soil rigidity.

As noted in the NRC staff's review of the Louisiana Energy Services, Inc.'s, National Enrichment Facility (NRC, 2005), the USGS results for the eastern New Mexico region may be conservative relative to other site-specific probabilistic seismic hazard assessments for southeastern New Mexico. For example, the USGS 2,500-year return period peak ground acceleration of 0.12 g for the proposed IIFP site is comparable to the 10,000-year return period ground motions of 0.15 g for the National Enrichment Facility, based on the site-specific seismic hazard assessment (LES, 2005) and the calculated 10,000-year return period peak ground acceleration of 0.15 g for the Waste Isolation Pilot Plant (DOE, 2003). Conservatism in the USGS model arises because the USGS hazard calculation used larger maximum magnitudes and seismicity rates than are generally supported by the historical or geological record. For example, the USGS seismic source model includes regional earthquakes with magnitudes of 6.0. These are large in contrast to the historical earthquake catalogs for New Mexico and west Texas, such as the one compiled by the applicant, which show that the largest recorded earthquakes have magnitudes of 5.5 or less.

The NRC staff reviewed the seismic hazard calculation for the IIFP site in accordance with Section 1.3 of NUREG-1520, Revision 0 (NRC, 2002). The NRC staff finds that the applicant adequately described and summarized the geological information needed to support evaluations of the seismic hazards. The information relied on by the applicant is consistent with the NRC staff's understanding of the seismic hazard potential in eastern New Mexico and western Texas based on information provided by the applicant, as well as information available from the geologic literature. The review verified that the seismic hazard calculation is consistent with the information used as a basis for the LA, ER, and ISA Summary. Based on this review, the NRC staff finds that the applicant has met the regulatory requirements in 10 CFR 70.65(b)(1).

1.3.3.6.7   Development of Site-Specific Spectra

As described above, the applicant used published soil characterization data from the region of the proposed facility to determine the geotechnical properties for the site response calculation at the time of this LA review. The applicant's commitment to use a methodology consistent with ASCE/SEI 7-05 and DOE-STD-1020-2002, as described in the following paragraphs, provides an adequate basis for regulatory compliance.

The NRC staff requested through RAIs (IIFP, 2011b) that the applicant commit to use conservative values for the soil-specific spectra and commit to a geotechnical and geophysical investigation and analysis plan for how the applicant would verify and refine the soil data. IIFP provided commitments to implement the soil characterization plan in the LA, Section 1.7.4.1 (IIFP, 2012a), and provided additional information on the sources for the soil data in LA, Section 1.7.4.2 (IIFP, 2012a). Thus, the NRC staff has confidence that the soil properties being used for the site-specific spectra are conservative and will be further refined during the detailed design.

In response to NRC staff's RAIs, the applicant provided a preliminary description of the site soil conditions (IIFP, 2011b). Based on its preliminary visits to the proposed IIFP facility site and discussions with Pettigrew Associates (a licensed engineering firm) located in Hobbs, NM, the applicant stated that the soil conditions at the site consist of approximately a 0.6 m- (2-ft) thick layer of soft caliche over hard caliche rock. Based on this soil profile, the conceptual design site soil class is either class B or class C. Because the USGS hazard information uses a site soil class of B, this class was used by IFFP. Therefore, no modification to the site hazard values will be needed if the final site geotechnical investigations confirm site soil class B conditions. If the geotechnical investigations to which the applicant committed reveal soil site class C conditions, then small amplifications of the hazard spectra are anticipated. In the RAI response, the applicant provided an analysis of the amplification of the design spectra, following the methodology in ASCE/SEI 7-05, (ASCE, 2006). This analysis indicated spectral amplification factors of between 1.2 and 1.7. In the LA, Section 3.2.4.3 (IIFP, 2012a), the applicant committed to follow the applicable portions of DOE-STD-1020-2002 for a PC-3 facility in order to develop the amplification factors that will be used to develop the seismic inputs for the seismic design. This approach to site response is consistent with the applicant's approach to determine seismic design criteria for IROFS, including the process buildings, following the guidance in DOE-STD-1020-2002 (DOE, 1995).

The NRC staff reviewed the seismic hazard calculation for the IIFP site, in accordance with NUREG-1520, Revision 0, Section 1.3 (NRC, 2002). The NRC staff finds that the applicant adequately described and summarized the site soil information, based on its preliminary assessment, to support evaluations of the seismic site response. The applicant has committed to keep the site soil data updated and incorporate new information into the detailed design as appropriate. The applicant also committed to developing a site response analysis consistent with their approach to determine the seismic design criteria per the guidance in DOE-STD-1020-2002 (DOE, 2002). The information on which the applicant relied is consistent with the NRC staff's understanding of the seismic hazard potential in eastern New Mexico and western Texas based on information the applicant provided, as well as information available from the geologic literature. The review verified that the seismic site response calculation is consistent with the information used as a basis for the LA (IIFP, 2012a), ISA Summary

(IIFP, 2012b), and ER (IIFP, 2011a). Based on this review of commitments made by the applicant to implement a written, site-specific plan to verify the soil characteristics; implement the soil characterization plan consistent with industry standards; and use the measurements to design and build the facility in accordance with accepted industry standards, the NRC staff finds that the applicant meets the requirements in 10 CFR 70.65(b)(1).

1.3.3.6.8   Surface Faulting

There is no geologic, geophysical, or seismological evidence of active surface faulting in the vicinity of the proposed facility site. As stated in the ISA Summary (IIFP, 2012b), the nearest recent faulting is located more than 161 km (100 mi) west of the site. Therefore, the applicant did not identify surface faulting as a credible disruptive event for the proposed IIFP facility. Based on this review, the NRC staff finds that the applicant meets the requirements in 10 CFR 70.65(b)(1).

1.3.3.6.9   Slope Stability

The applicant discussed the site topography in Section 1.1.1 of the LA (IIFP, 2012a) and Section 1.1.1 of the ISA Summary (IIFP, 2012b). The applicant stated that the IIFP site is relatively flat with slight undulations in elevation (IIFP, 2012a). In general, the topography of the IIFP site gradually slopes down to the east and south directions for several miles (IIFP, 2012b).

Based on the applicant's description of the IIFP site topography, the NRC staff finds that slope stability should not be a safety concern at the IIFP facility. Therefore, the NRC staff finds that the applicant meets the requirements in 10 CFR 70.65(b)(1).

1.3.3.6.10   Liquefaction and Soil Settlement

At the time of this LA review, the soil liquefaction, settlement, and bearing capacity for the proposed IIFP site were not currently available. The applicant developed a geotechnical and geophysical investigation and analysis plan and commits to acquiring necessary information to determine the liquefaction potential, soil settlement potential, and allowable bearing capacity of the soil for the IIFP site (see ISA Summary, Section 1.5.4 [IIFP, 2012b]). The field activities are to include dilatometer soundings, seismic cone penetration tests, crosshole seismic tests, standard penetration tests, auger borings, and soil sampling. The applicant indicated that it will conduct the geotechnical and geophysical investigation in accordance with appropriate ASTM standards listed in the LA, Section 3.1.5 (IIFP, 2012a).

The NRC staff reviewed the information the applicant provided regarding its plan to verify soil liquefaction, settlement, and bearing capacity for the proposed IIFP site. The NRC staff finds that the activities the applicant proposed for the geotechnical and geophysical investigation and analysis are reasonable and will enable the applicant to collect relevant data to verify the site liquefaction potential, soil settlement potential, and allowable bearing capacity of the soil for the IIFP site during implementation of the geotechnical and geophysical investigation plan. The NRC staff also finds that the ASTM standards (IIFP, 2012a) the applicant plans to use for conducting the geotechnical and geophysical investigation are accepted standards in the industry and appropriate for determining site class, seismic site response, soil liquefaction, settlement, and bearing capacity. Therefore, the NRC staff finds that the geotechnical and geophysical investigation and analysis plan will provide accurate data regarding the soil liquefaction, settlement, and bearing capacity for the IIFP site. The NRC staff finds that the applicant's commitments to incorporate the actual site soil characterization into the facility design, consistent with industry standards, provides adequate assurance of safety. Implementation of this plan will be monitored and verified by the NRC's construction inspection staff. Consequently, the NRC staff finds that the applicant meets the requirements in 10 CFR 70.65(b)(1).

1.3.3.6.11   Site Geology Conclusions

As discussed in Sections 1.3.3.5.1 through 1.3.3.5.10 of this SER, the NRC staff finds that the applicant provides adequate information on soil, faulting, and seismicity of the proposed site as presented in the LA (IIFP, 2012a) and ISA Summary (IIFP, 2012b). In addition, the NRC staff finds that the site soil characteristics are based on conservative values for their region in southeastern New Mexico. The applicant also provided appropriate seismic design basis information. The applicant provided sufficient information on the site geotechnical and geophysical investigation program that will be conducted to confirm the design of the IIFP and a commitment to update the design, as needed, once the measured site soil characterization becomes available. The applicant's descriptions are consistent with the more detailed information in the ISA Summary (IIFP, 2012b) and the ER (IIFP, 2011a). The information the applicant provided is consistent with the guidance in Sections 1.3.4.3 of NUREG-1520, Revision 0 (NRC, 2002), and is, therefore, acceptable.

1.3.4   Site Description Conclusions

The NRC staff has reviewed the site description for IIFP according to Section 1.3 of the Standard Review Plan. The applicant has adequately described and summarized general information pertaining to (1) the site geography, including its location relative to prominent natural and man-made features such as mountains, rivers, airports, population centers, schools, and commercial and manufacturing facilities; (2) population information on the basis of the most current available census data to show population distribution as a function of distance from this proposed facility; (3) meteorology, hydrology, and geology for the site; and (4) applicable design basis events. The reviewer verified that the site description is consistent with the information used as a basis for the ER, Emergency Management Plan, and ISA Summary.

The NRC staff finds that the site description, as discussed above, is consistent with the acceptance criteria in NUREG-1520 Chapter 1 (NRC, 2002) and meets the regulatory requirements in 10 CFR 40.32 and 10 CFR 70.65.

1.4   References

(ASCE, 2006) American Society of Civil Engineers, "Minimum Design Loads for Buildings and Other Structures," ASCE/SEI 7–05, Reston, VA: American Society of Civil Engineers, 2006.

(Bonnin, et al., 2011) Bonnin, M., D. Martin, B. Lin, T. Parzybok, M. Yekta, and D. Riley, "Precipitation Frequency Atlas of the United States: NOAA Atlas 14, Volume 1: Semiarid Southwest (Arizona, Southeast California, Nevada, New Mexico, Utah)," Version 5.0, Silver Spring, MD: U.S. Department of Commerce, National Oceanic and Atmospheric Administration, National Weather Service, 2011.

(Coats and Murray, 1985) Coats, D.W., and R.C. Murray, "Natural Phenomena Hazards Modeling Project: Extreme Wind/Tornado Hazard Models for Department of Energy Sites," Revision 1, CA: Lawrence Livermore National Laboratories, 1985.

(DOE, 2006) U.S. Department of Energy, "Accident Analysis for Aircraft Crash into Hazardous Facilities," DOE STD 3014 2006, Reaffirmed May 2006.

(DOE, 2003) U.S. Department of Energy, "Waste Isolation Pilot Plant Contact Handled (CH) Waste Safety Analysis Report," DOE/WIPPS95S2065, Rev. 7, 2003.

(DOE, 2002) U.S. Department of Energy, "Natural Phenomena Hazards Design and Evaluation Criteria for Department of Energy Facilities," DOE STD 1020 2002, 2002.

(DOE, 1995) U.S. Department of Energy, "Natural Phenomena Hazards Assessment Criteria," DOE 1023 95, 1995, Reaffirmed with Errata April 2002.

(Dumas, et al., 1980) Dumas, B.B., H.J. Dorman, and G.V. Latham, "A Reevaluation of the August 16, 1931 Texas Earthquake," Bulletin of the Seismological Society of America, Vol. 70, pp. 1,171–1,180, 1980.

(FEMA, 2011) Federal Emergency Management Agency, "Definitions of FEMA Flood Zone Designations," http://www.msc.fema.gov, 2011.

(Frankel, et al., 2002) Frankel, A.D., M.D. Petersen, C.S. Mueller, K.M. Haller, R.I.L. Wheeler, E.V. Leyendecker, R.L. Wesson, S.C. Harmsen, C.H. Cramer, D.M. Perkins, and K.S. Rukstales, "Documentation for the 2002 Update of the National Seismic Hazard Maps," U.S. Geological Survey Open File Report 02 420, Reston, VA: U.S. Geological Survey, 2002.

(ICRP, 1977) International Commission on Radiological Protection, "Recommendations of the Internaltional Commission on Radiological Protection," Publication 26, 1977.

(ICRP, 1982) International Commission on Radiological Protection, "Limits for Intakes of Radionuclides by Workers," Publication 30, 1982.

(ICRP, 1991) International Commission on Radiological Protection, "1990 Recommendations of the International Commission on Radiological Protection," Publication 60, 1991.

(ICRP, 1994a) International Commission on Radiological Protection, "Dose Coefficients for Intakes by Workers," Publication 68, 1994.

(ICRP, 1994b) International Commission on Radiological Protection, "Human Respiratory Tract Model for Radiological Protection," Publication 66, 1994.

(ICRP, 1994c) International Commission on Radiological Protection, "Dose Coefficients for Intakes by Workers," Publication 68, 1994.

(ICRP, 1995a) International Commission on Radiological Protection, "Age dependent Doses to Members of the Public from Intake of Radionuclides   Part 4 Inhalation Dose Coefficients," Publication 71, 1995.

(ICRP, 1996) International Commission on Radiological Protection, "ICRP Publication 72: Age dependent Doses to the Members of the Public from Intake of Radionuclides Part 5, Compilation of Ingestion and Inhalation Coefficients," Publication 72, 1996.

(ICRP, 1997) International Commission on Radiological Protection, "Individual Monitoring for Internal Exposure of Workers (preface and glossary missing)," Publication 78, 1997.

(IIFP, 2012a) International Isotopes Fluorine Products, Inc., "Fluorine Extraction Process and Depleted Uranium Deconversion Plant (FEP/DUP) License Application, Revision B," May 2012, Agencywide Documents Access and Management System (ADAMS) Accession No. ML12123A245.

(IIFP, 2012b) International Isotopes Fluorine Products, Inc., "ISA Summary Rev. B for IIFP," May 2012, Agencywide Documents Access and Management System (ADAMS) Accession No. ML12123A245.

(IIFP, 2012c) International Isotope Fluorine Products, Inc., "Emergency Plan Rev B of IIFP License Application," May 2012, Agencywide Documents Access and Management System (ADAMS) Accession No. ML12123A245.

(IIFP, 2011a) International Isotopes Fluorine Products, Inc., "Fluorine Extraction Process and Depleted Uranium Deconversion Plant (FEP/DUP) Environmental Report," Rev. A, 2011, Agencywide Documents Access and Management System (ADAMS) Accession No. ML100120758.

(IIFP, 2011b) International Isotopes Fluorine Products, Inc., "Official Response to Seismic and Structural RAIs," Rev. C, September 2011, Agencywide Documents Access and Management System (ADAMS) Accession No. ML11263A259.

(IIFP, 2009) International Isotopes Fluorine Products, Inc., "Memorandum of Agreement Between the International Isotopes, Inc., and the New Mexico Environment Department." October 2009. Agencywide Documents Access and Management System (ADAMS) Accession No. ML093080290

(LES, 2005) Louisiana Energy Services, "Integrated Safety Analysis Summary," Rev. 4, Albuquerque, NM, 2005.

(NFPA, 2004) National Fire Protection Association, "Standard for the Installation of Lightning Protection Systems," NFPA 780 2004, Quincy, MA: National Fire Protection Association, 2004.

(NRC, 2011) U.S. Nuclear Regulatory Commission, "Environmental Impact Statement for the Proposed Fluorine Extraction Process and Depleted Uranium Deconversion Plant in Lea County, New Mexico – Draft Report for Comment," NUREG–2113, 2002 (ADAMS) Accession No. ML093080290
(Note: Final NUREG 2113 will be published after this SER is finalized in fall 2012)

(NRC, 2010) U.S. Nuclear Regulatory Commission, "Standard Review Plan for the Review of Safety Analysis Reports for Nuclear Power Plants: LWR Edition," Chapter 3, "Design of Structures, Components, Equipment, and Systems," Section 3.5.1.6, Rev. 4, "Aircraft Hazards," NUREG 0800, 2010.

(NRC, 2006) U.S. Nuclear Regulatory Commission, "Tornado Climatology of the Contiguous United States," NUREG/CR 4461, Rev. 2, 2006.

(NRC, 2005) U.S. Nuclear Regulatory Commission, "Safety Evaluation Report for the National Enrichment Facility in Lea County, New Mexico," NUREG 1827, 2005.

(NRC, 2002) U.S. Nuclear Regulatory Commission, "Standard Review Plan for the Review of a License Application for a Fuel Cycle Facility," NUREG 1520, Revision 0, 2002.

(NRC, 1978) U.S. Nuclear Regulatory Commission, "Evaluations of Explosions Postulated To Occur on Transportation Routes Near Nuclear Power Plants," Regulatory Guide 1.91, Rev. 1, February 1978.

(The New York Times, 1992) "Light Earthquake Sends Tremor through Texas and New Mexico," The New York Times, http://www.nytimes.com/1992/01/03/us/light earthquake sends tremor through texas and new mexico.html, January 3, 1992.

(Trimble, 1980) Trimble, D.E., "The Geologic Story of the Great Plains," Geological Survey Bulletin 1493, Washington, DC: U.S. Geological Survey, 1980.

(Yarger, 2009) Yarger, F., "Seismic Probability in Lea County, NM: A Brief Analysis." Socorro, NM: New Mexico Institute of Mining and Technology, New Mexico Center for Energy Policy, 2009.

2.0   ORGANIZATION AND ADMINISTRATION

The purpose of the U.S. Nuclear Regulatory Commission's (NRC's) review of the International Isotopes Fluorine Products, Inc.'s (IIFP or the applicant), organization and administration is to ensure that the proposed management hierarchy and policies will provide reasonable assurance that the licensee plans, implements, and controls site activities in a manner that ensures the safety of workers and the public and protects the environment. The review also ensures that the applicant has identified and provided adequate qualification descriptions for key management positions.

2.1   Regulatory Requirements

The requirements in Title 10 of the *Code of Federal Regulations* 10 CFR 40.32, "General Requirements for Issuance of Specific Licenses," pertain to the applicant's training, experience and its proposed equipment, facilities and procedures for the effective implementation of health, safety, and environment functions. Effectively implementing these functions will better ensure adequate health and safety for workers and the public and protection of the environment, and not be inimical to the common defense and security.

2.2   Regulatory Guidance and Acceptance Criteria

Chapter 2 of NUREG-1520, Revision 0, "Standard Review Plan for the Review of a License Application for a Fuel Cycle Facility" (NRC, 2002), contains the guidance applicable to NRC's review of the organization and administration section of the license application (LA) (IIFP, 2012a). Although this guidance was written to address fuel cycle facilities regulated under 10 CFR Part 70, the requirements of 10 CFR Part 40, "Domestic Licensing of Source Material," and 10 CFR Part 70 are sufficiently similar that the majority of the criteria are applicable to the IIFP project. Section 2.4.3, "Regulatory Acceptance Criteria," of NUREG-1520, Revision 0 (NRC, 2002), lists acceptance criteria for new and existing fuel cycle facilities. Section 2.4.3 of NUREG-1520, Revision 0 (NRC, 2002), contains the acceptance criteria used for this review.

2.3   Staff Review and Analysis

Chapter 2 of the applicant's LA (IIFP, 2012a) describes the facility's organization and administration policies. These policies are designed to maintain a safe work place, protect the public, and comply with applicable regulations. The safety and quality assurance (QA) organizations are run independently to ensure that safety remains a priority. The International Isotopes Inc.'s (INIS), parent corporation has a quality management system which seeks to comply with the International Organization for Standardization's international standard for quality management, issued in 2000, ISO 9001, "Quality Management System Standard" (ISO, 2000). In addition, IIFP has a dedicated QA Program, defined in Appendix A to the LA (IIFP, 2012a). The purpose of these programs is to ensure that IIFP operations, products, and services comply with the regulatory requirements and are safe and reliable.

This chapter provides an overview of the facilities management structure and describes the responsibilities and qualifications required for key management functions. Figures 2-1 and 2-2 of the LA (IIFP, 2012a) provide flow charts to illustrate the chain of command both during the design and build (DB) phase of construction and the startup and operations phase. For both phases, the top tier of authority consists of the INIS Board of Directors, followed by the INIS

President, the IIFP President, and the IIFP Chief Operations Officer (COO). The LA describes the roles and qualifications of these and subordinate managers, which are summarized below.

2.3.1 Organization

As described in Chapter 2 of the LA (IIFP, 2012a), the INIS President and Chief Executive Officer (CEO) establish corporate policy and provide direction for all corporate activities. Under the INIS President, the IIFP President establishes policy, directs overall operations, and ensures that QA and safety procedures are disseminated throughout the entire facility. The IIFP COO serves under the IIFP President and oversees the Environmental Safety and Health (ESH) Manager and a QA Manager. During construction activities, the COO will oversee the facility design, engineering, and contractors. The COO will assign an Engineering Manager to oversee the day-to-day responsibility for design and construction. Once activities transition to operations, the COO will delegate authority for day-to-day operations to a Plant Manager.

During the design and construction phase of operations, the INIS corporate QA and ESH management will assist to establish the IIFP QA and ESH Programs, under the direction of the IIFP COO. The IIFP QA and ESH Managers will be hired before construction to ensure that a period of overlap exists to facilitate a smooth transition from corporate QA and ESH management to IIFP QA and ESH management.

During transition from construction to operations, the top level of the organizational structure will continue to be led by the IIFP President at the top, followed by the COO, followed by the independent positions of an Administration Manager, QA Manager, ESH Manager, Plant Manager, Engineering Manager and Training Manager (see LA Figure 2-2 [IIFP, 2012a]). Additional supervisors will serve under these managers to oversee operations.

The NRC staff concludes that the IIFP administration has a clearly defined management structure, which stretches from the IIFP President down to the supervisor level. The IIFP President has ultimate responsibility for the facility's safety, security, and the environment. The applicant described the management structure for both the DB and operations phases. The organizations have clear, unambiguous management structure and line communication, as illustrated in the LA Figures 2-1 and 2-2 (IIFP, 2012a). The description of the organizational groups responsible for facility oversight and management meet NRC requirements in 10 CFR 40.32(b) and comply with the guidance in Sections 2.4.3 of NUREG-1520, Revision 0 (NRC, 2002). Therefore, the NRC staff finds them acceptable.

2.3.2 Pre-operational Testing and Initial Startup

In Section 2.1.3 of the LA (IIFP, 2012a), the applicant described the plans for transition from the DB phase to the operations phase. The applicant will update the management structure for operations to include an Administration Manager, QA Manager, ESH Manager, Plant Manager, Engineering Manager, and Training Manager (see the LA Figure 2-2 [IIFP, 2012a]). Additional subordinate supervisors will oversee operations.

The applicant will ensure that these key management positions are filled in advance of startup to facilitate a smooth transition from construction to operations. Before operations, the COO and the Plant Manager will develop a transition plan for subordinate departments. The plan will address the turnover of physical systems, design information, records, and as-built drawings from the design build team (see the LA Figure 2-1 [IIFP, 2012a]) to the operations team (see the LA Figure 2-2 (IIFP, 2012a).

Before startup, the facility systems and equipment will undergo an acceptance review and testing in accordance with the QA Program (IIFP, 2012b). The testing will include verification of items relied on for safety (IROFS). Once the systems have successfully passed the QA review, the transition plan will be implemented to place the system into operations under the direction of the Plant Manager and the "Plant Operation Organization," as described in Section 2.1.4 of the LA (IIFP, 2012a). The acceptance review will cover the system design, function, records, and as-built drawings.

The NRC staff's review determined that the applicant committed to develop specific plans to transition systems from the construction phase to startup and operations. The transition plan will be reviewed consistent with the QA Program (IIFP, 2012b). The description of the transition plan provides adequate management and organizational controls to ensure that operations will meet NRC regulatory requirements. The information the applicant provided meets the guidance in Section 2.4.3 of NUREG-1520, Revision 0 (NRC, 2002), and is, therefore, acceptable.

2.3.3 Organizational Responsibilities and Qualifications

Section 2.2 of the LA (IIFP, 2012a) provides information concerning the minimum qualifications, functions, and responsibilities for key staff positions. This information is summarized below.

Corporate Employees

The President/CEO is responsible to the INIS Board of Directors and oversees all corporate activities. The President/CEO ensures that ESH and QA Programs are established and operated consistently throughout all INIS corporate operations. The President/CEO must have, as a minimum, a bachelor's degree in a scientific field or business and 8 years of experience in management in a related field (chemical, radiological, or nuclear facilities).

The corporate ESH and QA Managers support the development of the ESH and QA Programs at the IIFP facility, in addition to ensuring consistent programs throughout the parent and subsidiary corporate organizations. These corporate managers provide advice, oversight, and consultation to assist the IIFP QA and ESH Managers in implementing regulatory compliance. These individuals must have a minimum of a bachelor's degree in engineering or a scientific field and at least 5 years of related experience.

IIFP Facility Employees

The IIFP President reports directly to the INIS President and is ultimately responsible for all operations, safety, and security of the IIFP facility. Although corporate ESH and QA managers may provide support to the IIFP managers, the IIFP President is responsible for ensuring that the IIFP ESH and QA Programs implement, maintain, and distribute sound policies and procedures throughout the facility. The IIFP President must have, as a minimum, a bachelor's degree in a scientific field and at least 5 years of management experience in a related field.

The IIFP President appoints the COO to manage the overall safety and operations of the facility. This includes oversight of design, construction, and startup. Once operations begin, the COO will oversee hiring of key management, oversight of operations, and proper implementation of the ESH and QA Programs, among others. The COO has plant shutdown authority and grants this authority to appropriate line managers. The COO must review and approve restart activities. The COO must have, as a minimum, a bachelor's degree in a scientific field and 5 years of relevant experience.

The Engineering Manager oversees all construction and engineering activities during the DB phase of operations and works under the direction of the COO. The Engineering Manager oversees the engineering projects, reviews design changes, implements the engineering policies and procedures, and tests the operability of systems. During startup and operations, this individual will oversee all engineering-related activities and configuration management. The Engineering Manager must have, as a minimum, a bachelor's degree in an engineering field and a minimum of 5 years of relevant experience.

A Plant Engineering and Maintenance Manager will work under the Engineering Manager to oversee maintenance support for the facility. This individual will be responsible for supervising day-to-day engineering services and corrective maintenance. The individual must have, as a minimum, a bachelor's degree in engineering or science and at least 5 years of experience in a related field. Additional support staff, including the Maintenance Supervisors and Configuration Management Lead, will work under the direction of the Engineering Manager.

The INIS President and Board of Directors will select a DB contractor to design and build the IIFP facility. The DB contractor will report to the COO during construction. The DB contractor, under the direction of the IIFP managers, will be responsible for implementing all Federal, State, and local regulatory requirements. The DB contractor will have experience in work at related facilities (chemical, radiological or nuclear facilities).

The QA Manager is independent of other organization structures and reports directly to the COO. The QA Manager is responsible for establishing and maintaining the IIFP QA Program. The individual provides support to line managers and staff to ensure proper implementation of the QA Program throughout the facility. This individual has the authority to report issues directly to the IIFP President and to shut down operations, if needed. The QA Manager must have, as a minimum, a bachelor's degree in engineering or science, with at least 5 years of relevant experience in the implementation of a QA Program at a related facility.

The ESH Manager is responsible for overseeing plant safety, compliance with regulatory requirements, and environmental protection. This is accomplished through oversight of five separate organizations, described below. The ESH Manager reports directly to the COO. This individual has the authority to report issues directly to the IIFP President and to shut down operations, if needed. The ESH Manager oversees the Configuration Management Program to ensure that facility changes are reviewed by the safety and licensing organization and incorporated into facility documentation. The ESH Manager must have, as a minimum, a bachelor's degree in engineering, science, or a related field and 5 years of relevant experience in a related field.

The ESH Manager oversees five subordinate organizations, including the Fire Protection, Radiation Protection, Emergency Preparedness/Security, Industrial Hygiene/Safety and Environmental Programs. The ESH Manager is responsible for the effective coordination of these subordinate organizations:

- The Fire Protection Manager is responsible for developing and maintaining fire protection plans and procedures and ensuring their implementation in day-to-day operations.

- The Radiation Protection Manager ensures development and implementation of the Radiation Protection Program, including training staff and radiation monitoring.

- The Emergency Preparedness/Security Lead develops the Emergency Plan and Security Plan, oversees implementation of emergency procedures, plans exercises, and assesses the effectiveness of those plans.

- The Industrial Hygiene/Safety Lead ensures industrial safety through, training, audits, monitoring work place conditions, inspections, and compliance with the Occupational Safety and Health Administration's regulations.

- The Environmental Program Lead oversees effluent monitoring, regulatory compliance, an audit program, and recordkeeping.

Each of these submanagers is required to have, as a minimum, a bachelor's degree in an engineering or scientific field with between 2 and 5 years of applicable experience, as defined in Section 2.2 of the LA.

The Plant Manager will report directly to the COO and oversee the daily operations. These responsibilities will include day-to-day compliance with regulatory requirements and implementation of internal procedures. This individual will have the authority to shut down operations and must approve the restart of the facility. The Plant Manager must have, as a minimum, a bachelor's degree and 5 years of experience in a related field. Several types of subordinate managers will serve under the Plant Manager:

- The Facility Safety Engineer will oversee implementation of the integrated safety analysis (ISA), including IROFS.

- The Shift Superintendents will coordinate operations across the facility and direct facility operations on the backshift (e.g., evenings, weekends, holidays, etc.).

- The Area Shift Supervisors will oversee individual crews and implementation of procedures.

A Training Manager will report to the COO during the design and construction phase of the facility to assist the engineering department in developing operation procedures and a training program for new employees. Once operations begin, the Training Manager will oversee the training program as well as the document control system. The Training Manager must have, as a minimum, a bachelor's degree in science or engineering and a minimum of 5 years of relevant experience.

The Facility Safety Review Committee (FSRC) provides oversight and review of proposed changes to procedures and facility systems, including items in the ISA. The FSRC also oversees the audit programs and receives input from the As Low As Reasonably Achievable (ALARA) Safety Committee in matters related to radiation protection. The Committee reports directly to the COO, and the COO appoints the chairperson. The members of the FSRC are drawn from the safety and operations groups throughout the facility and have as a minimum a bachelor's degree in engineering or science and at least 3 years of applicable experience at a chemical or nuclear facility. Seven years of experience may be substituted for the bachelor's degree.

Based on its review, the NRC staff finds that the applicant has provided clear, unambiguous management controls and lines of communication and authority within the organization for managing the design, construction, and operation of the facility. The qualifications and responsibilities of key individuals involved in health and safety oversight are defined. The individuals have authority to shut down operations if conditions warrant. The proposed organization summarized in the LA, Chapter 2, provides an acceptable management system for ensuring that the design, construction, and operation of the facility will meet NRC regulatory requirements. The information the applicant provided meets the applicable acceptance criteria in Section 2.4.3 of NUREG-1520, Revision 0 (NRC, 2002), and is, therefore, acceptable.

2.3.4 Procedures

Managers are responsible for developing and implementing written procedures for their respective areas of responsibility. Procedures are used to define management measures and highlight operations which may be impacted by IROFS. Individuals are required to be trained in accordance with the written procedures. Non-routine tasks will be conducted under temporary orders or radiation work permits. The commitment to use procedures, as described throughout the LA and, in particular, Section 2.3.4, "Procedures" (IIFP, 2012a), meets the applicable acceptance criteria in Section 2.4.3 of NUREG-1520, Revision 0 (NRC, 2002), and is, therefore, acceptable.

2.3.5 Incidents and Items of Concern

All individuals working on the IIFP site have the responsibility and right to report to a supervisor conditions that are a safety or health concern. The supervisor is responsible for evaluating the report and suspending work on the task, if appropriate, or addressing the concern. Supervisors are encouraged to seek assistance from the ESH organization. The NRC staff has the right and responsibility to raise unresolved concerns to higher levels of management, up through the ESH Manager. As described in Section 2.3.7 "Employee Concerns," of the LA (IIFP, 2012a), additional methods of reporting potential unsafe conditions include the following:

- unusual event report to immediate supervisor

- direct contact with the ESH organization

- notification of facility supervisors and management

- notification of a member of a safety committee

- contacting IIFP facility managers

The ESH organization will conduct audits and inspections to self-identify potential safety concerns. Walk downs and inspections will be conducted frequently based on the safety concerns. The ESH organization will conduct a formal audit focused on safety and regulatory compliance quarterly.

Incidents or accidents will be formally investigated in accordance with the QA Program. Management will organize a qualified team of personnel from the plant to evaluate the root causes and recommend corrective actions. The corrective actions will be implemented, followed through to resolution, and documented.

Based on its review, the NRC staff finds that the applicant provided several methods for IIFP employees to report safety concerns to management. In addition, the applicant will conduct regular reviews of the facility, including audits, to self-identify potential problems. Items identified and accidents will be followed up with corrective actions. The information the applicant provided meets the applicable acceptance criteria in Section 2.4.3 of NUREG-1520, Revision 2 (NRC, 2002), and is, therefore, acceptable.

2.4   Evaluation Findings

The NRC staff has reviewed the organization and administration for the IIFP facility according to Chapter 2 of NUREG-1520, Revision 0 (NRC, 2002). The application provides (1) clear responsibilities and associated resources for the design, construction, operations, and modifications of the facility and (2) a plan for managing the project. The application adequately describes the management structure during both construction and operation of the facility. The flow charts promote a basic understanding of the chain of command. The qualifications required for key managers are also addressed. The NRC staff has reviewed these plans and commitments and concludes that they provide reasonable assurance that an acceptable organization, detailed administrative policies, and sufficient qualified resources have been established or are committed to satisfy the applicant's commitments for the design, construction, operations, and modifications of the facility. Therefore, the NRC staff concludes that the LA demonstrates compliance with the requirements in 10 CFR 40.32(b) and (c) and provides a reasonable assurance of safety.

2.5   References

(IIFP, 2012a) International Isotopes Fluorine Products, Inc., "Fluorine Extraction Process and Depleted Uranium Deconversion Plant (FEP/DUP) License Application, Revision B," May 2012, Agencywide Documents Access and Management System (ADAMS) Accession No. ML12123A245.

(IIFP, 2012b) International Isotopes Fluorine Products, Inc., "Appendix A QAPD Rev B of IIFP License Application," May 2012, Agencywide Documents Access and Management System (ADAMS) Accession No. ML12123A245.

(ISO, 2000) International Organization for Standardization, ISO 9001 2000, "Quality Management System Standard," 2000.

(NRC, 2002) U.S. Nuclear Regulatory Commission, "Standard Review Plan for the Review of a License Application for a Fuel Cycle Facility," NUREG 1520, Rev. 0, 2002.

3.0 INTEGRATED SAFETY ANALYSIS AND INTEGRATED SAFETY ANALYSIS SUMMARY

International Isotopes Fluorine Products, Inc. (IIFP or the applicant) is required to meet the requirements of Title 10 of the *Code of Federal Regulations* (10 CFR) Part 40, "Domestic Licensing of Source Material," which the U.S. Nuclear Regulatory Commission (NRC) is revising to require an integrated safety analysis (ISA) and ISA Summary for large facilities that are or would be licensed to possess significant quantities of uranium hexafluoride or uranium tetrafluoride. Since the NRC expected to finalize the revisions to 10 CFR Part 40 after IIFP submitted its application, the Commission directed the NRC staff, in the staff requirements memorandum to SECY-07-0146, "Regulatory Options for Licensing New Uranium Conversion and Depleted Uranium Deconversion Facilities," dated October 24, 2007 (NRC, 2007), to impose the performance requirements of Subpart H, "Additional Requirements for Certain Licensees Authorized To Possess a Critical Mass of Special Nuclear Material," of 10 CFR Part 70, "Domestic Licensing of Special Nuclear Materials," as part of the licensing basis for the application review. In anticipation of a final revised 10 CFR Part 40, the applicant conducted an ISA in accordance with the regulations in 10 CFR Part 70, Subpart H, and submitted the ISA Summary along with its license application (LA) to the NRC.

The NRC's review of the applicant's ISA and ISA Summary determined whether IIFP addressed appropriate hazards through independent analysis or qualitative evaluations of accepted engineering practices. The review also determined whether the applicant designated acceptable items relied on for safety (IROFS), management measures, and likelihoods and consequences for higher risk accident sequences and, whether, with IROFS, the applicant met the requirements of 10 CFR 70.61," Performance Requirements," which the NRC is incorporating into 10 CFR Part 40.

In particular, the review described in this chapter considered information provided by the applicant that is related to the following:

- commitments regarding the applicant's safety program, including the ISA, in accordance with the requirements of 10 CFR 70.62, "Safety Program and Integrated Safety Analysis," which the NRC is incorporating into 10 CFR Part 40

- the ISA Summary submitted in accordance with the requirements of 10 CFR 70.62(c)(3)(ii) and 10 CFR 70.65, "Additional Content of Applications," which the NRC is incorporating into 10 CFR Part 40

3.1 Regulatory Requirements

The following regulatory requirements of 10 CFR Part 70, which the NRC is incorporating into 10 CFR Part 40, apply to the ISA and ISA Summary content:

- 10 CFR 70.62(a), which requires the establishment and maintenance of a safety program, including performance of an ISA, that demonstrates compliance with the performance requirements of 10 CFR 70.61

- 10 CFR 70.62(c), which specifies requirements for conducting an ISA, including a demonstration that credible high-consequence and intermediate-consequence events meet the safety performance requirements of 10 CFR 70.61

- 10 CFR 70.64, "Requirements for New Facilities or New Processes at Existing Facilities," which outlines requirements for baseline design criteria (BDC) and facility and system design and facility layout

- 10 CFR 70.65(b), which describes the contents of an ISA Summary including identification of IROFS and management measures.

The regulations in 10 CFR 70.62 require an applicant to establish and maintain a safety program that demonstrates compliance with the performance requirements of 10 CFR 70.61. The safety program is required to contain (1) process safety information, (2) an ISA, and (3) management measures. The applicant must conduct and maintain the ISA, which must identify the following, in accordance with 10 CFR 70.62(c):

- radiological hazards related to possessing or processing licensed material at the facility

- chemical hazards of licensed material and hazardous chemicals produced from licensed material

- facility hazards that could affect the safety of licensed materials and thus present an increased radiological risk

- potential accident sequences caused by process deviations or other events internal to the facility and credible external events, including natural phenomena

- consequences and likelihood of occurrence of each potential accident sequence identified and the methods used to determine the consequences and likelihoods

- each IROFS identified in accordance with 10 CFR 70.61(e); the characteristics of its preventative, mitigative, or other safety function; and the assumptions and conditions under which the item is relied upon to support compliance with the performance requirements of 10 CFR 70.61

As required by 10 CFR 70.61, the ISA must evaluate compliance with performance requirements. The requirements in 10 CFR 70.61(b) specify that the risk of each credible, high-consequence event must be limited such that the likelihood of occurrence is highly unlikely. In addition, the requirements in 10 CFR 70.61(c) specify that the risk of each credible, intermediate-consequence event must be limited such that the likelihood of occurrence is unlikely.

The LA must describe the safety program developed under 10 CFR 70.65(a). In addition, the applicant is required to submit to the NRC an ISA Summary. As outlined in 10 CFR 70.65(b), the ISA Summary is required to contain the following elements:

- a general description of the site with emphasis on those factors that could affect safety

- a general description of the facility with emphasis on those areas that could affect safety, including an identification of the controlled area boundaries

- a description of each process analyzed in the ISA in sufficient detail to understand the theory of operation, and, for each process, the hazards that were identified in the ISA in accordance with 10 CFR 70.62(c)(1)(i)–(iii) and a general description of the types of accident sequences

- information that demonstrates the licensee's compliance with the performance requirements of 10 CFR 70.61, including a description of the management measures, and, if applicable, the requirements of 10 CFR 70.64, "Requirements for New Facilities or New Processes at Existing Facilities"

- a description of the team, qualifications, and the methods used to perform the ISA

- a list briefly describing each IROFS identified under 10 CFR 70.61(e) in sufficient detail to understand their functions in relation to the performance requirements of 10 CFR 70.61

- a description of the proposed quantitative standards used to assess the consequences to an individual from acute chemical exposure to licensed material or chemicals produced from licensed materials which are onsite, or expected to be onsite, as described in 10 CFR 70.61(b)(4) and 10 CFR 70.61(c)(4)

- a descriptive list that identifies all IROFS that are the sole item preventing or mitigating an accident sequence that exceeds the performance requirements of 10 CFR 70.61

- a description of the definitions of unlikely, highly unlikely, and credible, as used in the evaluations in the ISA

3.2   Regulatory Guidance and Acceptance Criteria

Chapter 3 of NUREG-1520, "Standard Review Plan for the Review of a License Application for a Fuel Cycle Facility," (NRC, 2002), contains the guidance applicable to the NRC's review of the applicant's ISA and ISA Summary (IIFP, 2012b). These sections are applicable in their entirety, except for the sections dealing with criticality safety and NUREG-1520,Section 3.4.3.2(5)(b)(i-ix) (NRC, 2002) regarding process hazard analysis methods. Because the applicant used the What-if method described in NUREG-1513, "Integrated Safety Analysis Guidance Document," (NRC, 2001), IIFP does not have to address the conditions in Section 3.4.3.2(5)(b)(i–ix). Sections 3.4.3.1 and 3.4.3.2 of NUREG-1520 (NRC, 2002) contain the acceptance criteria applicable to this review.

3.3   Staff Review and Analysis

This section contains the NRC staff's programmatic review of the applicant's proposed safety program, proposed ISA commitments, proposed ISA methodology, proposed BDC, and proposed defense-in-depth practices.

3.3.1   Safety Program and Integrated Safety Analysis Commitments

The NRC staff reviewed the applicant's proposed safety program commitments identified in Section 3.1 of the LA (IIFP, 2012a) to determine whether the three safety program elements (process safety information, ISA, and management measures) demonstrate compliance with the requirements of 10 CFR 70.62(a)–(d). The NRC staff also reviewed the applicant's proposed commitments to maintain records that demonstrate compliance with the requirements of 10 CFR 70.62(b)-(d) and records of IROFS failures, in accordance with 10 CFR 70.62(a)(2) and 10 CFR 70.62(a)(3).

3.3.1.1   *Process Safety Information*

In Section 3.1.1 of the LA (IIFP, 2012a), the applicant committed to maintaining process information addressing the following:

- the hazards of materials used or produced in the process, including information on chemical and physical properties

- technology of the process, including block flow diagrams or simplified process flow diagrams; brief outline of process chemistry; range of operating parameters (e.g., temperature, pressure, flow, and concentration); and evaluation of the health and safety consequences of potential process accidents

- equipment used in the process, which includes general information on topics such as the materials of construction, piping and instrumentation diagrams, ventilation requirements, design codes and standards employed, material and energy balances, IROFS, electrical classification, and relief system design and design basis

In Section 3.1.1 of the LA (IIFP, 2012a), the applicant commits to maintain process safety information up-to-date in accordance with the Configuration Management Program described in Chapter 11 of the LA (IIFP, 2012a). As discussed in Section 11.3.1 of this Safety Evaluation Report (SER), the applicant will use its configuration management system to control documentation and review design changes.

IIFP will submit changes to the ISA Summary to the NRC in accordance with 10 CFR 70.72(d).

Therefore, the NRC staff concludes that the above-mentioned program elements provide reasonable assurance of the following:

- Consistent with Section 3.4.3.1(1)(a) of NUREG-1520 (NRC, 2002), the LA (IIFP, 2012a) contains commitments to compile and maintain an up-to-date database of process safety information and is, therefore, acceptable.

- Consistent with Section 3.4.3.1(1)(b) of NUREG-1520 (NRC, 2002), the applicant will implement a safety program that includes procedures and criteria for changing the ISA, in accordance with 10 CFR 70.72, "Facility Changes and Change Process," and is, therefore, acceptable.

As described in Section 3.2.7 of the LA (IIFP, 2012a), the ISA (IIFP, 2012b) was performed and will be maintained by a team with expertise in engineering, process safety, safety analysis, and facility process operations. The NRC staff concludes that, consistent with 10 CFR 70.62(c)(2) and Section 3.4.3.1(1)(c) of NUREG-1520 (NRC, 2002), the ISA element of the applicant's safety program engaged personnel with appropriate experience and expertise in engineering and process operations and is, therefore, acceptable.

3.3.1.2 Integrated Safety Analysis Commitments

In Section 3.1.2 of the LA (IIFP, 2012a), the applicant stated that IIFP has conducted an ISA for each process. These analyses identified the radiological hazards, chemical hazards, potential accident sequences, consequences and likelihood of each accident sequence, and IROFS, including the assumptions and conditions under which they support compliance with the performance requirements of 10 CFR 70.61. IIFP evaluated the entire facility as part of a plant wide process hazards analysis with respect to radiological hazards. The NRC staff finds this acceptable since the ISA performed by IIFP addresses the hazards related to the safety of licensed materials and the purely chemical hazards of the separated fluoride compounds would be addressed under the Occupational Safety and Health Administration's Process Safety Management Program.

Section 3.3.2 of this SER describes the NRC staff's evaluation of the applicant's methods and criteria for implementing the ISA methodology.

In Section 3.1.2 of the LA (IIFP, 2012a), the applicant has committed to implementing programs to maintain the ISA and supporting documentation so that it is accurate and up to date. The applicant will submit changes to the ISA Summary to the NRC in accordance with 10 CFR 70.72(d)(1)–(3). Based on this commitment, the NRC staff has determined the ISA update process to be adequate and in accordance with the requirements of 10 CFR 70.62(a) for adequately maintaining the ISA Summary up to date.

Section 3.1.2 of the LA states that personnel used to update and maintain the ISA and ISA Summary (IIFP, 2012b) are trained in the ISA methods. Section 11.3.3, "Training and Qualification," of this SER describes the general training and qualification of personnel used to update or maintain the ISA. The ISA Summary (a nonpublic document) lists the ISA team members and their areas of expertise, qualifications, and experience. The NRC staff finds these commitments to be consistent with the requirements of 10 CFR 70.62(a) for adequately maintaining the ISA Summary up to date.

In Section 3.1.2 of the LA (IIFP, 2012a), the applicant commits to evaluating proposed changes to the facility or its operations using the ISA methods described in the ISA Summary. New or additional IROFS and appropriate management measures are designated as required. The adequacy of existing IROFS and associated management measures are promptly evaluated by IIFP staff to determine whether they are impacted by changes to the facility or its processes. The NRC staff finds compliance with this commitment along with the requirements of 10 CFR 70.72(d)(1)-(3) will result in IROFS being sufficiently available and reliable to ensure compliance with performance requirements of 10 CFR 70.61.

In Section 3.1.2 of the LA (IIFP, 2012a), the applicant commits to addressing unacceptable performance deficiencies associated with IROFS. The NRC staff finds compliance with this commitment along with the requirements of 10 CFR 70.62 will result in IROFS being sufficiently available and reliable to ensure compliance with performance requirements of 10 CFR 70.61.

Section 3.1.1 of the LA (IIFP, 2012a) contains a commitment to maintain the ISA and its supporting documentation by the Configuration Management Program described in Chapter 11 of the LA. The NRC staff finds this commitment to be consistent with Section 3.4.3.1(2)b of NUREG-1520 (NRC, 2002).

Section 3.2.7 of the LA (IIFP, 2012a) contains a commitment to train personnel in the facility's ISA methods and use suitably qualified personnel to update and maintain the ISA and ISA Summary. The NRC staff finds this commitment to be consistent with Section 3.4.3.1(2)c of NUREG-1520 (NRC, 2002).

Section 3.1.2 of the LA (IIFP, 2012a) contains a commitment to evaluate proposed changes to the facility or its operations by means of the ISA methods and to designate new or additional IROFS and appropriate management measures. The NRC staff finds this commitment to be consistent with Section 3.4.3.1(2)d of NUREG-1520 (NRC, 2002).

Section 3.1.2 of the LA (IIFP, 2012a) contains a commitment to address any IROFS's unacceptable performance deficiencies that are identified through updates to the ISA. The NRC staff finds this commitment to be consistent with Section 3.4.3.1(2)e of NUREG-1520 (NRC, 2002).

Section 3.1.2 of the LA (IIFP, 2012a) contains a commitment to maintain written procedures onsite. The NRC staff finds this commitment to be consistent with Section 3.4.3.1(2)f of NUREG-1520 (NRC, 2002).

Section 3.1.2 of the LA (IIFP, 2012a) contains a commitment to maintain IROFS so that they are available and reliable when needed. This commitment is consistent with Section 3.4.3.1(2)g of NUREG-1520 (NRC, 2002).

Based on the assessment provided above, the NRC staff finds the applicant's ISA commitments to be acceptable and consistent with the requirements of 10 CFR 70.62 and acceptance criteria contained in Sections 3.4.3(2)b, 3.4.3(2)c, 3.4.3(2)d, 3.4.3(2)e, 3.4.3(2)f, and 3.4.3(2)g of NUREG-1520 (NRC, 2002).

3.3.1.3   Management Measures

Section 3.1.3 of the LA (IIFP, 2012a) contains a commitment to utilize management measures to maintain the IROFS so that they are available and reliable to perform their safety functions when needed. Management measures are used to ensure the availability and reliability of IROFS on a continuous basis and for any planned changes made to the facility under 10 CFR 70.72 or any unplanned changes to the facility that result from events outside the control of the applicant, such as a seismic event. Management measures ensure compliance with the performance requirements assumed in the ISA documentation. The management measures are applied to particular structures, systems, components, equipment, and activities of personnel and may be graded commensurate with the reduction of risk attributable to that IROFS. Chapter 10 of the LA (IIFP, 2012a) further details the management measures, which are evaluated in Chapter 11 of this SER.

3.3.1.4   Baseline Design Criteria

Section 3.2.4.3 of the LA (IIFP, 2012a) and Section 4.4 of the ISA Summary (IIFP, 2012b) address BDC. BDC apply to all structures, systems, and components relied on for safety, including IROFS. BDC listed in 10 CFR 70.64 for new facilities or processes to which the applicant has committed include quality standards and records; natural phenomena hazards; fire protection; environmental and dynamic effects; chemical protection; emergency capability; utility services; inspection, testing, and maintenance; and instrumentation and controls. In addition, the applicant has committed to basing the IIFP facility and system designs on defense-in-depth practices. Preference is given to engineered controls over administrative controls. For engineered controls, preference is given to passive controls over active controls. IROFS are designed to perform their intended safety functions in the event of credible (1) natural external events, (2) fire and explosion conditions, (3) dynamic effects, such as the effects of missiles and discharging fluids, and (4) chemical releases that can affect radiological safety. Based on its review, the NRC staff finds that adequate implementation of the applicant's commitments contained in Section 3.2.4.3 of the LA (IIFP, 2012a) related to BDC in the areas of quality standards and records; natural phenomena hazards; fire protection; environmental and dynamic effects; chemical protection; emergency capability; utility services; inspection, testing, and maintenance; instrumentation and control; and defense-in-depth, will ensure the needed availability and reliability of IROFS and are, therefore, consistent with the requirements of 10 CFR 70.64.

3.3.1.5   *Safety Program and Integrated Safety Analysis Commitments Conclusion*

Based on the evaluations in Sections 3.3.1.1 through 3.3.1.3, above, the NRC staff concludes that the applicant meets the requirements of 10 CFR 70.62(a)(1)–(3) to establish and maintain a safety program that includes process safety information, ISA, management measures, and appropriate safety program records. The NRC staff also concludes that the applicant has an appropriate program for establishing and maintaining records of IROFS failures that will be retrievable for NRC inspection. IIFP has also appropriately addressed the BDC for new facilities and defense-in-depth requirements of 10 CFR 70.64.

3.3.2   Integrated Safety Analysis Methodology

Section 3.2.5 of the LA (IIFP, 2012a) and Section 5.2.1 of the ISA Summary (IIFP, 2012b) describe the ISA methodology used by IIFP. The IIFP ISA uses a hazard analysis method (i.e., the What-if method) to identify the hazards relevant to each node or the IIFP facility in general. IIFP's ISA team reviewed the hazards identified for the "credible worst-case" consequences. The team first assigned the credible high- or intermediate-severity consequence accident scenarios accident description identifiers, accident descriptions, frequency, or probability and then performed a risk index determination. The risk index was used by IIFP to evaluate unmitigated risk as unacceptable or acceptable.

For each accident scenario having an unacceptable unmitigated risk index, IIFP's ISA team defined IROFS and determined the mitigated likelihood for each accident scenario. Using the unmitigated initiating event frequency and the failure probability of each IROFS, the team also determined the mitigated scenario likelihood and mitigated risk. The risk index method is regarded as a screening method for indicating the adequacy or inadequacy of the IROFS for any particular accident. A risk analysis approach is used to evaluate the credible accidents that potentially exceed the levels identified in 10 CFR 70.61. Figure 3-1, "Integrated Safety Analysis Process Flow Diagram," of the LA (IIFP, 2012a) describes the ISA process steps. The following subsections correspond to the blocks in the flow diagram.

3.3.2.1   Define Nodes To Be Evaluated

The first step of the ISA is to systematically break down the process system, subsystem, facility area, or operation being studied into well-defined nodes. The ISA nodes establish the study area boundaries in which the various process systems and supporting systems entering or exiting the node, or activities occurring in the area, can be defined to allow interactions to be studied.

The applicant divided the plant site into four types of facilities as part of the process hazards analysis (PHA) effort: the depleted uranium tetrafluoride (DUF_4) facility, the silicon tetrafluoride (SiF_4) facility, the boron trifluoride (BF_3) facility, and support facilities. Specific process operations within these facilities are separated logically into "nodes" for PHA evaluation. The PHA is broken down in this manner to reduce the complexity of the facility to a manageable level and to organize the PHA process and results in a consistent format. These nodes define process boundaries for the PHA and are unique process steps within the facility. Equipment located outside the process boundary is not evaluated in the node, although interaction between systems and potential initiating events from other systems is considered. Operations were treated in this manner so that the entire IIFP facility was evaluated in a logical process flow approach. This approach is also used to evaluate the hazards associated with each process or operation and to identify any new hazards resulting from modifications made to an existing process or operation. The applicant identified boundaries that define the point of process separation of a hazardous chemical, as well as segregation points where the release of a hazardous chemical would not adversely affect licensed materials. The NRC staff finds that the ISA Summary lists the defined nodes for the IIFP facility.

3.3.2.2   Hazard Identification

3.3.2.3   As indicated in the LA, Section 3.2.5.2 (IIFP, 2012a), the applicant used the "What-if" analysis method to identify the hazards for the IIFP process. The NRC staff finds this to be appropriate as this method is consistent with the guidance provided in NUREG-1520 (NRC, 2002) and NUREG-1513 (NRC, 2001).

According to Section 3.2.5.2 of the LA (IIFP, 2012a), the hazard identification process addresses materials with the following characteristics:

- radioactive

- flammable

- explosive

- toxic

- reactive

According to Section 3.2.5.2 of the LA (IIFP, 2012a), the hazards identification process results in identification of radiological or chemical characteristics that have the potential for causing harm to workers, the public, or the environment. The hazards of concern for the IIFP facility relate to either a release (loss of confinement) of uranium hexafluoride (UF_6) or hydrogen fluoride (HF) or chemicals that may generate HF. In general, the loss of confinement of UF_6

would initially result in the moisture in the air reacting with the UF_6 to form uranyl fluoride (UO_2F_2) and HF as byproducts. UO_2F_2 is a significant inhalation problem because of its dispersible and small particle size. HF can also be released as the byproduct of uranium tetrafluoride (UF_4) or be generated by the exposure of SiF_4 or BF_3 to air. The HF, which is in a gaseous form, and UO_2F_2 could be transported through the IIFP facility and ultimately beyond the site boundary. Both HF and UO_2F_2 are toxic chemicals with the potential to cause harm to the workers or the public, as documented in Chapter 6 of the LA (IIFP, 2012a).

For licensed material or hazardous chemicals produced from licensed materials, chemicals of concern are those that, in the event of release, have the potential to exceed consequences defined in 10 CFR 70.61(b)(4) or 10 CFR 70.61(c)(4). In the LA (IIFP, 2012a), Tables 3-1, "Consequence Severity Categories Based on 10 CFR 70.61," and 3-2, "AEGL Thresholds from the EPA for Uranium Hexafluoride, Soluble Uranium, and Hydrogen Fluoride," present the criteria for evaluating potential releases and characterizing their consequence as either "high" or "intermediate" for members of the public and facility workers.

The NRC staff agrees with the applicant's assessment that UF_6, HF or chemicals that may generate HF are the hazards of concern for NRC- licensed activities at the IIFP facility.

The applicant assessed worker exposures based on a duration of 10 minutes, which according to the applicant represents a sufficient amount of time to evacuate an area of hazardous material leak. Public exposures were estimated to last for 30 minutes. The NRC staff finds this to be consistent with self-protective criteria for UF_6/HF plumes listed in NUREG-1140, "A Regulatory Analysis on Emergency Preparedness for Fuel Cycle and Other Radioactive Material Licensees," issued in 1988 and reprinted in August 1991 (NRC, 1988). The applicant used the acute exposure guideline level values of 1, 2, and 3 as threshold concentration levels to establish a low-, intermediate-, or high-severity consequence as shown in Table 3-1 of the LA (IIFP, 2012a).

The regulation in 10 CFR 70.61(b)(3) states, in part, for a high-consequence event, "An intake of 30 mg or greater of uranium in soluble form by any individual located outside the controlled area identified pursuant to Paragraph(f) of this section...." The UF_6 concentration in air is not directly equivalent to soluble uranium intake. Therefore, IIFP used a high-consequence threshold intake value of 75 milligrams (mg). This value corresponds to the threshold for life endangerment or "acute chemical exposure," as defined in 10 CFR 70.61(b)(4). It is noted that, several years ago, based on a Nuclear Energy Institute report addressing the health effects of soluble uranium, the NRC established a working group with industry representatives to address soluble uranium intake thresholds for life endangerment, long-lasting health effects, and mild transient health effects. The working group recommended that a regulatory guide be developed on this topic. The working group also recommended that this regulatory guide address inhalation and dermal exposures to HF. The NRC staff is currently considering developing such a regulatory guide that would recommend soluble uranium and HF ISA thresholds. If and when the NRC issues new or revised ISA thresholds for soluble uranium and HF, the NRC staff would anticipate that its licensees would adopt such thresholds at that time. However, until that time, the NRC staff finds the 75 mg intake value for soluble uranium and the qualitative HF thresholds provided in Table 3-1 of the LA (IIFP, 2012a) to be acceptable. The NRC staff finds the proposed rankings and all other thresholds, for offsite public, environment and workers, provided in Table 3-1 of , to be acceptable as they are consistent with the requirements in 10 CFR 70.61.

The applicant used the What-if analysis method, which is summarized in Table 5-5, "FEP/DUP Plan What-If Checklist," of the ISA Summary (IIFP, 2012b), to identify process hazards for the UF_6, UF_4, SiF_4, and BF_3 process systems at the IIFP facility. This PHA technique is used to identify and document items identified in the hazard analysis meetings. For identified single failure events (i.e., those accidents that result from the failure of a single control), the What-if method is the recommended approach.

IIFP used this approach for the process system hazard identification. The results of the unmitigated What-if scenarios are used directly as input to the development of the risk index. In addition, the hazard identification highlights potentially hazardous process conditions. As appropriate, IIFP assessed hazards individually for the potential impact on the discrete components of the process systems, as well as on a facility-wide basis for credible hazards from fires (e.g., external to the process system) and external events (e.g., seismic, severe weather).

The applicant's What-if techniques are an acceptable approach because it is consistent with NUREG-1513 (NRC, 2001). In addition, the accident sequences that were identified involve noncomplex systems and generally occur as a result of simple failure events. Therefore, the NRC staff concludes that the PHA method used by the applicant is acceptable for the identification of potential radiological, chemical, and facility hazards; potential accident sequences caused by process deviations; or other events internal to the facility, as well as credible external events, including natural phenomena that could lead to a loss of hazardous material confinement.

3.3.2.4   Identify Accident Scenarios

Credible accident scenarios or sequences are identified by analyzing initiating events. Using approved methods, the IIFP ISA team identified potential accident scenarios, including possible worst-case consequences; the causes (events that can initiate the accident); and controls to prevent the event or mitigate the consequences. Some controls are incorporated into the design features or administrative programs to provide defense-in-depth, in addition to the IROFS. Consequences of interest include radiological material releases; radiation exposures; chemical and toxic exposures from licensed material; or hazardous chemicals produced from licensed material, fires, and explosions. Hazards are defined to be materials, equipment, or energy sources with the potential to cause injury or illness to humans or adversely impact the environment.

When analyzing accident scenarios, the IIFP ISA team considered process deviations, human errors, internal facility events, and credible external events, including natural phenomena. The ISA team used the Division of Fuel Cycle Safety and Safeguards's (FCSS's) Interim Staff Guidance (ISG)-08, "Natural Phenomena Hazards," issued in 2005 (NRC, 2005), as guidance when evaluating natural phenomena hazards as initiating events. The team evaluated common mode failures and systems interactions that require preventive actions or control measures to prevent or mitigate accident scenarios. The team listed scenarios considered not credible. In addition to normal conditions, the team considered abnormal conditions, including startup, shutdown, maintenance, and process upsets.

For each accident scenario, the team identified enabling conditions and conditional events that affect the outcome of the accident scenario (e.g., conditions that affect the likelihood of the scenario or could mitigate the consequences to either workers or the public) where appropriate. An enabling condition does not directly cause the scenario, but must be present for the initiating event to proceed to the consequences described. Enabling conditions are expressed as probabilities and can reflect such things as the mode of operation (e.g., percent of operational online availability).

The team also identified conditional events that affect the probability of the undesired outcome. These include probabilistic consideration of individual or administrative actions that would not be considered IROFS but would affect the overall likelihood of the accident. For example, if a scenario involves personal injury hazards, at least one worker must be present in the affected area at the time of the event for the injury to occur. Thus, the presence of workers in the affected area is a conditional modifier for a consequence involving personal injury. Another example of a conditional event is the probability that a worker can successfully evacuate from an area given that a hazard is present.

In considering accident scenarios at the IIFP facility, it is necessary to determine which scenarios are considered not credible and which are credible. During the PHA, IIFP's ISA team considered each accident scenario as credible, unless the scenario could be determined to be not credible. Section 3.2.5.5 of the LA (IIFP, 2012a) describes the criteria IIFP used to determine whether an accident scenario is credible. The NRC staff finds the applicant's process for identifying accident scenarios consistent with the requirements of 10 CFR 70.61 and ISG-08 (NRC, 2005).

3.3.2.5   Determine Consequence Severity Level

Table 3-1 of the LA (IIFP, 2012a) presents the radiological and chemical consequences severity limits of 10 CFR 70.61 for each of the accident consequence categories. Table 3-2 of the LA (IIFP, 2012a) provides information on the chemical dose limits specific to the IIFP facility.

For each credible accident scenario identified, IIFP's ISA team assigned a severity ranking for the consequences using the consequence severity rankings provided in Table 3-1 of the LA (IIFP, 2012a). Assigning a severity ranking allowed each accident scenario to be categorized in terms of the performance requirements outlined in 10 CFR 70.61(b) and 10 CFR 70.61(c). The severity ranking system is listed below:

- A severity ranking of 3 corresponds to high consequences.

- A severity ranking of 2 corresponds to intermediate consequences.

- A severity ranking of 1 corresponds to low consequences.

When estimating the possible worst-case consequences of an accident scenario, IIFP's ISA team members used experience, guidance from NUREG/CR-6410, "Nuclear Fuel Cycle Facility Accident Analysis Handbook," issued March 1998 (NRC, 1998), and best judgment.

The severity of consequences is determined, both quantitatively and qualitatively, through a multitude of methods. Quantitative methods include source term and dispersion modeling. Qualitative methods may use as their basis worst-case assumptions or a comparison to similar events for which bounding conservative calculations have been made, or both. The consequence of concern is the chemo-toxic exposure to UF_6, UF_4, HF, uranium dioxide (UO_2), and UO_2F_2. The applicant evaluated the dose consequences for each of the accident descriptions were evaluated and compared them to the 10 CFR 70.61 criteria for high (severity level 3) and intermediate (severity level 2) consequences. For a severity level of 1 (low), there is "no safety consequence of concern," and no further analysis is required. The NRC staff finds the applicant's process to determine consequences of potential accidents consistent with Section 3.4.3.1(2)a of NUREG-1520 (NRC, 2002).

3.3.2.6   Determine Unmitigated Likelihood

IIFP determined the likelihood of an accident scenario occurring for the unmitigated case (unmitigated likelihood). Unmitigated likelihood is the likelihood or frequency that the initiating event or cause of the accident sequence occurs despite any actual or potential preventive or mitigating features. Therefore, this likelihood/frequency estimate assumes that none of the available safeguards or IROFS is available to perform its intended safety function. Table 3-4, "Unmitigated Likelihood Categories," of the LA (IIFP, 2012a) qualitatively demonstrates the likelihood of occurrence limits drawn from 10 CFR 70.61 for each of the three likelihood categories.

The IIFP's ISA team assigned a likelihood level for each accident scenario using the defined categories in Tables 3-5, "Event Likelihood Categories," and 3-6, "Determination of Likelihood Category," of the LA (IIFP, 2012a). When assigning a likelihood category, the team relied on process knowledge, accident scenario information, operating history, and manufacturer/product information to determine which category of likelihood was appropriate. For accident scenarios in which multiple initiating events were identified, the team estimated the likelihood for the most credible initiating event. This ensured that the accident scenario was screened using the most conservative estimate of risk. The NRC staff finds the applicant's process to determine unmitigated likelihoods of potential accidents consistent with Section 3.4.3.1(2)a of NUREG-1520 (NRC, 2002).

3.3.2.7   Determine Unmitigated Risk

The NRC staff finds that the scope of the ISA Summary includes the credible accident scenarios identified for the IIFP facility. These scenarios have the capability of producing conditions that fail to meet the performance requirements of 10 CFR 70.61(b), 10 CFR 70.61(c), or 10 CFR 70.61(d). For each credible accident scenario, IIFP's ISA team used the severity category ranking and unmitigated likelihood level to assign an unmitigated risk level. (The unmitigated risk level is determined from the product of the severity category and the unmitigated likelihood category.) The ISA team used the risk matrix in Table 3-7, "Unmitigated Risk Assignment Matrix," of the LA (IIFP, 2012a) to determine the unmitigated risk. The unmitigated risk associated with each accident scenario indicates the relative importance of the associated controls. Accident scenarios in which the consequences and likelihoods yielded an unacceptable risk index required further evaluation to determine the IROFS and mitigated risk, as described in Section 3.2.5.8 of the LA (IIFP, 2012a). If the unmitigated risk was less than or equal to 4, the unmitigated risk was acceptable and no further action was required. The applicant updated the What-if table to reflect the conclusion of no further action. The NRC staff finds the applicant's process to determine unmitigated risks of potential accidents consistent with Section 3.4.3.1(2)a of NUREG-1520 (NRC, 2002).

3.3.2.8   Items Relied on for Safety and Risk Development

For each accident scenario having an unacceptable unmitigated risk index (greater than 4), the ISA team defined the IROFS and determined the mitigated likelihood. The team used the unmitigated initiating event frequency and the failure probability of each IROFS to determine the mitigated likelihood.

The risk analysis presents an accident evaluation, including a detailed discussion concerning the selection of initiating events, IROFS, and the evaluation of the accident sequences. The risk analysis provides sufficient background and operational information to understand and examine accident scenarios that result in undesired outcomes for each initiating event. Each risk analysis provides details concerning an accident scenario's quantification, including (1) the method used, (2) the initiating-event frequency determination, (3) the IROFS credited to prevent or mitigate the initiating event or events being analyzed, (4) the failure probabilities for the credited IROFS, and (5) the overall likelihood estimates. The risk analyses are controlled documents, and the Configuration Management Program described in the LA, Chapter 10, ensures that they remain up to date. The ISA Summary presents the results from each risk analysis.

The mitigated likelihood of the accident scenario occurring with the preventive or mitigating IROFS in place must meet the provisions of 10 CFR 70.61, which require that unacceptable consequences be limited. The values of the index numbers for an accident scenario, depending on the number of events involved, are added to obtain a total likelihood index, T. Accident scenarios are then assigned to one of the three likelihood categories of the risk matrix, depending on the value of the likelihood index, in accordance with Table 3-5 of the LA (IIFP, 2012a).

Safeguards are design features or administrative programs that provide defense-in-depth, but are not IROFS and are not credited with preventing or mitigating accident scenarios.

IIFP committed to identify IROFS as a part of the ISA process and include the identification of the IROFS in the ISA Summary prepared and maintained for the IIFP facility. The IROFS are defined in such a way as to delineate their boundaries, describe the characteristics of the preventive/mitigating function, and identify the applicable assumptions and conditions.

The NRC staff finds the applicant's process to identify IROFS and assign reliability indices to these IROFS is consistent with Section 3.4.3.1(2)a of NUREG-1520 (NRC, 2002).

3.3.3   Integrated Safety Analysis Integration and Team

The ISA is intended to provide assurance that the potential failures, hazards, accident descriptions, scenarios, and IROFS have been investigated in an integrated fashion so as to adequately consider common-mode and common-cause situations. Included in this integrated review is the identification of IROFS functions that may simultaneously be beneficial and harmful with respect to different hazards, as well as interactions that the previously completed risk analysis might not have considered. This review is intended to ensure that the designation of one IROFS does not negate the preventive or mitigation function of another IROFS. The applicant's ISA team performed an integrated review during the process hazard review and an overall integration review after the nodes were completed. The following items warranted special consideration during the integration process evaluation:

- common-mode failures and common-cause situations

- support system failures (e.g., loss of electrical power or water), the failure of which could have a simultaneous effect on multiple systems

- divergent impacts of IROFS so as to ensure that the negative impacts of an IROFS, if any, do not outweigh the positive impacts (i.e., to ensure that the application of an

IROFS for one safety function does not degrade the defense-in-depth of an unrelated safety function)

- other safety and mitigating factors that do not achieve the status of IROFS that could impact system performance

- identification of scenarios, events, or event descriptions with multiple impacts, such as impacts on chemical, fire, or radiation safety (e.g., a flood which could cause both a loss of confinement and active safeguards)

- potential interactions between processes, systems, areas, and buildings; any interdependence of systems or potential transfer of energy or materials

- major hazards or events that tend to be common-cause situations leading to interactions between processes, systems, buildings, and the like

The NRC staff finds the applicant's process for integrating the ISA consistent with Section 3.4.3.1(2)a of NUREG-1520 (NRC, 2002).

The applicant's team with expertise in engineering, process safety, safety analysis, and facility process operations performed and will maintain the ISA. Specifically, the team that performed the ISA included individuals with experience in nuclear facility safety, radiological safety, PHA, safety analysis and risk assessment, fire safety, chemical process safety, operations and maintenance, and ISA methods. The NRC staff finds the applicant's ISA team used for performing and maintaining the ISA to be consistent with Section 3.4.3.2(5) of NUREG-1520 (NRC, 2002).

3.4   Integrated Safety Analysis Summary Review Conclusion

The NRC staff finds that the applicant's maintenance of process safety information is consistent with the requirements of 10 CFR 70.62(b) and the guidance of NUREG-1520 (NRC, 2002). The NRC staff further finds that the applicant's commitment to conduct and maintain an ISA is in accordance with the requirements of 10 CFR 70.62(c)(1) and the guidance in NUREG-1520 (NRC, 2002).

During construction of IIFP facility, the NRC staff will implement a construction inspection program to verify that IIFP complies with the LA. This inspection program will conclude with a readiness review prior to operations. The NRC staff have provided two license conditions in Section 1.2.3.5.3 of this SER that defines the scope of the construction inspection program and readiness review.

The NRC staff finds the ISA methodology to be complete because its use of the appropriate accident identification methodology from NUREG-1513 (NRC, 2001). The NRC staff finds the consequence determinations to be acceptable and in accordance with the guidance in NUREG/CR-6410 (NRC, 1998). The NRC staff also evaluated the consequence determination methodology and determined that it provides reasonable assurance that determination of consequences that meet the performance requirements of 10 CFR 70.61(b) and 10 CFR 70.61(c) will result from its use. The NRC staff considers the likelihoods to have been derived using acceptable methods and to comply with acceptable definitions of "not unlikely," "unlikely," and "highly unlikely," in accordance with the requirements of 10 CFR 70.61 and the guidance in NUREG-1520 (NRC, 2002). The NRC staff finds that these descriptions (see Section 3.3.2.4 of this SER) conform to the guidance provided in NUREG-1520 (NRC, 2002) and meet the requirements of 10 CFR 70.65(b)(5).

The NRC staff finds that the likelihood evaluation, including the definition of "not credible," conforms to the guidance provided in NUREG-1520 (NRC, 2002) and is acceptable (see Section 3.3.2.3 of this SER). The NRC staff also finds that the determination of chemical consequences (see Section 3.3.2.4 of this SER) conforms to the guidance provided in NUREG-1520 (NRC, 2002) and is acceptable.

On July 6–8, 2010, the NRC staff conducted a site visit in Oak Ridge, TN, with IIFP and its contractor. The purpose of the visit was to allow the NRC staff to conduct a horizontal and vertical slice review of the applicant's onsite ISA documents and to discuss draft requests for additional information (RAIs) regarding the safety review.

After the NRC staff reviewed the ISA Summary submitted with the LA and identified RAI questions and comments, the NRC staff conducted a horizontal and vertical slice review of the ISA and PHA. Following guidance in NUREG-1520 (NRC, 2002), for the horizontal review, the NRC staff reviewed the process used to identify and evaluate credible accidents. For the vertical slice review, the NRC staff examined the applicant's analysis of certain selected accidents and the adequacy of the IROFS. The reviewers confirmed that the applicant's methodology to determine the consequences of the accident sequences identified in the PHA is consistent with the information presented in the ISA Summary. The horizontal review revealed that, at the time of the site visit, the ISA continued to undergo updates, in part because of the continuing development of specific design information. The vertical review examined several accident sequences in depth, including the applicant's What-if analysis, accident sequences, methods for assigning likelihood and consequence levels, and designation of IROFS. The chemical, fire, and ISA reviewers selected several accident sequences which the applicant described in detail. The NRC staff requested updates to the ISA Summary, which IIFP provided in its RAI reply (IIFP, 2011b).

Many hazards and potential accidents can result in unintended exposure of persons to radiation, radioactive materials, or toxic chemicals incident to the processing of licensed materials. The NRC staff finds that the applicant has performed an ISA to identify and evaluate those hazards and potential accidents as required by the regulations. The NRC staff has reviewed the ISA Summary and other information, and finds that it provides reasonable assurance that the applicant has identified IROFS and established engineered and administrative controls to ensure compliance with the performance requirements of 10 CFR 70.61. Specifically, the NRC staff finds that the ISA results, as documented in the ISA Summary, provide reasonable assurance that the IROFS, the management measures, and the applicant's programmatic commitments will, if properly implemented, make all credible intermediate-consequence accidents unlikely, and all credible high-consequence accidents highly unlikely.

In accordance with 10 CFR 70.66(a), the NRC staff has determined that IIFP complied with the requirements of 10 CFR 70.21, "Filing"; 10 CFR 70.22, "Contents of Applications"; 10 CFR 70.23, "Requirements for the Approval of Applications"; and 10 CFR 70.60, "Applicability," through 10 CFR 70.65 relating to Subpart H.

3.5 References

(IIFP, 2012a) International Isotopes Fluorine Products, Inc., "Fluorine Extraction Process and Depleted Uranium Deconversion Plant (FEP/DUP) License Application, Revision B," May 2012, Agencywide Documents Access and Management System (ADAMS) Accession No. ML12123A245.

(IIFP, 2012b) International Isotopes Fluorine Products, Inc., "ISA Summary Rev. B for IIFP," May 2012, Agencywide Documents Access and Management System (ADAMS) Accession No. ML12123A245.

(NRC, 2007) U.S. Nuclear Regulatory Commission, Staff Requirements Memorandum to SECY 07 0146, "Regulatory Options for Licensing New Uranium Conversion and Depleted Uranium Deconversion Facilities," October 10, 2007.

(NRC, 2005) U.S. Nuclear Regulatory Commission, "Natural Phenomena Hazards," FCSS ISG-08, 2005.

(NRC, 2002) U.S. Nuclear Regulatory Commission, "Standard Review Plan for the Review of a License Application for a Fuel Cycle Facility," NUREG 1520, Rev. 0, March 2002.

(NRC, 2001) U.S. Nuclear Regulatory Commission, "Integrated Safety Analysis Guidance Document," NUREG-1513, May 2001.

(NRC, 1998) U.S. Nuclear Regulatory Commission, "Nuclear Fuel Cycle Facility Accident Analysis Handbook," NUREG/CR-6410, March 1998.

(NRC, 1988) U.S. Nuclear Regulatory Commission, "A Regulatory Analysis on Emergency Preparedness for Fuel Cycle and Other Radioactive Material Licensees," NUREG-1140, 1988 (reprinted August 1991).

4.0   RADIATION PROTECTION

The purpose of this review is to determine whether the International Isotopes Fluorine Products, Inc. (IIFP or the applicant), Radiation Protection (RP) Program is adequate to protect the radiological health and safety of workers and to comply with the associated regulatory requirements in the following parts of Title 10 of the *Code of Federal Regulations* (10 CFR):

- 10 CFR Part 19, "Notices, Instructions and Reports to Workers: Inspection and Investigations"

- 10 CFR Part 20, "Standards for Protection against Radiation"

- 10 CFR Part 40, "Domestic Licensing of Source Material"

- 10 CFR Part 70, "Domestic Licensing of Special Nuclear Material"

Chapter 9 of this Safety Evaluation Report (SER) discusses public and environmental protection.

4.1   Regulatory Requirements

The RP regulatory requirements are located in 10 CFR Part 20. The regulations in 10 CFR 40.32, "General Requirements for Issuance of Specific Licenses," apply to the organization, qualifications and written procedures for the RP staff. The regulations in 10 CFR 19.12, "Instructions to Workers," apply to the Radiation Safety Training Program. The regulations in 10 CFR 70.61, "Performance Requirements," apply to restricting radiological risk associated with potential process accidents involving radioactive materials.

4.2   Regulatory Guidance and Acceptance Criteria

Sections 4.4.1.3, 4.4.2.3, 4.4.3.3, 4.4.4.3, 4.4.5.3, 4.4.6.3, 4.4.7.3, and 4.4.8.3 of NUREG-1520, Revision 0, "Standard Review Plan for the Review of a License Application (LA) for a Fuel Cycle Facility" issued March 2002 (NRC, 2002), outlines the acceptance criteria for the U.S. Nuclear Regulatory Commission (NRC) staff's review of the RP Program.

4.3   Staff Review and Analysis

4.3.1   Radiation Protection Program Implementation

The NRC staff reviewed the applicant's RP Program implementation against the acceptance criteria in Section 4.4.1.3 of NUREG-1520, Revision 0 (NRC, 2002). The following discussion summarizes the NRC staff's analysis as to whether the information provided by the applicant in the LA (IIFP, 2012a) meets the criteria.

In Section 4.1 of the LA (IIFP, 2012a), the applicant stated that the RP Program will meet the requirements of 10 CFR Part 20, Subpart B, by committing that the program will be consistent with the guidance provided in Regulatory Guide 8.2, "Guide for Administrative Practice in Radiation Monitoring," issued February 1973 (NRC, 1973). The applicant further committed to maintaining occupational exposures as low as reasonably achievable (ALARA) through exposure monitoring consistent with the regulations in 10 CFR Part 20, frequent interactions between the Radiation Safety Committee and operations personnel, and annual RP Program assessments with senior management.

The applicant will establish annual exposure goals for occupationally exposed personnel to ensure that personnel doses received will be below the limits specified in 10 CFR 20.1201. The RP Program content and implementation will be reviewed annually, at a minimum. In addition, the applicant will establish controls such that no member of the public is expected to receive a total effective dose equivalent (TEDE) in excess of 0.25 millisieverts per year (mSv/yr) (25 millirems per year [mrem/yr]).

The applicant outlined the operations organizational structure in Section 2.1.4 of the LA (IIFP, 2012a) and described the RP Program structure and defined the responsibilities of key program personnel in Section 4.1.1 of the LA (IIFP, 2012a). The Chief Operations Officer (COO) will have the overall responsibility of ensuring that facility operations are conducted in a manner that protects the employee, the environment, and the public from radiological, chemical, and industrial hazards; and that these operations will be carried out in accordance with all applicable regulations, licenses, and permits. The COO will perform his or her duties in accordance with written policies and procedures. The COO provides for safety and control of operations and protection of the environment by delegating and assigning responsibility to qualified plant and line supervisors.

The Environmental, Safety and Health (ESH) Manager will report to the COO, and will interact with all IIFP managers on matters of EHS policies, regulatory requirements, plant safety and environmental compliance. The ESH Manager will be responsible for directing activities to ensure the facility complies with appropriate rules, regulations, and codes. This includes ESH activities associated with RP, chemical safety, environmental protection, industrial hygiene, industrial safety, emergency preparedness, security, regulatory affairs, and licensing. The ESH Manager will work with other managers and supervisors of the plant to ensure consistent interpretations of the requirements, perform independent reviews, and support facility and operations change control reviews. The ESH organization provides independent oversight of plant operations. The ESH manager will have the responsibility and authority to elevate any ESH or security-related issue to the IIFP President and Chief Executive Officer (CEO).

The Radiation Protection Manager (RPM) will be administratively independent of operations and will report directly to the ESH Manager. The RPM will also have the authority to report to the COO any unresolved concerns related to ESH and RP. The RPM will be responsible for effectively implementing the RP Program and will ensure that the facility is staffed with suitably trained RP personnel and that sufficient resources are provided to implement an effective program. The RPM will approve restart of any operation that gets shut down by the RP function or as a result of RP concerns.

The RP staff, including technicians and support personnel, will report to the RPM. Major responsibilities of the RPM and the RP staff will include, but not be limited to, the following:

- Establish and maintain the RP Program, procedures, and training.

- Conduct radiation and contamination monitoring and control programs.

- Evaluate radiation exposures of employees, contractor personnel, and visitors and ensure the maintenance records and reporting of results.

- Establish and maintain the ALARA Program, including being a key member of the ALARA Safety Committee.

- Evaluate the integrity and reliability of radiation detection instruments.

- Provide support for integrated safety analyses (ISAs) and configuration control.

Section 4.1.2 of the LA (IIFP, 2012a) addresses the independence of the RP Program. The RP Program will be independent of operations because both the RP and ESH Managers function independently of operations and have the authority to elevate concerns within the organization.

As stated in Section 4.1.3 of the LA (IIFP, 2012a), the applicant commits that the ALARA Committee will review, at least annually, the content and implementation of the RP Program. The review will consider facility changes, new technologies, and other process enhancements that could improve overall program effectiveness. The NRC staff finds that the commitments in the LA (IIFP, 2012a) satisfactorily address the acceptance criteria in Section 4.4.1.3 of NUREG-1520, Revision 0 (NRC, 2002).

4.3.2 As Low As Reasonably Achievable Program

The NRC staff reviewed the applicant's ALARA Program commitments against the acceptance criteria in NUREG-1520, Revision 0, Section 4.4.2.3 (NRC, 2002). The following discussion summarizes the NRC staff's analysis as to whether the information provided by the applicant in the LA (IIFP, 2012a) meets the criteria.

Section 4.2 of the LA (IIFP, 2012a) addresses the proposed ALARA Program, which will function as a subset of the RP Program. The objective of the program will be to make every reasonable effort to maintain facility exposures to radiation as far below the dose limits of 10 CFR 20.1201 as is practical and to maintain radiation exposures to members of the public below the dose limits of 10 CFR 20.1101(d). The design and implementation of the ALARA Program will be consistent with guidance provided in Regulatory Guide 8.2 (NRC, 1973); Regulatory Guide 8.13, Revision 3, "Instruction Concerning Prenatal Radiation Exposure," issued June 1999 (NRC, 1999a); Regulatory Guide 8.29, Revision 1, "Instruction Concerning Risks from Occupational Radiation Exposure," issued February 1996 (NRC, 1996); and Regulatory Guide 8.37, "ALARA Levels for Effluents from Materials Facilities," issued July 1993 (NRC, 1993a).

The ALARA Program will include the following features:

- management commitment, demonstrated through a written policy statement, procedures, other directives, and periodic management reviews

- formal program audits, conducted on at least an annual basis

- well-supervised and defined RP capability, including appropriate supervisors and technicians, with all personnel onsite having the authority to stop work as needed to ensure that appropriate safety precautions are observed

- appropriate training for the workforce, including training consistent with the requirements of 10 CFR 19.12 and incorporating appropriate portions of the guidance provided in

Regulatory Guides 8.13, Revision 3 (NRC, 1999a) and 8.29, Revision 1 (NRC, 1996)

- appropriate authority vested in RP personnel including stop work authority

- consideration of the need for plant modifications as warranted for reducing exposures and doses to personnel.

The applicant will implement documented RP Program policies to ensure that the ALARA goal is met. Procedures will incorporate the ALARA philosophy into routine operations and ensure exposures will be maintained below the levels identified in 10 CFR 20.1101 levels (i.e., 10 mrem/yr to a member of the public through airborne emissions). As stated previously, the applicant will also establish controls such that no member of the public will receive a TEDE in excess of 0.25 mSv/yr (25 mrem/yr). Radiological controlled areas (RCAs) will be established within the facility and identified through signs, ropes, gates, fences, or other visible means. Each zone will have specific entry requirements, survey requirements, and dosimetry requirements. The establishment of these areas will support the ALARA commitment to minimize the spread of contamination and reduce unnecessary exposure of personnel to radiation.

The applicant will have an ALARA Safety Committee that supports the Facility Safety Review Committee. The ALARA Safety Committee will consist of key members of facility management, supervisory staff, and workforce and will meet periodically on a frequency established in the RP ALARA Program. The ALARA Safety Committee will use the guidance provided in Regulatory Guide 4.21, "Minimization of Contamination and Radioactive Waste Generation: Life-Cycle Planning," issued June 2008 (NRC, 2008); Regulatory Guide 8.10, Revision 1, "Operating Philosophy for Maintaining Occupational Radiation Exposures as Low as Is Reasonably Achievable," issued May 1977 (NRC, 1977); and Regulatory Guide 8.37 (NRC, 1993a) to formulate facility operating philosophy in reducing exposures. Membership of the ALARA Safety Committee includes the following individuals:

- the COO

- the RP Manager

- selected department managers

- the ESH Manager

- selected supervisors and hourly personnel

The ALARA Program is intended/expected to facilitate interaction between RP and operations personnel. The ALARA Safety Committee, comprising staff members responsible for RP and operations personnel, including hourly workers, will be utilized in achieving this goal. The scope of the ALARA Safety Committee's activities will include at a minimum, annual review of the following:

- site radiological operating performance, including trends in airborne concentrations, personnel exposures, and environmental monitoring results

- operations and exposure records to determine where exposures may be reduced

- employee training and methods for utilizing information on the job to keep exposure ALARA

- potential modifications of procedures and equipment when changes will reduce exposure at reasonable cost

Specific goals of the ALARA Program will include maintaining occupational exposures, as well as environmental releases, as far below regulatory limits as is reasonably achievable. For environmental effluents, ALARA goals will be initially set to 20 percent of 10 CFR Part 20, Appendix B. The design and operation of the facility will also incorporate the ALARA concept. The size and number of areas with higher dose rates will be minimal. Written procedures will incorporate instruction such that time spent in higher dose rate areas will be controlled, and projects will be evaluated to ensure that workers receive minimum radiological exposure. Areas where personnel spend significant amounts of time will be designed to maintain the lowest dose rates reasonably achievable.

Based on the review summarized above, the NRC staff finds that the commitments in the LA (IIFP, 2012a) satisfactorily address the acceptance criteria in Section 4.4.2.3 of NUREG-1520, Revision 0 (NRC, 2002).

4.3.3 Organization and Personnel Qualifications

The NRC staff reviewed the applicant's organization and personnel qualifications against the acceptance criteria in NUREG-1520, Revision 0, Section 4.4.3.3 (NRC, 2002). The following discussion summarizes the NRC staff's analysis as to whether the information provided by the applicant in the LA (IIFP, 2012a) meets the criteria.

Several sections of the LA (IIFP, 2012a) discuss organization and personnel qualifications, although most of the information is repeated in Section 4.3. The applicant stated that the RP staff will be assigned responsibility for implementation of the RP Program functions. The facility will only employ suitably trained RP personnel. Staffing will be consistent with the guidance provided in Regulatory Guide 8.2 (NRC, 1973) and Regulatory Guide 8.10, Revision 1 (NRC, 1977).

The ESH Manager will be responsible for establishing and overseeing the RP, Industrial Safety and Industrial Hygiene, Environmental Protection, Fire Protection, and Emergency Preparedness/Security Programs to ensure compliance with applicable Federal, State, and local regulations and requirements. The ESH Manager will report to the COO, but also interacts with all managers on matters of ESH policies, regulatory requirements, plant safety, and environmental compliance. In addition, the ESH Manager has the authority and responsibility to elevate any ESH concerns to the corporate EHS manager and the IIFP President. The ESH Manager will have, as a minimum, a bachelor's degree in engineering, science, or a related field and 5 years of experience of ESH activities at chemical, radiological, or nuclear facilities.

The RPM will report directly to the ESH Manager and has the responsibility for establishing and implementing the RP Program. These duties will include the training of personnel in the use of equipment, control of radiation exposure of personnel, continuing evaluation and determination of the radiological status of the facility, and conducting the Radiological Environmental Monitoring Program. The RPM will have, as a minimum, a bachelor's degree in engineering or a scientific field and a minimum of 5 years of responsible experience that includes assignments involving responsibility for RP and the application and direction of RP programs.

Staff health physicists will have, as a minimum, a bachelor's degree in engineering or a scientific field and experience commensurate with health physics and RP duties. Staff radiation control technicians will have a high school diploma and experience commensurate with RP duties.

The facility organization chart, reproduced as Figure 4-1 of the LA (IIFP, 2012a), establishes clear organizational relationships among the RP staff and the other facility line managers. In matters involving radiological protection, the RPM will have the responsibility and authority to elevate any radiation safety or environmental issue to the COO. The RPM will be skilled in the interpretation of RP data and regulations and familiar with the operation of the facility and RP concerns relevant to the facility. The RPM will be a resource for radiation safety management decisions.

RP technicians, engineers, and supervisors will perform the functions of assisting and guiding workers in the radiological aspects of the job. These individuals will have the responsibility and authority to stop work or mitigate the effect of an activity, if it is suspected that the initiation or continued performance of a job, evaluation, or test will result in the violation of approved RP requirements.

Based on the review as summarized above, the NCR staff finds that the commitments in the LA (IIFP, 2012a) satisfactorily address the acceptance criteria in Section 4.4.3.3 of NUREG-1520, Revision 0 (NRC, 2002).

4.3.4 Written Procedures

The NRC staff reviewed the applicant's written procedure commitments against the acceptance criteria in NUREG-1520, Revision 0, Section 4.4.4.3 (NRC, 2002). The following discussion summarizes the NRC staff's analysis as to whether the information provided by the applicant in the LA (IIFP, 2012a) meets the criteria.

Sections 2.3.4, 4.4, 4.4.1, and 11.4 of the LA (IIFP, 2012a) discuss written procedure commitments. The applicant stated that it will use approved written procedures to conduct operations at the facility involving licensed materials. The applicant will develop or modify procedures through a formal process incorporating change controls, as described in Section 11.1 of the LA. The procedure process will utilize nine basic elements to accomplish procedure development, review, approval, and control. The LA, Section 11.4.2 (IIFP, 2012a), identifies these elements as (1) identification, (2) development, (3) verification, (4) review and comment resolution, (5) approval, (6) validation, (7) issuance, (8) change control, and (9) periodic review. Changes to work procedures, including safety requirements, will be reviewed with production personnel by their immediate supervisor or delegate.

The RP staff will prepare, review, and approve RP procedures to carry out activities related to the RP Program. Approved written procedures will control RP activities to ensure that the activities are implemented in a safe, effective, and consistent manner. RP procedures will be reviewed and revised, as necessary, to incorporate facility or operational changes or changes to the ISA. The RP staff will prepare draft procedures that affected personnel will review to ensure that the procedures are appropriate and reasonable to implement. The RPM (or designee) will review and approve final RP procedures, as well as proposed revisions to RP procedures.

The RWP system will administer non-routine activities, particularly those performed by non-IIFP employees and not covered by approved written procedures. The RWP will describe the work to be performed and define the authorized activities. The RWP will specify the necessary radiation safety controls, as appropriate, to include personnel monitoring devices, attendance of RP staff, protective clothing, respiratory protective equipment, special air sampling, and additional precautionary measures to be taken. The RWP will also describe the radiological conditions in the immediate work area covered by the RWP. The RPM or designee must approve the RWP. The designee must meet the qualification requirements of the RPM. RWPs will have a predetermined period of validity with a specified expiration or termination time. Standing RWPs may be issued for routinely performed activities, such as tours of the plant.

Before commencing work that requires an RWP, employees performing the job will have to review the RWP and document their review. An RP technician will monitor the work as required. RWPs will be available to workers for review at any time and will include expiration dates. An RP technician or the RPM (or designee) reviews the status of issued RWPs on a periodic basis. An RWP will be closed when the applicable work activity for which it is written is complete and terminated. Copies of RWPs and any associated records will be kept for the life of the facility.

Based on the review summarized above, the NRC staff finds that the commitments in the LA (IIFP, 2012a) satisfactorily address the acceptance criteria in Section 4.4.4.3 of NUREG-1520, Revision 0 (NRC, 2002).

4.3.5 Training

The NRC staff reviewed the applicant's training commitments against the acceptance criteria in NUREG-1520, Revision 0, Section 4.4.5.3 (NRC, 2002). The following discussion summarizes the NRC staff's analysis as to whether the information provided by the applicant in the LA (IIFP, 2012a) meets the criteria.

The applicant discussed its training commitments in Section 4.5 of the LA (IIFP, 2012a), as well as in Sections 11.3.7.1, 11.3.7.2, and 11.3.8. The applicant stated that RP training will comply with the requirements in 10 CFR 19.12 and 10 CFR 20.2110 and will take into consideration a worker's normally assigned work activities. Basic RP training will cover the following topics:

- radiation safety principles, policies, and procedures

- radiation hazards and health risks

- correct handling of radioactive materials

- location of and adherence to RP procedures

- minimization of exposures to radiation and radioactive materials
- contamination control

- access and egress controls

- monitoring for internal and external exposures

- ALARA and exposure limits

- exposure monitoring methods and instrumentation

- personal and area dosimetry

- donning and doffing of personal protective equipment

- emergency response.

Additional topics can be included if applicable to the individual's job responsibilities.

The applicant will establish training programs for the various types of job functions (e.g., production, maintenance, RP technician, and contractor personnel) commensurate with RP responsibilities associated with each such position. Trained personnel will escort visitors to the controlled access area (CAA).

Individuals attending training sessions must pass an initial examination covering the training contents to ensure the understanding and effectiveness of the training. The periodicity of refresher training will depend on the worker's responsibilities; however, the basic refresher training will occur annually (not to exceed 15 months) and will include an exam. Training requirements will be documented and tracked for employees. The applicant will manage and store training records in accordance with 10 CFR 20.2110. Audits and assessments of production and maintenance personnel responsible for following the requirements related to the topics in the training they have received will also be used to evaluate the effectiveness of the training programs.

Since contractor employees will perform diverse tasks in the CAA, training for these employees will be designed to address the type of work they perform. In addition to applicable radiation safety topics, training contents may include RWPs; special bioassay sampling; and special precautions for welding, cutting, and grinding in the CAA. The training manager will assign instructors to conduct these training programs who have the necessary knowledge to address chemical safety and RP. The applicant will maintain records of the training programs.

Individuals requiring unescorted access to the CAA will receive annual continuing education training. Production personnel will be further instructed in the specific safety requirements of their work assignments by qualified personnel during on-the-job training. Employees must demonstrate an understanding of work assignment requirements based on observations by qualified personnel before working without direct supervision. Changes to work procedures, including safety requirements, will be reviewed with production personnel by their immediate supervisor or delegate.

The RPM will review the contents of the Radiation Safety Training Program used by IIFP biannually. The review will address changes which have occurred in policies, procedures, and requirements, as well as changes to the ISA.

Periodically, the training program will be systematically evaluated to measure the program's effectiveness in producing competent employees. The trainees will be encouraged to offer feedback after completion of classroom training sessions to provide data for program improvements. Training program evaluations will identify program strengths and weaknesses, determine whether the program content matches current job needs, and determine whether corrective actions are needed to improve the program's effectiveness. The training organization will be responsible for leading the training program evaluations. The RPM will review the evaluation information and implement changes in the training program as necessary. Based on the review as summarized above, the NRC staff finds that the commitments in the LA (IIFP, 2012a) satisfactorily address the acceptance criteria in Section 4.4.5.3 of NUREG-1520, Revision 0 (NRC, 2002).

4.3.6 Ventilation and Respiratory Protection Programs

The NRC staff reviewed the applicant's Ventilation and Respiratory Protection Program commitments against the acceptance criteria in NUREG-1520, Revision 0, Section 4.4.6.3 (NRC, 2002). The following discussion summarizes the NRC staff's review as to whether the information provided by the applicant in the LA (IIFP, 2012a) meets the criteria.

The applicant discussed ventilation commitments in Section 4.6.1 of the LA (IIFP, 2012a) and stated that the confinement of uranium will be a design requirement for the facility. Areas where uranium will be processed with the potential for dusts, mists, or fumes and areas where toxic chemicals are processed or produced are provided with dust collection or scrubber systems, or both, to maintain exposures ALARA. The design of building ventilation systems in process areas and control rooms will be sized with adequate flows and pressure differentials to ensure that potential airborne concentrations of radioactivity do not exceed derived air concentration (DAC) values, calculated from the dose coefficients specified in the International Commission on Radiological Protection (ICRP) Publication 68, "Dose Coefficients for Intakes by Workers" (ICRP, 1994). The applicant has requested an exemption to utilize the ICRP-68 values as opposed to the values presented in Appendix B, "Annual Limits on Intake (ALIs) and Derived Air Concentrations (DACs) of Radionuclides for Occupational Exposure; Effluent Concentrations; Concentrations for Release to Sewerage," to 10 CFR Part 20, "Standards for Protection against Radiation," in Section 1.5 of the LA (IIFP, 2012a). The Commission has previously directed the NRC staff to approve the use of ICRP-68 on a case-by-case basis in the staff requirements memorandum, dated April 21, 1999, associated with SECY-99-0077, "To Request Commission Approval to Grant Exemptions from Portions of 10 CFR Part 20," dated March 12, 1999. The ventilation system will also be designed so that air will flow from areas of low contamination potential towards areas of higher contamination potential to minimize the spread of contamination.

The ventilation program, radiation detectors and alarms, process vents, and associated containment systems will be checked routinely as part of the operating process controls and Preventive Maintenance Program. Operations and maintenance relative to the Ventilation Program, including calibrations, change management, measurements, and analysis, will be performed using approved written procedures.

Several measures will be in place to ensure effective operation of the ventilation control systems. Differential pressure will be monitored and alarmed for high-efficiency particulate air filters used for control rooms in which uranium is processed. Operating procedures will specify limits and set points on differential pressure consistent with manufacturer's recommendations. Filters will be changed if they fail to function properly or if the differential pressure exceeds the manufacturer's ratings.

Dust collector units in the depleted uranium tetrafluoride (DUF_4) and fluorine extraction process buildings will be monitored and alarmed if the differential pressure exceeds established limits. Operating procedures will specify limits and set points for acceptable differential pressures and uranium sample results. Operating procedures will also specify that at least two dust collector units will be operated in series; if not, the process system being serviced by the dust collectors must be placed in a shutdown or standby mode.

Written procedures, approved by the Plant Engineering and Maintenance Manager and the RPM (or designated alternates), will specify filter and dust collector inspection, testing, maintenance, and change-out criteria. Change-out frequency will be based on considerations of filter loading, operating experience, differential pressure data, and any monitoring data that exceeds set administrative control limits.

Pressures will be continuously monitored and controlled for the plant offgas scrubbing system and across the process system vented to the scrubbing system. Limits will be set to ensure adequate safety margin of pressure controls for the vent gas plant scrubbing system. Operation procedures and operator aids will also provide for corrective response when alarms are received relative to the system pressure controls.

Airflow rates at exhausted enclosures and close-capture points related to uranium processing and handling areas, when in use, will be adequate to preclude escape of airborne uranium and minimize potential for intake by workers. Airflow rates will be checked routinely when in use and after modification of any hood-exhausted enclosure, close-capture point equipment, or ventilation system serving these barriers.

The applicant discussed the Respiratory Protection Program commitments in Section 4.6.2 of the LA (IIFP, 2012a). The applicant will conduct the Respiratory Protection Program in accordance with Subpart H of 10 CFR Part 20 and consistent with the guidance in Regulatory Guide 8.15, Revision 1, "Acceptable Programs for Respiratory Protection," issued October 1999 (NRC, 1999b).

Approved written procedures for using Respiratory Protective Equipment will be implemented consistent with 10 CFR 20.1703(c)(4) and will dictate the following:

- monitoring, including air sampling and bioassays

- supervision and training of respiratory protection users

- fit testing of respirators

- respirator selection

- breathing air quality

- inventory and control of respirators

- cleaning of respirators

- storage, issuance, maintenance, repair, testing, and quality assurance of respiratory protection equipment

- recordkeeping

- limitations on respirator use and relief from respirator use.

The applicant will maintain records of the Respiratory Protection Program (including training for respiratory use and maintenance) in accordance with the Records Management Program, as described in Chapter 11 of the LA (IIFP, 2012a).

Based on the analysis as summarized above, the NRC staff finds that the commitments in the IIFP's LA (IIFP, 2012a) satisfactorily address the application acceptance criteria in Section 4.4.6.3 of NUREG-1520, Revision 0 (NRC, 2002).

4.3.7 Radiation Survey and Monitoring Programs

The NRC staff reviewed the applicant's Radiation Survey and Monitoring Program commitments against the acceptance criteria in NUREG-1520, Revision 0, Section 4.4.7.3 (NRC, 2002). The following discussion summarizes the NRC staff's analysis as to whether the information provided by the applicant in the LA (IIFP, 2012a) meets the criteria.

The applicant discussed the Radiation Survey and Monitoring Programs in Section 4.7 of the LA (IIFP, 2012a). IIFP stated that it will conduct necessary and reasonable surveys, in accordance with 10 CFR 20.1501(a) and 10 CFR 20.1501(b), that satisfy the applicable regulations and are adequate to identify the extent of radiation levels, concentrations, or quantities of radioactive material and the potential radiological hazards. Instruments and equipment will be calibrated periodically in accordance with 10 CFR 20.1501(b). Personnel dosimeters will be processed by a vendor accredited by the National Voluntary Laboratory Accreditation Program (NVLAP), in accordance with 10 CFR 20.1501(c). IIFP will also monitor exposure to radiation and radioactive material to demonstrate compliance with occupational dose limits, in accordance with 10 CFR 20.1502, "Conditions Requiring Individual Monitoring of External and Internal Occupational Dose." Written procedures will address survey and monitoring objectives, sampling procedures and data analysis methods, types of equipment and instrumentation to be used, frequency of measurements, recordkeeping and reporting requirements, and actions to be taken in case measurements exceed administrative or regulatory limits.

The applicant will measure external occupational dose in accordance with 10 CFR 20.1501(a). Individually assigned dosimeters will be used to determine deep dose equivalent and shallow dose equivalent from external sources of radiation. In accordance with written procedures, personnel dosimeters will be distributed to individuals based on their job functions, commensurate with the amount of time an individual spends working with or near radioactive materials. An NVLAP-accredited vendor will process personnel dosimeters. The capability will exist to process dosimeters expeditiously if there is an indication of an exposure in excess of established action guides. Written procedures will establish action guides for external exposures.

IIFP will design and implement its Personnel Monitoring Program for internal occupational radiation exposures based on the requirements of 10 CFR 20.1201, 10 CFR 20.1204, "Determination of Internal Exposure," 10 CFR 20.1502(b), and 10 CFR 20.1704(i). Intakes will be assigned to individuals based on one or more types of measurements as follows: air sampling, in vitro bioassay (i.e., urinalysis or fecal); and in vivo bioassay (i.e., lung counting). The type and frequency of measurements for an individual will be determined by his or her job function and properties of the licensed material associated with a known or suspected intake. The measurements will be commensurate with the amount of time an individual spends working with or near radioactive material. Intakes will be converted to committed dose equivalent and committed effective dose equivalent for the purposes of limiting and recording occupational doses.

Approved written procedures will establish action levels to prevent an individual from exceeding the occupational exposure limits specified in 10 CFR 20.1201. Work activity restrictions will be imposed when an individual's exposure exceeds 80 percent of the 10 CFR 20.1201 limit. Control actions will include temporarily restricting the individual from working in an area containing airborne radioactivity, and actions are taken as necessary to prevent recurrence. Exposure to airborne radioactive material may also be controlled through limiting access to areas, limiting exposure time, and use of respiratory equipment.

Conduct of the *in vitro* bioassay program will primarily evaluate the intake of soluble uranium to ensure that the 10 CFR 20.1201(e) intake limit of 10 milligrams per week is not exceeded. Personnel assigned to work in areas where soluble airborne uranium compounds will be present in concentrations likely to result in intakes exceeding 10 percent of the applicable limits in 10 CFR 20.1201 will be monitored by urinalysis or fecal bioassay methods, or both. Written procedures will specify the minimum sampling frequency for these individuals. *In vitro* monitoring may will also be used for individuals involved in non-routine operations, perturbations, or incidents. The applicant will use an offsite laboratory that meets the performance standards specified in American National Standards Institute/Health Physics Society (ANSI/HPS) N13.22-1995, "Bioassay Program for Uranium" (ANSI/HPS, 1995), and ANSI/HPS N13.30-1996, "Performance Criteria for Radiobioassay" (ANSI/HPS, 1996), to process and analyze in vitro bioassay samples.

The applicant will conduct *in vivo* lung counting as necessary to supplement or verify in vitro bioassay results. Qualified contractors will perform *in vivo* lung counting in accordance with ANSI/HPS N13.35-2009, "Specifications for the Bottle Manikin Absorption Phantom" (ANSI/HPS, 2009) performance standards.

In accordance with procedure, the summation of external and internal occupational radiation exposure will be reported as a TEDE and will be calculated consistent with the guidance in Regulatory Guide 8.34, "Monitoring Criteria and Methods to Calculate Occupational Radiation Doses," issued July 1992 (NRC, 1992a).

The applicant will design and implement an Air Sampling Program in areas of the IIFP facility that are identified as potential airborne radioactivity areas. This program will include procedures to conduct air surveys and to calibrate and maintain RP airborne sampling equipment in accordance with the manufacturers' recommendations. Evaluations of air sampling effectiveness will be performed in accordance with the methods and acceptance criteria in Regulatory Guide 8.25, Revision 1, "Air Sampling in the Workplace," issued June 1992 (NRC, 1992c).

Air samples will be continuously taken from each main process area where airborne concentrations could potentially exceed 0.3 DAC when averaged over 40 hours to assess the concentrations of uranium in the air. In accordance with procedures, the air samples will be collected in such a way that the concentrations of uranium measured are representative of the air which workers breathe. Filters from air samplers will be changed each shift during normal operating periods or at more frequent intervals following the detection of an event that may have released airborne uranium, based upon knowledge of the particular circumstances. Filters will be changed less frequently during periods when no work is in progress. Grab samples will be obtained during maintenance activities that are known to have the potential to generate airborne radioactivity levels in excess of 1.0 DAC.

Each air sampler will be equipped with a flow meter to indicate flow rate of air sampled. These flow meters will be calibrated or replaced every 18 months, at a minimum. Air sampling results in excess of 2.5 DAC (8-hour sample), and not resulting from specific known causes, will be investigated to determine the probable cause. Operations or equipment will be shut down, and immediate corrective action will be taken at locations where an air sample exceeds 10 DAC without a specific known cause.

The IIFP facility will be designed and operated in accordance with 10 CFR 20.1406, "Minimization of Contamination," to minimize contamination, facilitate eventual decommissioning, and minimize where practicable the generation of radioactive waste. The following are examples of methods that will be employed for minimizing contamination:

- containment of radioactive material throughout the facility

- monitoring for equipment leaks

- training on proper techniques for handling radioactive material

- airflow from areas of low radioactivity to higher radioactivity.

Routine surveys will be performed in areas that are most likely to be contaminated or where contamination from licensed processes, radioactive decay products or other radionuclide contaminants may concentrate. RP technicians will perform routine contamination surveys in the change rooms, plant exit walkways, and the laboratory area. The RP staff will determine survey frequencies, compare the survey results to action guide values specified in procedures, and ensure that the appropriate responses are taken. If the results exceed the action guide values, the RPM (or designee) will be informed and will determine whether an investigation or corrective actions are necessary.

Personnel contamination surveys will be required for external contamination on clothing and the body by personnel exiting the change rooms. If contamination is found in excess of background levels, the individual will attempt self-decontamination (except for facial contamination) at the facilities provided in the change rooms. If decontamination attempts are not successful, or if facial contamination is detected, decontamination assistance will be provided by the RP function (typically an RP technician). If skin or personal clothing is still contaminated above background levels, the individual will not be permitted to leave the area without the prior approval (in accordance with procedure) of the RPM.

The applicant will implement and document corrective actions for airborne occupational exposure based on the frequency and magnitude of events causing releases of airborne uranium that exceed administrative limits. Portable air sample surveys will supplement routine air sampling as required to evaluate nonroutine activities or breaches in containment. RP and operations staff will investigate the cause of the release and implement recommended actions to prevent future releases.

Appropriate radiation detection instruments will be available in sufficient number to ensure that adequate radiation surveillance can be accomplished. Selection criteria for portable and laboratory counting equipment will be based on the types of radiation detected, maintenance requirements, ruggedness, interchangeability, and upper and lower limits of detection capabilities.

Portable instrumentation will be calibrated in accordance with manufacturing recommendations before initial use, after major maintenance, and on a routine basis specified by the manufacturer. Calibration will consist of a performance check on each range scale of the instrument with a radioactive source of known activity traceable to a recognized standard, such as the National Institute of Standards and Technology. Before each use, operability checks will be performed on monitoring and laboratory counting instruments. The background and efficiency of laboratory counting instruments will be determined on a daily basis when in use.

The applicant discussed clearance of materials from RCAs in Section 4.7.13 of the LA (IIFP, 2012a). When removing equipment and materials from RCAs, with the exception of hazardous chemicals produced from licensed operations, IIFP staff will follow the guidance in NRC's branch technical position, "Guidelines for Decontamination of Facilities and Equipment Prior to Release for Unrestricted Use or Termination of Licenses for Byproduct, Source, or Special Nuclear Material" (NRC, 1993b). Volumetrically contaminated materials will be released if the uranium concentration of the material does not exceed 30 picocuries per gram or the dose to a member of the public, taking into consideration the subsequent use of the material, does not exceed 1 mrem per year. The RP staff must approve the release of equipment or materials from RCAs. Written procedures will govern the equipment and material screening and evaluation process.

Hazardous chemicals produced from licensed materials, as defined in 10 CFR 70.4, "Definitions," will be considered "separated from licensed materials" by meeting the concentration levels described in 10 CFR 40.13(a) for "unimportant quantities of source material." The term "unimportant quantities of source material" is defined in 10 CFR 40.13(a) as "...source material in any chemical mixture, compound, solution, or alloy in which the source material is by weight less than one-twentieth of 1 percent (0.05 percent) of the mixture, compound, solution, or alloy." Written procedures will govern the analytical methods applied to determine the concentration of source material in hazardous chemicals. IIFP will conduct statistical sampling to provide confidence that the quantity of source materials remains below the "unimportant quantities" threshold.

The applicant discussed leak testing of sealed sources in Section 4.7.14 of the LA (IIFP, 2012a) and made commitments consistent with branch technical positions that are acceptable to staff ("License Condition for Leak-Testing Sealed Byproduct Material Sources" (NRC, 1993c); "License Condition for Leak-Testing Sealed Plutonium Sources" (NRC, 1993d); "License Condition for Plutonium Alpha Sources" (NRC, 1993e); "License Condition for Leak-Testing Sealed Source Which Contains Alpha and/or Beta-Gamma Emitters" (NRC, 1993f); and "License Condition for Leak-Testing Sealed Uranium Sources" (NRC, 1993g), all of which were issued April 1993). When not in use, sources will be stored in a closed container adequately designed and constructed to contain radioactive material that may otherwise be released during storage. The sources will be tested for leakage using the dry wipe test method in accordance with International Standardization Organization (ISO) 9978, "Radiation Protection—Sealed Radioactive Sources—Leakage Test Methods" (ISO, 1992). Sealed sources will be leak checked at 6-month intervals, not to exceed that specified on the sealed source and device registration certificate. The check will use a quantitative analysis capable of detecting

185 becquerels (Bq) (0.005 microcuries (μCi)) of radioactivity. Leak tests will not be required if any one of the following conditions is true:

- sources contain only tritium

- sources contain only licensed material with a half-life of less than 30 days

- sources contain only a radioactive gas

- sources contain 3.7 megabecquerels (100 μCi) or less of beta-emitting or gamma-emitting material or 370 kilobecquerels (10 μCi) or less of alpha emitting material

- sources are stored and are not being used (must be leak tested before use or transfer).

Sources that exhibit removable contamination in excess of 185 Bq (0.005 μCi) will be removed from service and disposed of in accordance with regulations.

Section 4.7.15 of the LA (IIFP, 2012a) discusses personnel access control to RCAs. For most RCAs, routine access points will be established through change rooms. Each change room will include a step-off area provided between the contamination controlled and noncontrolled areas. Instructions controlling entry and exit from RCAs will be posted at the entry points. Survey meters will be provided in the step-off area of each change room for use by personnel leaving the RCA. Posted instructions will address the use of the survey meters, donning and doffing of protective clothing, and appropriate decontamination methods. Alternate access points to RCAs will be established for specific activities not accommodated by the change rooms. Such access will be governed by approved written procedures or RWPs, which will establish controls to prevent the spread of contamination to noncontrolled areas.

RCAs that may pose a risk to employees will be identified and posted in compliance with the requirements in 10 CFR 20.1901, "Caution Signs"; 10 CFR 20.1902, "Posting Requirements"; and 10 CFR 20.1903, "Exceptions to Posting Requirements." Access to these areas will be controlled so that only appropriately trained individuals are allowed entry. Signs will be regularly inspected for conformance. The following areas will be identified and posted if applicable in accordance with definitions provided in 10 CFR 20.1003, "Definitions":

- radiation area

- high radiation area (unlikely to have but a sealed calibration source may require)

- airborne radioactivity area

- radioactive material area

In addition, contamination areas will be posted in accordance with approved written procedures. Signs will be posted at the entry points of areas requiring protective clothing. Radiation safety training and approved written procedures will instruct employees on requirements for entering and working in posted areas.

The applicant discussed the Radiation Reporting Program in Section 4.7.16 of the LA (IIFP, 2012a). IIFP will establish a Radiation Reporting Program to maintain records of the RP Program, radiation survey results, Corrective Action Program referrals, RWPs, and planned special exposures. The Radiation Reporting Program will be consistent with the guidance in Regulatory Guide 8.7, Revision 2, "Instructions for Recording and Reporting Occupational Radiation Exposure Data," issued November 2005 (NRC, 2005).

The Radiation Reporting Program will report to the NRC any event resulting in an occupational exposure to radiation exceeding the dose limits in 10 CFR 20.1201, within the time specified in 10 CFR 20.2202, "Notification of Incidents"; 10 CFR 40.60, "Reporting Requirements"; and 10 CFR 70.74, "Additional Reporting Requirements." The Radiation Reporting Program is also committed to preparing and submitting to the NRC an annual report of individual monitoring results, as required by 10 CFR 20.2206(b).

Radiation exposure data for an individual, and the results of any measurements, analyses, and calculations of radioactive material deposited or retained in the body of an individual, will be reported to the individual as specified in 10 CFR 19.13, "Notifications and Reports to Individuals." During basic radiation safety training, individuals will be advised of their right to request radiation exposure data. In accordance with 10 CFR 19.11, "Posting of Notices to Workers," IIFP management will post current copies at locations where they may be reviewed of the following documents:

- the regulations in 10 CFR Part 19 and 10 CFR Part 20

- the license, or documents incorporated into the license by reference and amendments thereto

- the operating procedures applicable to licensing activities

Based on the review as summarized above, the NRC staff finds that the commitments in the IIFP's LA (IIFP, 2012a) satisfactorily address the acceptance criteria in Section 4.4.7.3 of NUREG-1520, Revision 0 (NRC, 2002).

4.3.8 Additional Program Requirements

The NRC staff reviewed the applicant's additional program commitments against the acceptance criteria in NUREG-1520, Revision 0, Section 4.4.8.3 (NRC, 2002). The following discussion summarizes the NRC staff's analysis as to whether the information provided by the applicant in the LA (IIFP, 2012a) meets the criteria.

The applicant discussed additional program commitments in Section 4.8 of the LA (IIFP, 2012a). IIFP will maintain records of the RP Program (including program provisions, audits, and reviews of the program context and implementation); radiation survey results (air sampling, bioassays, external exposure data from monitoring individuals, and internal intakes of radioactive material); and results of Corrective Action Program referrals, RWPs, and planned special exposures. Records will be maintained consistent with the commitments in Section 11 of the LA (IIFP, 2012a).

Procedures will dictate that IIFP will report to the NRC, within the time specified by 10 CFR Part 20, Subpart M, "Reports," and 10 CFR 70.74, any event resulting in an occupational exposure to radiation exceeding the dose limits in 10 CFR Part 20.

IIFP will prepare and submit an annual report to the NRC of the results of individual monitoring, as required by 10 CFR 20.2206(b).

The IIFP Corrective Action Program will evaluate any radiation incident resulting in an occupational exposure that exceeds the dose limits in 10 CFR 20.1201, or is required to be reported in accordance with Subpart M of 10 CFR Part 20, 10 CFR 40.60, and 10 CFR 70.74. The applicant will report to the NRC the corrective actions taken (or planned) to protect against a recurrence and the proposed schedule to achieve compliance.

Based on the review as summarized above, the NRC staff finds that the commitments in the LA (IIFP, 2012a) satisfactorily address the acceptance criteria in Section 4.4.9.3 of NUREG-1520, Revision 0 (NRC, 2002).

4.3.9   Control of Radiological Risk Resulting from Accidents

The applicant submitted the ISA Summary (IIFP, 2012b), as well as the accident consequence evaluations (IIFP, 2010) supporting the ISA Summary. These documents sufficiently described the radiological hazards, accident sequences, consequences, and methodology for staff to initially evaluate compliance with the radiological provisions in 10 CFR Part 70, Subpart H. Descriptions and justifications of the initiating events, preventions, and mitigating factors were also presented to an acceptable extent. In addition, the applicant submitted an EP (IIFP, 2011) to address response to accidents at the facility. This plan generally conforms to guidance in Regulatory Guide 3.67, "Standard Format and Content for Emergency Plans for Fuel Cycle and Materials Facilities," issued January 1992 (NRC, 1992b).

As discussed in Appendix B to this SER, the NRC staff independently evaluated a sampling of the applicant's accident sequences to verify the methodologies employed and consequence categorizations. The NRC staff's results were consistent with consequence categories determined by the applicant. Overall, the applicant concluded in the accident consequence evaluations (IIFP, 2010) that radiological dose, soluble uranium exposures, and environmental releases could be categorized as "low" for all credible accident sequences evaluated consistent with categorizations in 10 CFR 70.61. The chemicals associated with the processes present the primary hazard expected at the facility, which the consequence analysis determined as having the greater impact relative to the performance criteria in 10 CFR 70.61. Because the applicant determined that radiological consequences from credible process accidents are "low," there were no IROFS identified to reduce or eliminate radiological exposures and no management measures identified to maintain the IROFS. The NRC staff also noted that the applicant identified IROFS and associated management measures to prevent or mitigate chemical exposures resulting from process accidents, as addressed in Chapter 6 of this SER.

Based on the NRC staff's evaluation of the analysis as summarized above, the NRC staff finds that the radiological evaluations the applicant performed to support the ISA, as well as the EP developed for the facility, demonstrate compliance with 10 CFR Part 70, Subpart H.

4.4   Evaluation Findings

The applicant has committed to establish and maintain an acceptable RP Program that includes the following:

- an effective, documented program to ensure that occupational radiological exposures are ALARA

- an organization with adequate qualification requirements for the RP personnel

- approved, written RP procedures and RWPs for RP activities

- RP training for all personnel who have access to restricted areas

- a program to control airborne concentrations of radioactive material with engineering controls and respiratory protection

- a radiation survey and monitoring program that includes requirements for controlling radiological contamination within the facility and monitoring external and internal radiation exposures

- other programs implement corrections at the facility, maintain records, and report to the NRC in accordance with 10 CFR Part 20, 10 CFR Part 40, and 10 CFR Part 70.

In addition, the applicant evaluated, in the ISA Summary (IIFP, 2012b), accident sequences with potential intermediate and high radiological consequences. Because none of the accident sequences resulted in intermediate or high radiological consequences, the radiological portion of the ISA Summary meets the performance criteria of 10 CFR 70.61.

The NRC staff concludes that the applicant's RP Program is adequate and meets the requirements of 10 CFR Part 19, 10 CFR Part 20, 10 CFR Part 40, and Subpart H of 10 CFR Part 70. Conformance to the analysis and commitments to the LA will ensure safe operation.

4.5   References

(ANSI/HPS, 2009) American National Standards Institute/Health Physics Society, "Specifications for the Bottle Manikin Absorption Phantom," ANSI/HPS N13.35-2009, 2009.

(ANSI/HPS, 1996) American National Standards Institute/Health Physics Society, "Performance Criteria for Radiobioassay," ANSI/HPS N13.30-1996, 1996.

(ANSI/HPS, 1995) American National Standards Institute/Health Physics Society, "Bioassay Program for Uranium," ANSI/HPS N13.22-1995, 1995.

(ICRP, 1994) International Commission on Radiological Protection, "Dose Coefficients for Intakes by Workers," Publication 68, 1994.

(IIFP, 2012a) International Isotopes Fluorine Products, Inc., "Fluorine Extraction Process and Depleted Uranium Deconversion Plant (FEP/DUP) License Application, Revision B," May 2012, Agencywide Documents Access and Management System (ADAMS) Accession No. ML12123A245.

(IIFP, 2012b) International Isotopes Fluorine Products, Inc., "ISA Summary Rev. B for IIFP," May 2012, Agencywide Documents Access and Management System (ADAMS) Accession No. ML12123A245.

(IIFP, 2011) International Isotope Fluorine Products, Inc., "(NonPublic) Emergency Plan Rev B of IIFP License Application," May 2012, Agencywide Documents Access and Management System (ADAMS) Accession No. ML12123A245.

(IIFP, 2010) International Isotope Fluorine Products, Inc. (IIFP), "Letter from INIS, RE: Integrated Safety Analysis Supporting Documents Including: Consequence Evaluation Calculations and Accident Consequence Evaluations TAC L32739," June 2010. Agencywide Documents Access and Management System (ADAMS) Accession No. ML102560265

(ISO, 1992) International Organization for Standardization, "Radiation Protection—Sealed Radioactive Sources—Leakage Test Methods," ISO 9978, 1992.

(NRC, 2008) U.S. Nuclear Regulatory Commission, "Minimization of Contamination and Radioactive Waste Generation: Life-Cycle Planning," Regulatory Guide 4.21, June 2008.

(NRC, 2005) U.S. Nuclear Regulatory Commission, "Instructions for Recording and Reporting Occupational Radiation Exposure Data," Regulatory Guide 8.7, Rev. 2, November 2005.

(NRC, 2002) U.S. Nuclear Regulatory Commission, "Standard Review Plan for the Review of a License Application for a Fuel Cycle Facility," NUREG-1520, Rev. 0, March 2002.

(NRC, 1999a) U.S. Nuclear Regulatory Commission, "Instruction Concerning Prenatal Radiation Exposure," Regulatory Guide 8.13, Rev. 3, June 1999.

(NRC 1999b) U.S. Nuclear Regulatory Commission. "Acceptable Programs for Respiratory Protection," Regulatory Guide 8.15, Rev. 1, October 1999.

(NRC, 1996) U.S. Nuclear Regulatory Commission, "Instruction Concerning Risk from Occupational Radiation Exposure," Regulatory Guide 8.29, Rev. 1, February 1996.

(NRC, 1993a) U.S. Nuclear Regulatory Commission, "ALARA Levels for Effluents from Materials Facilities," Regulatory Guide 8.37, July 1993.

(NRC, 1993b) U.S. Nuclear Regulatory Commission, Branch Technical Position, "Guidelines for Decontamination of Facilities and Equipment Prior to Release for Unrestricted Use or Termination of Licenses for Byproduct, Source, or Special Nuclear Material," 1993.

(NRC, 1993c) U.S. Nuclear Regulatory Commission, Branch Technical Position, "License Condition for Leak-Testing Sealed Byproduct Material Sources," April 1993.

(NRC, 1993d) U.S. Nuclear Regulatory Commission, Branch Technical Position, "License Condition for Leak-Testing Sealed Plutonium Sources," April 1993.

(NRC, 1993e) U.S. Nuclear Regulatory Commission, Branch Technical Position, "License Condition for Plutonium Alpha Sources," April 1993.

(NRC, 1993f) U.S. Nuclear Regulatory Commission, Branch Technical Position, "License Condition for Leak-Testing Sealed Source Which Contains Alpha and/or Beta-Gamma Emitters," April 1993.

(NRC, 1993g) U.S. Nuclear Regulatory Commission, Branch Technical Position, "License Condition for Leak-Testing Sealed Uranium Sources," April 1993.

(NRC, 1992a) U.S. Nuclear Regulatory Commission, "Monitoring Criteria and Methods to Calculate Occupational Radiation Doses," Regulatory Guide 8.34, July 1992.

(NRC, 1992b) U.S. Nuclear Regulatory Commission, "Standard Format and Content for Emergency Plans for Fuel Cycle and Materials Facilities," Regulatory Guide 3.67, January 1992.

(NRC, 1992c) U.S. Nuclear Regulatory Commission, "Air Sampling in the Workplace," Regulatory Guide 8.25, Rev. 1, June 1992.

(NRC, 1977) U.S. Nuclear Regulatory Commission, "Operating Philosophy for Maintaining Occupational Radiation Exposures as Low as Is Reasonably Achievable," Regulatory Guide 8.10, Rev. 1, May 1977.

(NRC, 1973) U.S. Nuclear Regulatory Commission, "Guide for Administrative Practice in Radiation Monitoring," Regulatory Guide 8.2, February 1973.

5.0 NUCLEAR CRITICALITY SAFETY

The IIFP facility is a source material facility licensed under Title 10 of the *Code of Federal Regualtions* (10 CFR) Part 40 and the Integrated Safety Analysis (ISA) requirements in 10 CFR Part 70, Subpart H. As such the facility does not process or possess special nuclear material (SNM) other than check sources used for calibration licensed by the Agreement state. International Isotopes Fluorine Products, Inc. (IIFP), does not have a Nuclear Criticality Safety (NCS) program or procedures as described in Chapter 5, "Nuclear Criticality Safety" of the license application (IIFP, 2012a).

5.1 Regulatory Requirements

Title 10 CFR Part 40 does not contain criticality requirements for source material due to the absence of enriched material. IIFP must demonstrate compliance with the 10 CFR Part 70, Subpart H, including requirements involving criticality.

As part of the ISA, IIFP did identify one criticality accident scenario which involves the inadvertent receipt and processing of a shipment of fissile material. IIFP has identified controls which will prevent this type of accident. The cylinders typically processed at the facility consist of the 14-ton uranium hexafluoride tails cylinders, not the typical 2½-ton enriched cylinders. In addition, an item relied on for safety (IROFS).

5.2 Evaluation of Findings

Based on this review, the U.S. Nuclear Regulatory Commission (NRC) staff finds acceptable the applicant's assessment that IIFP is not authorized to possess SNM under the NRC license and, therefore, is not required to implement an NCS program. Consistent with the ISA requirements in 10 CFR 70.61(d) and 10 CFR 70.64(a), the applicant has identified adequate controls and IROFS to prevent a criticality involving an inadvertent shipment of fissile material.

Therefore, the NRC staff concludes that the applicant's information regarding NCS meets the requirements of 10 CFR Part 70, Subpart H, and provides reasonable assurance for the protection of public health and safety, including workers and the environment.

5.3 References

(IIFP, 2012a) International Isotopes Fluorine Products, Inc., "Fluorine Extraction Process and Depleted Uranium Deconversion Plant (FEP/DUP) License Application, Revision B," May 2012, Agencywide Documents Access and Management System (ADAMS) Accession No. ML12123A245.

(IIFP, 2012b) International Isotopes Fluorine Products, Inc., "ISA Summary Rev. B for IIFP," May 2012, Agencywide Documents Access and Management System (ADAMS) Accession No. ML12123A245.

6.0   CHEMICAL PROCESS SAFETY

The purpose of the U.S. Nuclear Regulatory Commission's (NRC's) review of the International Isotopes Fluorine Products, Inc. (IIFP or applicant), Chemical Safety Program and the facility's design is to evaluate whether the applicant will adequately protect workers, the public, and the environment during normal operations against chemical hazards of licensed material and its byproducts. The Chemical Safety Program and the facility's design must also protect against facility conditions and operator actions that can affect the safety of licensed materials and thus present an increased chemical and radiological risk.

6.1   Regulatory Requirements

In Staff Requirements Memorandum SECY-07-0146, "Regulatory Options for Licensing New Uranium Conversion and Depleted Uranium Deconversion Facilities," dated October 10, 2007 (NRC, 2007a), the Commission stated, "If new license applications are submitted before the completion of the rulemaking [to Title 10 of the Code of *Federal Regulations* (10 CFR) Part 40], the NRC staff shall impose 10 CFR Part 70, Subpart H, performance requirements as part of the licensing basis for the application review." Therefore, since revisions to 10 CFR Part 40 are not completed, the regulatory bases for the review must address the chemical process safety requirements in 10 CFR 70.65, "Additional Content of Applications." In addition, the chemical process safety review should provide reasonable assurance of compliance with 10 CFR 70.61, "Performance Requirements"; 10 CFR 70.62, "Safety Program and Integrated Safety Analysis"; and 10 CFR 70.64, "Requirements for New Facilities or New Processes at Existing Facilities."

6.2   Regulatory Guidance and Acceptance Criteria

Chapter 6 of NUREG-1520, "Standard Review Plan for the Review of a License Application for a Fuel Cycle Facility," issued March 2002 (NRC, 2002), contains the guidance applicable to NRC's review of chemical process safety for the proposed facility. This chapter is applicable in its entirety. The NRC staff also used the following as guidance documents for this review: NUREG-1601, "Chemical Process Safety at Fuel Cycle Facilities," issued August 1997 (NRC, 1997), and NUREG-1513, "Integrated Safety Analysis Guidance Document," issued May 2001 (NRC, 2001). Section 6.4.3 of NUREG-1520 (NRC, 2002) provides the acceptance criteria applicable to this review.

6.3   Staff Review and Analysis

The NRC staff reviewed the license application (LA) (IIFP, 2012a) and the Integrated Safety Analysis (ISA) Summary (IIFP, 2012b) submitted by the applicant and considered the following eight areas as they relate to chemical process safety:

(1)   process description
(2)   chemical accident sequences
(3)   chemical accident consequences
(4)   chemical process items relied on for safety (IROFS)
(5)   chemical process management measures
(6)   emergency management
(7)   baseline design criteria (BDC)
(8)   defense-in-depth

The NRC staff reviewed the applicant's responses to requests for additional information (RAI) (IIFP, 2011) and ISA documents during an in-office review (NRC, 2010), as necessary, to gain a better understanding of the process and safety requirements. The NRC staff evaluated the information to determine whether the facility's design complied with the chemical protection BDC and defense-in-depth requirements as they relate to chemical process safety specified in 10 CFR 70.64(a)(5) and 10 CFR 70.64(b), respectively. Section 6.3.6 of this Safety Evaluation Report (SER) discusses compliance with these regulations in more detail. The following sections summarize the NRC staff's evaluation and general information about the proposed fluorine extraction process and depleted uranium deconversion plant (FEP/DUP).

6.3.1 Process Description

The applicant described the proposed FEP/DUP in Section 1.1.3 of the LA (IIFP, 2012a) and provided a more detailed description in Chapter 3 of the ISA Summary (IIFP, 2012b). The process will begin at the proposed depleted uranium hexafluoride (DUF_6) to depleted uranium tetrafluoride (DUF_4) deconversion plant. The proposed DUF_6 to DUF_4 deconversion plant, which is described in Section 6.3.1.1 of this SER, will produce DUF_4 and anhydrous hydrogen fluoride (AHF). DUF_4 will be transferred to one of two proposed fluorine extraction processes (FEPs), which will produce fluoride products and uranium oxides. AHF will be sold as a chemical commodity. The two proposed processes in which DUF_4 will be transferred are the silicon tetrafluoride (SiF_4) process and boron trifluoride (BF_3) process. The products that will be obtained from the proposed FEP are fluoride gases (SiF_4 and BF_3), and uranium oxides (uranium dioxide [UO_2] or triuranium octoxide [U_3O_8]). The applicant plans to sell the fluoride gases to several industries. The uranium oxides are more stable forms of uranium and are suitable for disposal at licensed waste facilities. Sections 6.3.1.2 and 6.3.1.3 of this SER describe the proposed SiF_4 and BF_3 processes, respectively. The proposed IIFP facility will also consist of support processes, which include process offgas treatment (i.e., potassium hydroxide [KOH] scrubbing system), environmental protection process, fluoride products storage and packaging, AHF staging containment, and fluoride products trailer loading. This section describes these support processes.

The applicant provided engineering drawings as part of the LA submittal. The engineering drawings are process flow diagrams (PFDs) of each of the facility processes. The PFDs are publicly available and can be obtained through the NRC's Agencywide Documents Access and Management System (ADAMS) using the Accession Number ML100120768. The descriptions provided below are based on the applicant's description and these PFDs.

6.3.1.1 *DUF₆ to DUF₄ Deconversion Plant*

The DUF_6 will arrive at the proposed IIFP facility in 14-ton, type 48-Y cylinders owned by the IIFP deconversion customer. The DUF_6 in these cylinders will be in the solid state. These cylinders will be built to the standards of the American National Standards Institute (ANSI) (ANSI, 2001) and will be transported by truck trailers approved by the U.S. Department of Transportation (DOT).

Upon receipt, IIFP staff will visually inspect full cylinders of DUF_6 for damage and survey them for radiation and removable contamination. Documents that accompany the shipment and contain information regarding cylinder identification, weight, and uranium assay will be reviewed and verified for accuracy. Uranium assay will be qualitatively verified by performing a nondestructive gamma survey measurement. Once accepted for receipt, the facility cylinder hauler vehicle will unload the cylinder and place it in the full DUF$_6$ storage pad area until it is scheduled for feed to the deconversion process.

Once the cylinder is scheduled to be fed into the deconversion process, the cylinder hauler vehicle will retrieve the cylinder from the full DUF$_6$ storage pad area and place it onto a cylinder scale cart for movement into the autoclave room. The cylinder bridge crane will lift the cylinder from the scale cart and place it on the accountability scale where the cylinder weight will be measured, recorded, and compared against the applicable weight limits (i.e., 27,560 pounds [lb] of DUF$_6$ for type 48-Y cylinders). If the cylinder weight is within safety limits, the cylinder bridge crane will lift it and place into one of two steam-heated containment type autoclaves. These autoclaves will be cylindrical pressure vessels designed to meet the requirements of Section VIII, Division 1, of the American Society of Mechanical Engineers (ASME) Boiler and Pressure Vessel Code (ASME, 2004). The autoclaves will have a maximum allowable working pressure of 200 pounds per square inch gauge (psig) at 250 degrees Fahrenheit (F).

The cylinder will be connected to the feed manifold using a connection line (pigtail) compatible with uranium hexafluoride (UF$_6$). The pigtail will be visually inspected and tested for leaks. Before heating the cylinder, IIFP staff will perform tests to ensure that it is safe to heat the cylinder. These tests will include opening the cylinder valve to ensure valve clarity and checking the cylinder pressure (cold pressure check) to ensure that the cylinder does not contain significant amounts of noncondensable or impurity gases. Once all requirements for feeding have been met, the autoclave hydraulic closing system will close the autoclave door.

The DUF$_6$ cylinder will be heated using saturated steam and, as a result, the gaseous DUF$_6$ will flow through the heat-traced and insulated piping of the feed header to the DUF$_6$ surge tank. When the quantity of DUF$_6$ in a cylinder has been reduced to the extent that the desired flow of DUF$_6$ cannot be maintained, the residual DUF$_6$ will be removed from the cylinder using a process called heeling. Heeling will be accomplished by connecting the cylinder to a low pressure source, such as the purge and evacuation system. The residual DUF$_6$ will be transferred to another 48-Y cylinder located inside a cylinder cold box. The cylinder cold box will be located in the autoclave building. After the heeling process is complete, the cylinder weight will be measured to verify that it is empty. Once verified, the cylinder hauler vehicle will transport the cylinder to the empty cylinder area where it is stored until it is returned to the deconversion customer.

The DUF$_6$ will be fed from the surge tank to a high-temperature reaction vessel where it will react with hydrogen (H$_2$) to produce solid DUF$_4$ and gaseous AHF. The reaction is exothermic and is shown in Equation 6-1 below. The gaseous H$_2$ supply will be produced onsite by steam reforming of natural gas followed by purification using pressure swing adsorption. Steam reforming of natural gas is one of the most developed and commercially used technologies (Kutz, 2007) for hydrogen production.

$$UF_{6\,(gas)} + H_{2\,(gas)} \rightarrow UF_{4\,(solid)} + 2HF_{(gas)} \hspace{2cm} Eq.\ 6\text{-}1$$

The solid DUF$_4$ powder will be continuously withdrawn from the bottom of the reaction vessel using a cooling screw mechanism and transferred to storage hoppers. The DUF$_4$ in the storage hoppers will be transferred to the FEP plant for use as raw material feed in the production of SiF$_4$ and BF$_3$. A two-stage dust collector system will be provided to control and recycle DUF$_4$ dusts that are internal to the solids-handling equipment and generated by air or gas flow associated with the handling equipment. The collected fines from the dust collector system will be discharged into the vacuum transfer system to the DUF$_4$ feed hopper located in the proposed

FEP plant. The applicant will continuously monitor the vent stack flow to the environment for uranium emissions.

The cooling screw mechanism will have an offgas plenum where the entrained DUF_4 particles, unreacted DUF_6, gaseous AHF, unreacted H_2, and nitrogen (N_2) (used as a buffer on the cooling screw bearing seals) will flow and pass through a filter system to remove entrained DUF_4 particles. The filter system will comprise primary and secondary stages configured in series. The primary stage will be a cyclone separator with a section of sintered metal filter tubes. Cyclone separators have been used in the United States for about 100 years and are one of the most widely used of all industrial gas-cleaning devices (Benítez, 1993). The cyclone separator will remove the larger entrained particles, while finer particles will be screened by the sintered metal filter tubes and fall down to the cyclone separator. The collected solids will be discharged to a fines hopper for transfer to the DUF_4 powder stream. The secondary stage is a backup to the first-stage filter. It will capture the particles that may pass the first-stage filter. The solids from the secondary stage will be discharged into the vacuum transfer system to the DUF_4 feed hopper located in the proposed FEP plant.

Following the filter system, the offgas will flow through one of the two parallel banks of carbon-bed traps. The carbon-bed trap material will be activated carbon. The two parallel banks will provide operational flexibility, and each bank will have three carbon-bed traps configured in series. The carbon-bed traps will remove any trace quantities of unreacted DUF_6 left in the offgas. Once a carbon-bed trap is determined to be full, the flow to the carbon-bed trap will be isolated; and the carbon will be removed and transferred to the decontamination building for removal, replacement, and eventual approved disposal of uranium-contaminated activated carbon. After exiting the carbon-bed traps, the offgas will enter a series of two heat exchangers. The first heat exchanger will be a partial hydrogen fluoride (HF) condenser. The offgas from the partial HF condenser will flow into the second heat exchanger, the total HF condenser. The condensed HF from both condensers will be drained to the HF storage tanks. The AHF staging containment building will house the HF storage tanks; this building will be separated from licensed materials. Section 6.3.1.7 of this SER describes the AHF staging containment building. The HF will be temporarily stored and then loaded into truck trailers inside the AHF staging containment building for shipment to customers. DOT will approve the trailers for shipment.

Residual offgases (mainly unreacted H_2, N_2, and small amounts of HF) will leave the total HF condenser to a hydrogen burner system to combust any unreacted H_2. The small amounts of HF in the offgas leaving the hydrogen burner system will flow to the KOH scrubbing system. The KOH scrubbing system is designed to remove the small amounts of HF before venting the offgases to the atmosphere. Section 6.3.1.4 of this SER describes this system.

The NRC staff finds that the ISA Summary (IIFP, 2012b) adequately describes the DUF_6 to DUF_4 deconversion plant because it provides information that allows the NRC staff to understand the process and the chemical hazards that could result from potential accident sequences.

6.3.1.2 SiF_4 Process

The proposed SiF_4 process will contain two equal production trains. For simplification, this section will only describe one of these production trains.

DUF$_4$ will be transferred from the proposed DUF$_6$ to DUF$_4$ deconversion plant to a hopper in the proposed SiF$_4$ process. Silicon dioxide (SiO$_2$) will be stored in another hopper. The contents from the DUF$_4$ and SiO$_2$ hoppers will be mixed and then fed to a rotary calciner. In the rotary calciner, DUF$_4$ and SiO$_2$ will react to form gaseous SiF$_4$ and solid UO$_2$. Equation 6-2, below, shows this reaction. Small amounts of moisture (i.e., water vapor) may be contained in the powder mixture and, because of the high temperatures in the rotary calciner DUF$_4$ and water, will react to form gaseous HF and solid UO$_2$. Equation 6-3 shows this reaction.

$$UF_{4\ (solid)} + SiO_{2\ (solid)} \rightarrow SiF_{4\ (gas)} + UO_{2\ (solid)}$$ Eq. 6-2

$$UF_{4\ (solid)} + 2H_2O_{(vapor)} \rightarrow 4HF_{(gas)} + UO_{2\ (solid)}$$ Eq. 6-3

The resulting SiF$_4$ gas and trace impurities (mainly HF) will exit the rotary calciner as an offgas , while the UO$_2$ powder will be discharged at the end of the rotary calciner through a cooling screw mechanism and transferred to storage hoppers. A gas plenum that provides a redundant flow path for the offgases will be located at the top of the cooling screw mechanism. A two-stage dust collector system will control and recycle uranium oxide dusts that are internal to the solids-handling equipment and generated by air or gas flows associated with the handling equipment. The proposed SiF$_4$ and BF$_3$ processes will share the two-stage dust collector system, and the collected solids will be discharged to an uranium oxide hopper. The vent stack flow to the environment will be continuously monitored for uranium emissions. The uranium oxide in the storage hoppers will be packaged into DOT-approved shipping containers and transported to an offsite licensed disposal facility.

Offgas leaving the rotary calciner and the cooling screw mechanism gas plenum will flow through two-stages of filters to capture entrained particles. The two-stage filter system will be similar to the filter system in the proposed DUF$_6$ to DUF$_4$ deconversion plant, described in Section 6.3.1.1 of this SER. The collected solids from both filters will be returned to the fines hopper to become part of the powder mix fed to the rotary calciner. After exiting the filter system, the offgas will flow to one of two precondensers to remove HF and other trace impurities from the SiF$_4$ product stream. The collected HF and other trace impurities will be batch transferred to the KOH scrubbing system. Two precondensers will be provided for operational flexibility (i.e., one precondenser can transfer its contents to the KOH scrubbing system while the other is operating normally).

The SiF$_4$ product offgas will leave the precondenser and flow into a series of two product cold traps. The series of product cold traps will be configured into two parallel banks (i.e., two primary product cold traps and two secondary product cold traps). Again, this arrangement will be provided for operational flexibility. The primary product cold trap will solidify SiF$_4$, and the secondary product cold trap will solidify any SiF$_4$ remaining in the product offgas after passing the primary product cold trap. The offgas from the primary and secondary product cold traps will be sent to the KOH scrubbing system. After a product cold trap signals that it is full of solid SiF$_4$ product, the cold trap will be set to an unloading mode by increasing the coolant temperature, thus causing the solid SiF$_4$ to turn into liquid or gaseous SiF$_4$. The liquid or gaseous SiF$_4$ will flow to an evaporator vessel using a transfer compressor. The applicant stated that the evaporator vessel and transfer compressor will be located in the proposed FEP product storage and packaging area. The proposed FEP product storage and packaging area will be physically separated from the SiF$_4$ process and licensed material. Section 6.3.1.6 of this SER describes the FEP product storage and packaging area.

The NRC staff finds that the ISA Summary (IIFP, 2012b) adequately describes the SiF_4 process because it provides information that allows the NRC staff to understand the process and the chemical hazards that could result from potential accident sequences.

6.3.1.3 BF_3 Process

DUF_4 will be transferred from the proposed DUF_6 to DUF_4 deconversion plant to a hopper in the BF_3 process. Diboron trioxide (B_2O_3) will be stored in another hopper. The contents from the DUF_4 and B_2O_3 hoppers will be mixed, fed to a preheater, and then fed to a rotary calciner. The temperature in the preheater will be controlled to cause a reaction of small amounts of DUF_4 with moisture (shown in Equation 6-3 above) that may be contained in the powder mixture. This will minimize the amount of HF impurities in the product gas stream. The HF will leave the preheater, move through a series of filters, and then enter the KOH scrubbing system. The filter system will be similar to the two-stage filter system of the proposed DUF_6 to DUF_4 deconversion plant. The collected solids will be transferred to the fines hopper. The contents of the fines hopper will mix with the DUF_4 and B_2O_3 powders before entering the preheater. In the rotary calciner, DUF_4 and B_2O_3 will react to form gaseous BF_3 and solid UO_2. Equation 6-4 shows this reaction.

$$UF_{4\,(solid)} + B_2O_{3\,(solid)} \rightarrow BF_{3\,(gas)} + UO_{2\,(solid)} \qquad \text{Eq. 6-4}$$

The resulting BF_3 gas and trace impurities (mainly HF) will exit the rotary calciner as an offgas , while the UO_2 powder will discharge at the end of the rotary calciner through a cooling screw mechanism and then transfer to storage hoppers. A gas plenum that will provide a redundant flow path for the offgases will be located at the top of the cooling screw mechanism. A two-stage dust collector system will control and recycle uranium oxide dusts internal to the solids handling equipment and generated by air or gas flows associated with the handling equipment. The SiF_4 and BF_3 processes will share this two-stage dust collector system, which is described in Section 6.3.1.2 of this SER. The uranium oxide in the storage hoppers will be packaged into DOT approved shipping containers and transported to an offsite licensed disposal facility.

Offgas leaving the rotary calciner and the cooling screw mechanism gas plenum will flow through two-stages of filters to capture entrained particles. The two-stage filter system will be similar to the filter system in the proposed DUF_6 to DUF_4 deconversion plant. The collected solids from both filters will be returned to the fines hopper to become part of the powder mix fed to the preheater and then to the rotary calciner. After exiting the filter system, the offgas will flow to one of two precondensers to remove HF and other trace impurities from the BF_3 product stream. The collected HF and other trace impurities will be batch transferred to the KOH scrubbing system. As in the SiF_4 process, the two precondensers will be provided for operational flexibility (i.e., one precondenser can transfer its contents to the KOH scrubbing system while the other is operating normally).

The BF_3 product offgas will leave the precondenser to a series of two product cold traps. The series of product cold traps will be configured into two parallel banks (i.e., two primary product cold traps and two secondary product cold traps). Again, this arrangement will be provided for operational flexibility. The primary product cold trap will solidify BF_3, and the secondary product cold trap will solidify any BF_3 remaining in the product offgas after passing the primary product cold trap. The offgas from the primary and secondary product cold traps will be sent to the KOH scrubbing system. After a product cold trap signals that it is full of solid BF_3 product, the cold trap will be set to an unloading mode by increasing the coolant temperature, thus causing the solid BF_3 to turn into liquid or gaseous BF_3. The liquid or gaseous BF_3 will flow to an evaporator vessel using a transfer compressor. The applicant stated that the evaporator vessel and transfer compressor will be located in the FEP product storage and packaging area. The FEP product storage and packaging area will be physically separated from the BF_3 process and licensed material. Section 6.3.1.6 of this SER describes the FEP product storage and packaging area.

The NRC staff finds that the ISA Summary (IIFP, 2012b) adequately describes the BF_3 process because it provides information that allows the NRC staff to understand the process and the chemical hazards that could result from potential accident sequences

6.3.1.4   KOH Scrubbing System

The KOH scrubbing system will comprise two parallel lines, and each line will have three scrubber stages. One of the lines will process HF from the proposed DUF_6 to DUF_4 deconversion plant and from the SiF_4 and BF_3 precondensers. The other line will process the final offgas coming from the SiF_4 and BF_3 processes, containing uncollected SiF_4 and BF_3 and trace quantities of other fluorides. KOH will be used as the scrubber liquor in each of the three scrubber stages. The three scrubber stages will be (1) a primary wet venturi scrubber, (2) a secondary countercurrent flow gas liquid-packed tower, and (3) a bed of sized coke.

The system equipment will consist of a KOH storage tank, KOH pump tank, regenerated KOH tank, two venturi scrubbers, two spare venturi scrubbers, two packed towers, and two coke boxes. Redundant pumps will be available for each scrubber, pump tank, and storage tank. To remove the fluoride components from the offgas stream, the following reactions will occur:

$$HF_{(anhydrous)} + H_2O \rightarrow HF_{(aqueous)} \qquad \text{Eq. 6-5}$$

$$HF + KOH \rightarrow KF + H_2O \qquad \text{Eq. 6-6}$$

$$3SiF_4 + 4KOH \rightarrow 2K_2SiF_6 + 2H_2O + SiO_2 \qquad \text{Eq. 6-7}$$

$$SiO_2 + 2KOH \rightarrow K_2SiO_3 + H_2O \qquad \text{Eq. 6-8}$$

$$4BF_3 + 3KOH \rightarrow 3KBF_4 + B(OH)_3 \qquad \text{Eq. 6-9}$$

The KOH scrubbing system will vent the treated offgases through a single stack, and it will be designed to remove fluoride-bearing components in the offgas stream at approximate efficiencies of greater than 80 percent, 95 percent, and 99 percent for the first, second, and third stages, respectively. The overall system removal efficiency will be designed at greater than 99.9 percent. The system stack will be continuously sampled and routinely analyzed to measure for traces of fluorides or uranium in the vent gas.

The KOH solution will be used and recycled within each of the scrubbers until the KOH concentration needs replenishment. The concentration of KOH in the scrubber equipment will be maintained at a safe margin to ensure that it effectively reacts with fluoride components in the offgas stream. When the KOH concentration needs replenishment, some of the spent scrubbing solution containing potassium fluoride (KF), water, and some excess KOH will be pumped from the scrubber recycle tanks to the environmental protection process (EPP). Section 6.3.1.5 of this SER describes the EPP.

6.3.1.5   Environmental Protection Process

The EPP will primarily be a means of treating two types of solutions that result from the production processes—KF solutions (KOH regeneration process) and weak aqueous HF (HF neutralization process). Each of these materials will originate from scrubbing systems designed to prevent the emission of fluoride compounds into the air. The KF solution will be a byproduct of using KOH as a scrubbing medium (see Equation 6-6 above). In the KOH regeneration process of the EPP, the KF, water, and excess KOH spent solution from the plant's KOH scrubbing system will be reacted with a lime slurry. Calcium fluoride (CaF_2) and regenerated KOH solution will be produced. The regenerated KOH will be recycled and reused in the plant's scrubbing process. The CaF_2 will be filtered, dried, and packaged for shipment to an approved commercial waste burial site, to an HF producer, or to other potential users.

The other stream treated in the EPP will be weak aqueous HF solution, water, or KOH solution that may contain a low concentration of fluorides. Also, small spills that potentially occur and require cleanup from spill control containment areas may contain weak fluoride concentrations. In this case, the fluoride-bearing liquids may have too much water to send to the KOH regeneration and recycle system. The HF neutralization process will use lime slurry to react with weak HF to produce CaF_2 and water.

HF Neutralization

The HF neutralization process will be designed to operate intermittently, as needed. A lime silo will be provided, including an installed dust collector. The silo will hold an inventory of hydrated lime. Lime will be fed to a mix tank where it will be mixed with harvested water. The slurry generated will be approximately 30 percent solids. Dilute HF solution will be transferred from the weak HF solution tank to an agitated acid reaction vessel. The lime slurry from the mix tank will be also transferred to the acid reaction vessel. The materials in the acid reaction vessel will require a retention time of about 1 hour or greater for reaction completion. With the reaction complete, materials from the acid reaction vessel will be transferred to a thickener tank for settling. After thickening, CaF_2 and excess lime will be transferred by a slurry-type pump from the bottom of the thickener to a rotary drum vacuum filter. Solids will be discharged from the filter to a dryer capable of removing excess water. Liquors from the rotary vacuum filter will be recycled to the weak HF solution tank for recycling. After drying, CaF_2 will be packaged suitable for sale or disposal in an appropriate offsite licensed Resource Conservation Recovery Act (RCRA) disposal facility. Equation 6-10 shows the primary chemical reactions:

$$2HF + Ca(OH)_2 \rightarrow CaF_2 + 2H_2O \qquad\qquad\qquad \text{Eq. 6-10}$$

KOH Regeneration

Lime will be fed to an agitated mix tank where it will mix with harvested water. The slurry generated will contain approximately 30 percent solids. Spent KOH solution (KF solution containing a weak concentration of KOH) will be transferred from a spent KOH storage tank to an agitated reaction vessel. The lime slurry from the mix tank will be also transferred to the reaction vessel. The materials in the reaction vessel tank will be given a retention time of about 1 hour or greater for reaction completion. With the reaction complete, materials from the reaction vessel will be transferred to a thickening tank for settling. Excess lime and CaF_2 will be transferred by a slurry pump from the bottom of the thickener to a rotary drum vacuum filter. Solids will be discharged from the filter to a dryer capable of processing excess water. Liquors will be transferred to a clarifier where trace solids will be settled. Regenerated KOH will be removed from the top of the clarifier and passed through a set of filters to the regenerated KOH storage tank. The regenerated KOH solution will be pumped to the plant's KOH scrubbing system as needed for reuse by the scrubbers. Solids will be transferred via a slurry pump from the bottom of the clarifier to the rotary drum vacuum filter and subsequently transferred to the dryer. The dried material will be packaged and stored for sale or sent to an approved offsite licensed RCRA disposal facility. Equation 6-11 shows the primary chemical reaction:

$$2KF + Ca(OH)2 \rightarrow CaF2 + 2KOH \qquad \text{Eq. 6-11}$$

6.3.1.6 *Fluorine Extraction Process Storage and Packaging Area*

As described in Sections 6.3.1.2 and 6.3.1.3 of this SER, the FEP storage and packaging area will be physically separated from licensed material. Therefore, the applicant stated that this process is outside the scope of NRC's licensing review. However, the applicant stated that it conducted a process hazard analysis in accordance with the Occupational Safety and Health Administration (OSHA) regulations in 29 CFR 1910.119, "Process Safety Management of Highly Hazardous Chemicals." For completeness, this section will briefly describe the FEP storage packaging area.

From the evaporator and transfer compressor, the SiF_4 will flow into a cooler and then into the storage tubes. There will be 15 storage tubes for SiF_4 storage and one dump tube for accepting SiF_4 gas in the event a relief valve from one of the other storage tubes actuates. Five of the 15 storage tubes will be spares to be used when necessary. For BF_3, the storage process will be essentially the same as that for SiF_4. The difference is that there will be 10 storage tubes instead of 15. Five of the 10 tubes will be spares. The storage tubes for both products will be ASME Code rated pressure vessels.

The SiF_4 or BF_3 gas will be packaged into small container cylinders or large container tube trailers. The small container cylinders and large container tube trailers will arrive at the applicant's proposed facility from IIFP's customers. When packaging is scheduled, the cylinder or tube trailer will be connected to the respective SiF_4 or BF_3 packaging manifold. Each packaging manifold will be enclosed in a containment-type packaging station that will be connected via a hood and ducts to the emergency KOH scrubbing system. Detectors will be located inside and outside the packaging station that will alarm and automatically shut off the flow of product gas at the storage tube manifold from which the package is being filled, and it will shut off the flow at the packaging manifold to the cylinder or tube trailer being filled. If a leak is detected, the packaging station will open valves on the station to vent to the emergency KOH scrubbing system.

The small container cylinders and large container tube trailers will be ASME Code rated pressure vessels and will require DOT's approval for shipment.

6.3.1.7 *AHF Staging Containment and Fluoride Products Trailer Loading Buildings*

Similar to the FEP storage and packaging area, the AHF staging containment and fluoride products trailer loading buildings will be separated from licensed material. Safeguards and operational controls will be provided to meet the requirements of OSHA regulations (29 CFR 1910.119) or Federal and State of New Mexico environmental permit requirements. Again, for completeness, this section will briefly describe the processes in these buildings.

As described in Section 6.3.1.1 of this SER, AHF from the partial and total HF condensers will be transferred to the AHF storage tanks located in the AHF staging containment building. Dikes will be provided to each AHF storage tank. Each dike will be sized to hold the contents of the entire tank and minimize the surface area of evaporation of AHF in the case of a leak from an AHF storage tank. The AHF will be temporarily stored, and when the inventory in a storage tank reaches a level for shipment, the AHF will be loaded into a customer-owned tank trailer staged in the fluoride products trailer product building. The tank trailers will be DOT approved for AHF shipment to customers.

The fluoride products trailer loading building will serve two purposes. As mentioned above, it will serve as a loading place for AHF tank trailers from storage. The second purpose will be to load gas-tube trailers with SiF_4 or BF_3 transferred from the FEP product storage and packaging building (briefly described in Section 6.3.1.6 of this SER).

The AHF staging containment and fluoride products trailer loading buildings will be connected and will have a fluoride detection system and water-spray deluge system. If a leak of AHF or fluoride product gas occurs, the fluoride detection system will automatically initiate an alarm and isolate the transfer of products at the storage tanks and tank trailer fill lines. The system will also automatically initiate the water-spray deluge system to knock down the fluoride vapors within the buildings. The water from the water-spray deluge system is gravity drained to a holding tank that vents to the KOH scrubbing system. If the holding tank contains aqueous HF, it would be sent to the EPP (described in Section 6.3.1.5 of this SER).

6.3.1.8   Hazardous Chemicals and Chemical Interactions

The hazardous chemicals present in significant quantities in the proposed IIFP facility will be UF_6, uranium tetrafluoride (UF_4), UO_2, HF, SiF_4, and BF_3. Any UF_6 that is released to the environment will react exothermically with moisture in the air producing solid uranyl fluoride (UO_2F_2) and HF gas. Equation 6-12 shows the reaction of gaseous UF_6 with water vapor at elevated temperatures:

$$UF_{6\,(gas)} + 2H_2O_{(vapor)} \rightarrow UO_2F_{2\,(solid)} + 4HF_{(gas)} + heat \qquad \text{Eq. 6-12}$$

At room temperature, depending on the relative humidity of the air, the products of this reaction are UO_2F_2, hydrates, and HF-H_2O fog, which will be seen as a white cloud. Equation 6-13 shows a typical reaction with excess water:

$$UF_{6\,(gas)} + (2+4x)H_2O_{(vapor)} \rightarrow UO_2F_2 \cdot 2H_2O_{\,(solid)} + 4\,HF \cdot xH_2O_{(fog)} + heat \qquad \text{Eq. 6-13}$$

These reactions, if occurring in the gaseous phase at ambient or higher temperatures, are very rapid and near instantaneous.

The chemical characteristics of UF_4 do not have the same reactivity as UF_6 when released to the environment; that is, it reacts slowly with moisture at ambient temperature (DOE, 1999). UO_2 does not react with water (Benedict, 1981), and it will oxidize to U_3O_8 in air at ambient temperature. Nonetheless, uranium is a heavy metal that, in addition to being radioactive, can have toxic chemical effects, primarily on the kidneys, if it enters the blood stream by means of ingestion or inhalation. HF is an extremely corrosive gas that can damage the lungs and cause death if inhaled at sufficiently high concentrations.

Appendix A, "List of Highly Hazardous Chemicals, Toxics and Reactives (Mandatory)," to 29 CFR 1910.119 does not list SiF_4 as a highly hazardous chemical. However, the applicant stated that SiF_4 is corrosive to the eyes, skin, and respiratory tract (IIFP, 2012a). It may react violently and decompose exothermically with water or moisture in the air to form HF and silicic acid (NOAA, 2011a). Equation 6-14 shows the SiF_4 decomposition by water:

$$SiF_4 + 4H_2O \rightarrow 4HF + Si(OH)_4 + heat \qquad Eq. 6\text{-}14$$

Appendix A to 29 CFR 1910.119 lists BF_3 as a highly hazardous chemical with a threshold quantity of 250 lb. The applicant stated that BF_3 is corrosive and can cause irritation of the eyes, nose, throat, and skin (IIFP, 2012a). It hydrolyzes in moist air to form toxic and corrosive hydrogen fluoride, fluoroboric acid (HBF_4), and boric acid (H_3BO_3) (IIFP, 2012a; 2012b, and NOAA, 2011b). Equation 6-15 shows the BF_3 decomposition by water:

$$2BF_3 + 3H_2O \rightarrow 2HF + HBF_4 + B(OH)_3 \qquad Eq. 6\text{-}15$$

6.3.1.9   Materials of Construction

The applicant considered the interactions between process equipment and process fluids/gases in the design of the proposed IIFP facility. IIFP stated in the LA (IIFP, 2012a) that the materials used for construction are compatible with the process operational physical parameters of temperature and pressure and corrosion resistant to UF_6, AHF, SiF_4, and BF_3. The applicant commits to use materials for construction in accordance with applicable codes and standards, such as the 2009 edition of the International Building Code (IBC, 2009).

The cylinders used to store and transport DUF_6 consist of carbon steel and are standard DOT-approved containers, designed and fabricated in accordance with ANSI N14.1, "Uranium Hexafluoride Packaging for Transport" (ANSI, 2001). The cylinders are painted to resist corrosion caused by being stored outside in open air where they are exposed to atmospheric conditions. Also, the cylinders are routinely inspected to assess corrosion and corrosion rates. Packages of SiF_4 and BF_3 include small container cylinders and tube trailers; packages for AHF include tube trailers. The small cylinders and tube trailers are ASME Code rated pressure vessels and require DOT's approval for transportation.

6.3.1.10   Process Description Conclusion

Based on the review and verification of the information provided in the LA (IIFP, 2012a) and the ISA Summary (IIFP, 20012b), including the consistency between the description of the process and the associated hazards, the NRC staff finds that the applicant has provided detailed process descriptions sufficient to enable an understanding of the chemical process hazards and the chemical hazards that could result from potential chemical interactions. The rationale for these findings is that the applicant provided expected process operating conditions, interactions between chemicals, and interactions between process chemicals and the equipment material that will be used to contain these process chemicals. In addition, the applicant's information allowed the development of potential accident sequences. Therefore, the information that the applicant provided, as described above, meets the guidance in Section 6.4.3.1, bullets (1) and (2), of NUREG-1520 (NRC, 2002) and is acceptable.

6.3.2   Chemical Accident Sequences

Tables 3-7, 3-8, 3-9, and 3-10 of the ISA Summary (IIFP, 2012b) describe the chemical accident sequences in the proposed DUF_6 to DUF_4 deconversion plant, SiF_4 process, BF_3 process, and auxiliary process systems, respectively. Tables 4-3, 4-4, 4-5, and 4-6 summarize the accident sequences and assign the indices used to determine the overall risk of each accident sequence in the proposed DUF_6 to DUF_4 deconversion plant, SiF_4 process, BF_3 process, and auxiliary process systems, respectively. The consequences of the chemical accident sequences identified in these tables have the potential to exceed the performance requirements of 10 CFR 70.61. Table 5-1 shows the areas/systems for which the applicant identified chemical accident sequences with the potential to have intermediate or high consequences. The auxiliary process systems are located in the FEP oxide staging building, which is a common area that receives uranium oxide drums from both the SiF_4 and BF_3 processes.

The regulations in 10 CFR 70.61(b) and 10 CFR 70.61(c) define high- and intermediate-consequence events, respectively. A high-consequence event is the release of licensed material or hazardous chemicals produced from licensed material that, if an individual were to be exposed, could endanger the life of a worker or lead to irreversible or other serious, long-lasting health effects to a member of the public. A high-consequence event also includes an intake of 30 milligrams (mg) of soluble uranium by a member of the public. An intermediate-consequence event is the release of licensed material or hazardous chemicals produced from licensed material that, if an individual were to be exposed, could lead to irreversible or other serious, long lasting health effects to a worker or could cause mild transient health effects to a member of the public. The NRC staff performed a risk-informed evaluation of the chemical accident sequences by reviewing the ISA Summary which contains the intermediate- and high-consequence events postulated by the applicant. Also, the NRC staff performed a vertical slice review of certain accident sequences.

The NRC staff concludes that IIFP has identified appropriate chemical accident sequences based on the applicant's use of a recommended process hazards analysis method (i.e., the What-if) contained in NUREG-1513 (NRC, 2001). The information provided by the applicant, as described above and in Chapter 3 of this SER, includes a list of the accident sequences with their respective consequence and likelihood identified in the ISA Summary (IIFP, 2012b) that involve hazardous chemicals produced from licensed material and chemical risks of plant conditions that affect the safety of licensed material. The applicant also described the postulated high-consequence events, how they will be detected, and the mitigative measures in Sections 4.1, 5.3, and 8.3, respectively, of the Emergency Plan (EP) (IIFP, 2012c). The actions described on the EP (IIFP, 2012c) are consistent with the consequences of the accident sequences identified in the ISA Summary (IIFP, 2012b). Based on the above, the information provided by the applicant meets the guidance in Section 6.4.3.1, bullet (2), of NUREG-1520 (NRC, 2002) and is, therefore, acceptable.

6.3.3   Chemical Accident Consequences

The chemical exposure limits proposed by the applicant for UF_6, UF_4, HF, SiF_4, and BF_3 are based on the acute exposure guideline limit (AEGL) values. Specifically, for AHF, SiF_4, and BF_3 these values apply, from an NRC licensing perspective, before these compounds are separated from licensed material because these compounds have licensed material as precursors. Once separated from licensed material, OSHA regulates these compounds. It should be noted that HF is under NRC's regulatory jurisdiction if it is produced from the reaction of UF_6 with water (see Equations 6-12 and 6-13 above) and from the reaction of SiF_4 and BF with water (see

6-12

Equations 6-14 and 6-15 above) if it forms because of a release of SiF_4 and BF_3 in the FEP process building. The conversion to HF could affect the safe handling of licensed material in this building (i.e., UF_4 and UO_2). The applicant proposed to use the 10-minute AEGL values for exposures to workers with a duration of 10 minutes or less. The worker is trained to take proper actions (i.e., escape) upon sensing the initial effects of released hazardous material. For the public, the exposure duration was assumed to be 30 minutes. This is consistent with one of the conservative factors listed in NUREG-1140, "A Regulatory Analysis on Emergency Preparedness for Fuel Cycle and Other Radioactive Material Licensees," issued January 1988 (NRC, 1988).

There are no AEGL values for UO_2. For this reason, the applicant derived the chemical exposure limits for UO_2 from the emergency response planning guidelines (ERPGs). ERPG values are based on exposures for up to 60 minutes, and the applicant has scaled the exposures to 10 minutes and 30 minutes for the worker and public, respectively.

For soluble uranium (i.e., UF_6 or UO_2F_2) intakes, the applicant proposed to use 75 mg of soluble uranium intake as the high-consequence threshold for the worker. The NRC staff finds this value acceptable because it is based on the model in ICRP 66 (ICRP-66), "Human Respiratory Tract Model for Radiological Protection," that derives the systematic burned value of 0.3 milligram per kilogram of body weight (ICRP, 1994). The NRC staff approved the use of the ICRP-66 model and determined that using 75 mg of soluble uranium as the threshold for permanent renal damage is consistent with a high-consequence event to a worker, as defined in 10 CFR 70.61(b)(4)(i) (NRC, 2007b). For the public, the high-consequence threshold for a soluble uranium intake is 30 mg. This value is acceptable because it is based on the performance requirement in 10 CFR70.61(b)(3).

The comparison of the AEGLs or ERPGs with the calculated airborne concentrations allows the applicant to classify a high, intermediate, or low chemical consequence level. Table 5-2, reproduced from the LA (IIFP, 2012a), presents the chemical consequence levels and values. For worker and public chemical exposures, Tables 5-3 and 5-4, taken from the ISA Summary (IIFP, 2012b), present more detailed chemical consequence values for licensed material and hazardous chemicals produced from licensed material, respectively.

Table 6-1 Chemical Consequence Levels and Values (IIFP, 2012a)

Consequence	Workers	Offsite Public
High	CE > AEGL-3 (10-minute exposure)	CE > AEGL-2 (30-minute exposure)
Intermediate	AEGL-2 < CE ≤ AEGL-3 (10-minute exposure)	AEGL-1 CE ≤ AEGL-2 (30-minute exposure)
Low	Lower than CE above	Lower than CE above

Note: CE = chemical exposure

Table 6-2 Worker Chemical Exposure Values (10 Minutes) (IIFP, 2012b)

Chemical	High Consequences AEGL-3/(mg/m^3)	Intermediate Consequences AEGL-2/(mg/m^3)
UF_6	216	28
UF_4	216	28
UO_2F_2	216	28

HF	139	77.7
SiF$_4$	81	27
BF$_3$	140	47
UO$_2$ (ERPG)	300	201

Table 6-3 Public Chemical Exposure Values (30 Minutes) (IIFP, 2012b)

Chemical	High Consequences AEGL-2/(mg/m^3)	Intermediate Consequences AEGL-1/(mg/m^3)
UF$_6$	19	3.6
UF$_4$	19	3.6
UO$_2$F$_2$	19	3.6
HF	28	0.82
SiF$_4$	18	0.21
BF$_3$	47	2.5
UO$_2$ (ERPG)	32	0.68

The NRC staff's review of the ISA and supporting documentation found that the source term values are reasonable because the applicant provided the material at risk values for each building and followed the methodology described in NUREG/CR-6410, "Nuclear Fuel Cycle Facility Accident Analysis Handbook," issued March 1998 (NRC, 1998). The applicant used a computer code named HGSYSTEM, Version 3.0, to model the site boundary atmospheric dispersion of HF. The applicant also used modeling methods for source term determination, release fraction, dispersion factors, and meteorological conditions. The NRC staff performed an independent evaluation of the consequences of potential accidents identified in the ISA Summary. The NRC staff evaluated a representative selection of the types of accidents postulated for the proposed IIFP facility. Appendix B to this SER discusses this evaluation. The information provided by IIFP meets the guidance of Section 6.4.3.1, bullets (3) and (5), of NUREG-1520 (NRC, 2002) because the applicant (1) identified and used appropriate techniques in estimating the concentration of hazardous chemicals produced from licensed material, (2) used the performance requirements criteria of 10 CFR 70.61, and (3) ensured that the consequence analysis conformed to the guidance in NUREG/CR-6410 (NRC, 1998). Therefore, the NRC staff finds the applicant's proposed methodology for source term determination and consequence analysis to be acceptable.

6.3.4 Items Relied on for Safety and Management Measures

6.3.4.1 Chemical Process Items Relied on for Safety

Table 6-1 of the ISA Summary (IIFP, 2012b) is a descriptive list of all IROFS. It includes (1) the type of control of the IROFS (i.e., administrative control, enhanced administrative control, active or passive-engineered control); (2) the safety function and how it is implemented; (3) the management measures applied to ensure that the IROFS is available and reliable when needed; and (4) the index used to determine the overall risk of a particular accident sequence and its basis. Table 8-1 of the ISA Summary (IIFP, 2012b) identifies the sole IROFS credited to prevent a high-consequence event, and Table 8-2 identifies the sole IROFS credited to prevent an intermediate-consequence event. All of the chemical process sole IROFS are passive-engineered controls. The applicant stated in the ISA Summary that a passive-engineered control does not require human intervention to carry out its safety function; and it relies on natural forces, such as gravity or natural convection, to maintain safe process conditions. This description is consistent with the term as used in NUREG-1520, Glossary (NRC, 2002). Specifically for chemical process safety, the passive-engineered control is the structural design criteria of certain process equipment in the DUF_6 to DUF_4 deconversion plant and SiF_4 and BF_3 processes. The NRC reviewed the IROFS and management measures described above and finds they provide reasonable assurance that process equipment will remain available and reliable. In addition, NRC staff finds that the process equipment will not suddenly fail, leading to a release of licensed material or hazardous chemicals produced from licensed material, without a detectable process upset or external impact. The applicant has adequate detection systems, procedures, and shutdown capabilities to ensure that the structural design criteria of certain process equipment is not challenged.

Relying on multiple IROFS for an accident sequence provides the protection against releases of licensed material or hazardous chemicals produced from licensed material. For example, to prevent a release of licensed material caused by heating an overfilled UF_6 cylinder, the applicant proposed an IROFS procedure that prevents an overfilled cylinder to be placed in the autoclave. The IROFS procedure is complemented by an IROFS accountability scale and by the IROFS autoclave integrity.

The applicant postulated accident sequences that rely on sole IROFS. The chemical process's sole IROFS prevent intermediate-consequence events, and it is the structural design criteria of certain process equipment (discussed in the first paragraph of this section).

The NRC staff reviewed the listed IROFS, process descriptions, and process flow diagrams provided in Chapters 6 and 3 of the ISA Summary (IIFP, 2012b) and the engineering drawings, respectively, to identify where each IROFS would be used and how the IROFS would function to prevent the accident sequence or mitigate its consequences. The identified IROFS protect against a loss of confinement of licensed material or hazardous chemicals produced from licensed material during operation of the facility.

Based on this system-level review, the NRC staff concludes that the applicant meets the guidance of Section 6.4.3.2, bullet (2), of NUREG-1520 (NRC, 2002) because the ISA Summary (IIFP, 2012b) identified chemical process IROFS to prevent or mitigate the consequences of accident sequences that involve the chemical hazards of licensed material and hazardous chemicals produced from licensed material. In addition, the ISA Summary (IIFP, 2012b) identified the hazards being mitigated and the risk category of each accident sequence.

6.3.4.2   Management Measures

The following sections briefly discuss the integration of chemical safety into the elements of management measures. The applicant identified management measures to ensure that chemical safety IROFS would be available and reliable to perform their safety function when needed. The applicant proposed to implement a graded approach based on the level of protection needed by each IROFS to meet the performance requirements. Chapter 11 of this SER provides an overall evaluation of the management measures applied to the proposed IIFP facility.

6.3.4.2.1   Configuration Management

The Configuration Management Program's proposal in Section 11.1 of the LA for the proposed IIFP facility, includes those elements that ensure that the facility's technical baseline is thoroughly documented and maintained. The Configuration Management Program also ensures that all safety, security, and licensing organizations review and approve changes to the technical baseline. The technical baseline consists of facility drawings, procedures, specifications, and other technical documents—including the ISA (IIFP, 2012b).

The applicant stated that it will submit changes to the technical baseline that require prior NRC approval in a license amendment request (as required by 10 CFR 70.72(d)(1)) and will not implement such changes without prior NRC's approval. The regulations in 10 CFR 70.72(c) include the criteria for changes that require NRC pre-approval. The applicant also stated that it will submit changes that do not require prior NRC approval annually, as well as the revised ISA Summary pages, as required by 10 CFR 70.72(d)(2) and 10 CFR 70.72(d)(3), respectively (IIFP, 2012b).

6.3.4.2.2   Maintenance

The applicant outlined the maintenance and functional testing programs in Section 11.2 of the LA. Maintenance activities provide reasonable assurance that IROFS will be available and reliable to perform their safety functions when needed. Appropriate plant management is responsible for ensuring operational readiness of IROFS under this control. For this reason, the maintenance function is administratively closely coupled to the engineering function (IIFP, 2012a). The applicant will evaluate the impact of maintenance activities on the specific systems and on other nearby systems (IIFP, 2012b).

Maintenance activities generally fall into the following categories:

- surveillance and monitoring

- testing

- preventive maintenance

- corrective maintenance

Surveillance and monitoring identify conditions that require corrective maintenance and ensure that preventive maintenance is effective. The chemical process safety discipline within the applicant's organization will evaluate the results of surveillance and monitoring activities to determine any impact on the ISA and any updates needed. The applicant stated that testing includes functional tests, performance tests, software checks and updates, and instrument calibration. In addition, the applicant will develop testing plans for IROFS (including chemical safety) before the IROFS installation, and IROFS will be tested following maintenance and on an annual basis (unless otherwise noted in Chapter 6 of the ISA Summary [IIFP, 2012b]). The applicant stated that preventive maintenance includes periodic refurbishment or like-kind replacement of IROFS at a frequency based on the expected life of the IROFS along with the relative importance of the IROFS in meeting the performance requirements and the results of surveillance and monitoring. Sole IROFS will be refurbished or replaced at a greater frequency. Corrective maintenance includes repair of like-kind replacement of equipment that has failed to perform or is performing outside of desired safety and process parameter limits (IIFP, 2012b).

6.3.4.2.3 Training and Qualifications

Section 11.3 of the LA describes the training program for operations of the facility, including preoperational functional testing and initial startup testing. The applicant proposed to train and qualify employees to ensure safe operation of the IIFP facility. The training program requirements are applicable, but not limited, to employees who perform activities that affect IROFS or items that may affect the function of IROFS (IIFP, 2012a). Technical training is provided to assist employees in gaining an understanding of applicable fundamentals, procedures, and practices related to IROFS. Employees are provided with formal classroom training along with specific on-the-job training. IIFP will use a job task analysis, as needed, to supplement training when tasks associated with IROFS are involved (IIFP, 2012b).

Chemical safety (hazard communication) training is provided to all employees as one of the multiple topics included in the general employee training. This general employee training also covers general industrial safety. The applicant also conducts more specific training in various aspects of industrial and chemical safety protection to train new employees in specific job duties and to provide refresher training topics to workers depending on employee job responsibilities. In particular, any employee or contractor using hazardous materials is trained to ensure safe handling, use, and disposal of such materials.

IIFP provides annual training to emergency response personnel, as discussed in Section 10.2.5 of the IIFP EP (IIFP, 2012c). The applicant stated that the emergency response personnel are prepared to respond to various emergency conditions, including a chemical accident.

6.3.4.2.4 Procedures

The applicant proposed that all activities involving IROFS are conducted in accordance with approved procedures. Procedures are used to ensure that activities involving IROFS are carried out in a safe manner and in accordance with regulatory requirements (IIFP, 2012a). Section 11.4 of the LA (IIFP, 2012a) describes and outlines the types of procedures, including (1) operating procedures, (2) administrative procedures, (3) maintenance procedures, and (4) emergency procedures.

The operating procedures are used to directly control production and to clearly identify applicable safety limits and IROFS. Administrative procedures are used to perform activities that support production, including management measures such as, but not limited to, chemical safety. Maintenance procedures are used to maintain, test, and calibrate facility IROFS. Finally, the emergency procedures outline duties, responsibilities, action levels, and actions to be taken by responders pertinent to specific accident scenarios and other categorized non-routine operational events. In addition, administrative procedures ensure that individuals and groups with assigned responsibilities in an emergency have easy access to a current copy of each procedure that pertains to their functions (IIFP, 2012a).

6.3.4.2.5 Audits and Assessments

Chemical safety is one of the areas in which the applicant proposed to conduct audits and assessments. Section 11.5 of the LA (IIFP, 2012a) describes audits and assessments. Audits are focused on verifying compliance with regulatory and procedural requirements and licensing commitments. Assessments are focused on evaluating the effectiveness of activities and ensuring that IROFS, and any items that affect the function of IROFS, are reliable and available to perform their intended safety functions (IIFP, 2012a).

The results of audits and assessments are documented and reported as specified in plant procedures, which include the managers responsible for the activities audited or assessed (IIFP, 2012a). The Corrective Action Program tracks audit and assessment results (IIFP, 2012a).

6.3.4.2.6 Incident Investigation and Corrective Actions

The applicant described the incident investigations and corrective action process in Section 11.6 of the LA (IIFP, 2012a). The applicant proposed to use an incident investigation process to report deficiencies, abnormal events, and potentially unsafe conditions or activities (IIFP, 2012a). When an incident occurs, facility management will form a qualified team to determine root causes of the event and develop recommendations to reduce the likelihood of recurrence. Lessons learned will be developed, and unaffected organizations can review their operations for similar type initiators. Written incident investigation and corrective action procedures address the process of incident identification, investigation, root-cause analysis, recording, and followup; IIFP staff use these procedures when performing incident investigations and corrective actions. These procedures also address hazardous chemical safety requirements (IIFP, 2012a). Corrective actions are assigned and tracked programmatically to ensure that timely and adequate corrections to deficiencies are incorporated.

6.3.4.2.7 Records Management

The applicant will maintain the following records, which are described in Section 11.7 of the LA (IIFP, 2001a), for the proposed IIFP facility:

- results of surveys to determine the dose from external sources and used in the assessment of individual dose equivalents

- results of measurements and calculations used to determine individual intakes of radioactive material and used in the assessment of internal dose

- results of air sampling, surveys, and bioassays

- results of measurements and calculations used to evaluate the release of radioactive effluents to the environment

- records of spills or other unusual occurrences involving the spread of contamination in and around the facility, equipment, or site

- as-built drawings and modifications of structures and equipment in restricted areas where radioactive materials are used or stored

- IROFS design specifications and maintenance records

- training and qualification records

- audit, assessment, and inspection results

- incident investigation reports

- quality assurance records.

6.3.4.3 *Items Relied on for Safety and Management Measures Conclusion*

The information IIFP provided, as described above, meets the guidance in Section 6.4.3.2, bullet (2), of NUREG-1520 (NRC, 2002) and is acceptable because the applicant has adequately identified the administrative and engineered controls (IROFS) to prevent chemical accident sequences or mitigate their consequences at the proposed facility. The applicant also identified the hazards being mitigated and the risk category.

In addition, the information IIFP provided, as described above and in Chapter 11 of this SER, meets the guidance in Section 6.4.3.2, bullet (3), of NUREG-1520 (NRC, 2002) and is acceptable because the applicant sufficiently described its procedures to ensure the reliable operation of engineered controls and the correct implementation of administrative controls.

6.3.5 Emergency Management

As described further in Chapter 8 of this SER and consistent with the requirements in 10 CFR 40.31(j)(1)(ii), IIFP has submitted an EP and program which includes response to mitigate the potential impact of any process chemical release, including requirements for notification and reporting of accidental chemical releases. The emergency response team is outfitted, equipped, and trained for hazardous material response; and local agencies can supplement the response with additional response teams. The applicant has an agreement with the Fire Department of the City of Hobbs, NM, to supplement the IIFP facility's emergency response team with additional response teams, if needed. The applicant stated in Section 10.2.4 of the EP (IIFP, 2012c) that IIFP facility personnel meet at least every 2 years with each offsite assistance group to accomplish training and review items of mutual interest, including relevant changes to the Emergency Management Program.

In Chapter 13 of the EP (IIFP, 2012c) for the proposed IIFP facility, the applicant committed to meeting the requirements of the Emergency Planning and Community Right-to-Know Act of 1986 (Title III, Public Law 99-499), as required by 10 CFR 40.31(j)(3)(xiii). The applicant identified the material safety data sheets (MSDSs) that will be available to local agencies. The applicant will provide the MSDSs for UF_6, UF_4, UO_2, U_3O_8, HF, SiF_4, and BF_3 to local agencies. The applicant also committed to providing additional MSDSs to local agencies if any significant quantities of hazardous chemicals are added to the facility inventory. Table 9 of the EP (IIFP, 2012c) provides the major chemicals used at the IIFP facility and their locations.

The NRC staff finds that the applicant has provided reasonable assurance that measures used to mitigate the consequences of accident sequences identified in the ISA Summary (IIFP, 2012b) are consistent with actions described in Chapter 8 of NUREG-1520 (NRC, 2002). The information IIFP provided, as described above, meets the guidance in Section 6.4.3.1, bullet (2), of NUREG-1520 (NRC, 2002), and is acceptable because the applicant described the postulated high-consequence events, how they will be detected, and the mitigative measures in Sections 4.1, 5.3, and 8.3, respectively, of the EP (IIFP, 2012c). The actions described in the EP (IIFP, 2012c) are consistent with the consequences of the accident sequences identified in the ISA Summary (IIFP, 2012b).

6.3.6 Baseline Design Criteria and Defense-in-Depth

The applicant provided design-basis information for chemical process safety IROFS for the proposed facility in the LA (IIFP, 2012a) and ISA Summary (IIFP, 2012b). For chemical protection, 10 CFR 70.64(a)(5) states the following:

> Chemical protection. The design must provide for adequate protection against chemical risks produced from licensed material, facility conditions which affect the safety of licensed material, and hazardous chemicals produced from licensed material.

Applicable to new facilities, the regulations in 10 CFR 70.64(b) require the facility and system design and facility layout to be based on defense-in-depth practices. The regulations also require that the design must incorporate, to the extent practicable, preference for the selection of engineered controls over administrative controls to increase overall system reliability and features that enhance safety by reducing the challenges to IROFS. The applicant provided information in the process description of the ISA Summary (IIFP, 2012b) regarding non-IROFS safeguards to be used as defense-in-depth to reduce the challenge to IROFS.

The chemicals of concern are UF_6, UF_4, UO_2, HF, SiF_4, BF_3, and H_2. Chapters 3 and 6 of the ISA Summary (IIFP, 2012b), respectively, provide details of the design and safety features of all chemical process systems. The applicant's design of the chemical process systems includes numerous controls, in addition to the IROFS, for maintaining safe conditions during operation. The applicant accomplishes this through several means, including the following:

- managing the arrangement and size of material containers and processes

- selecting and using materials compatible with process chemicals

- providing inherently safer operating conditions (e.g., limiting steam temperature in the autoclave, ensuring a slight vacuum or near atmospheric pressure in the rotary calciners)

- providing process interlocks, controls, and alarms within the process.

The NRC staff reviewed the applicant's proposed design of the proposed FEP/DUP contained in Section 3.1 of the ISA Summary and the process hazards description in Section 3.2 (IIFP, 2012b). The NRC staff notes that the applicant preferred engineered controls over administrative controls in the selection of IROFS. This is demonstrated by the fact that no accident sequence is prevented or mitigated only by administrative controls. Usually the accident sequences are prevented or mitigated by a combination of administrative and engineered controls. As discussed in Section 6.3.4 of this SER, when sole IROFS are credited to prevent an accident sequence, the selected IROFS is an engineered control. The NRC staff also notes that the applicant preferred to prevent an accident sequence rather than mitigate it. The majority of IROFS are preventive. In contrast, mitigative IROFS are credited in a few of the accident sequences identified by the applicant.

The NRC staff notes that the applicant's design of the UF_6 feeding system uses a cylinder qualified under ANSI N14.1 (ANSI, 2001) as the primary confinement vessel and an ASME Code pressure vessel as a secondary confinement system. The applicant proposed a limit of

235 degrees F for the temperature of the steam used to vaporize the UF_6 in the cylinder inside the autoclave. This temperature limit provides a margin of safety from the limit used in the ANSI N14.1 standard (i.e., 250 degrees F) (ANSI, 2001). The rationale for the limit of 250 degrees F is, if a full UF_6 cylinder (i.e., 27,560 lb for a 48-Y cylinder) is heated to this temperature, the cylinder would have 5 percent of its total volume available. The NRC staff concludes that this design approach for the feeding portion of the process is acceptable because it uses recognized nuclear fuel cycle industry codes and standards and provides an adequate safety margin.

The NRC staff reviewed the results of the applicant's What-if analysis, as discussed in Chapter 3 of this SER. This method is widely used in the chemical industry during the design phase to identify operability and safety issues and is identified as an acceptable method in Section 2.4 of NUREG-1513 (NRC, 2001). As applied to the proposed depleted uranium deconversion and fluorine extraction processes, the What-if analysis considered a variety of internal process, facility, and external hazards that could breach the process and release licensed material and hazardous chemicals produced from licensed material. Tables 4-3, 4-4, 4-5, and 4-6 of the ISA Summary (IIFP, 2012b) present the results of the applicant's ISA. The tables contain information concerning the accident sequences identified as a result of the What-if analysis, the unmitigated risk of each applicant-identified accident sequence, and the IROFS applied to prevent the accident sequence or mitigate its consequences. The NRC staff also reviewed selected high-consequence and intermediate-consequence accident scenarios to confirm that the applicant addressed chemical events that could exceed the performance requirements of 10 CFR 70.61.

Based on the above, the NRC staff concludes that the information IIFP provided and its proposed design meet the guidance in Section 6.4.3.3 of NUREG-1520 (NRC, 2002) and provide for adequate protection against chemical risks produced from licensed materials, facility conditions which affect the safety of licensed material, and hazardous chemicals produced from licensed material, thus meeting the requirements of 10 CFR 70.64(a)(5) and 10 CFR 70.64(b).

6.4   Evaluation Findings

The NRC staff evaluated the application using the criteria previously listed. Based on the review of the LA (IIFP, 2012a) and ISA Summary (IIFP 2011b), the NRC staff concludes that the applicant has described and assessed accident sequences that can result from the handling, storage, or processing of licensed materials and that can potentially have significant chemical consequences and effects. The applicant has prepared a hazard analysis that identified and evaluated those chemical process hazards and potential accidents and established safety controls providing reasonable assurance of safe facility operation. To ensure that the performance requirements in 10 CFR Part 70 are met, the applicant stated that controls are maintained, available, and reliable to perform their safety-related functions when needed. The NRC staff has reviewed these safety controls and the applicant's plan to managing chemical process safety and finds them acceptable.

The NRC staff concludes that the applicant's plan for managing chemical process safety and chemical process safety controls meets the requirements of 10 CFR Part 70, Subpart H, and 10 CFR Part 40 and therefore provides reasonable assurance that the public health and safety and the environment will be protected.

6.5   References

(ASME, 2004) American Society of Mechanical Engineers, "Boiler and Pressure Vessel Code," Section VIII, Division 1, 2004

(ANSI, 2001) American National Standards Institute, "Uranium Hexafluoride—Packaging for Transport," ANSI N14.1, February 2001.

(Benedict, 1981) Benedict, Mason, Thomas H. Pigford, and Hans Wolfgang Levi, *Nuclear Chemical Engineering* (2nd edition), (pp: 223). McGraw-Hill, 1981.

(Benítez, 1993) Benítez, Jaime, *Process Engineering and Design for Air Pollution Control*, (pp: 333). Prentice-Hall. 1993.

(ICRP, 1994) International Commission on Radiological Protection, "Human Respiratory Tract Model for Radiological Protection," Publication 66, 1994.

(IBC, 2009) International Code Council, International Building Code, 2009.

(IIFP, 2012a) International Isotopes Fluorine Products, Inc., "Fluorine Extraction Process and Depleted Uranium Deconversion Plant (FEP/DUP) License Application, Revision B," May 2012, Agencywide Documents Access and Management System (ADAMS) Accession No. ML12123A245.

(IIFP, 2012b) International Isotopes Fluorine Products, Inc., "ISA Summary Rev. B for IIFP," May 2012, Agencywide Documents Access and Management System (ADAMS) Accession No. ML12123A245.

(IIFP, 2012c) International Isotope Fluorine Products, Inc., "Emergency Plan Rev B of IIFP License Application," May 2012, Agencywide Documents Access and Management System (ADAMS) Accession No. ML12123A245.

(IIFP, 2011) International Isotopes Fluorine Products, Inc., "Official Responses to Chemical Process Safety RAIs," March 2011 Agencywide Documents Access and Management System (ADAMS) Accession No. ML110950330.

(Kutz, 2007) Kutz, Myer, *Environmentally Conscious Alternative Energy Production*, John Wiley & Sons. 2007.

(NOAA, 2011a) National Oceanic and Atmospheric Administration, "Chemical Datasheet—Silicon Tetrafluoride," http://cameochemicals.noaa.gov/chemical/1449, Access Date: January 21, 2011.

(NOAA, 2011b) National Oceanic and Atmospheric Administration, "Chemical Datasheet—Boron Trifluoride," http://cameochemicals.noaa.gov/chemical/255, Access Date: January 21, 2011.

(NRC, 2010) U.S. Nuclear Regulatory Commission, "July 6–8, 2010, Summary of Site Visit to Conduct Integrated Safety Analysis Horizontal and Vertical Slice and Discussion Draft Requests for Additional Information with International Isotopes Fluorine Products, Inc.," July 2010. ADAMS Accession No. ML102070210

(NRC, 2007a) U.S. Nuclear Regulatory Commission, Staff Requirements Memorandum to SECY-07-0146, "Regulatory Options for Licensing New Uranium Conversion and Depleted Uranium Deconversion Facilities," October 10, 2007.

(NRC, 2007b) U.S. Nuclear Regulatory Commission, "Reply to Proposed Usage of International Commission on Radiological Protection 66 in Determination of Soluble Uranium Intake Threshold," July 2007. ADAMS Accession No. ML072010285

(NRC, 2002) U.S. Nuclear Regulatory Commission, "Standard Review Plan for the Review of a License Application for a Fuel Cycle Facility," NUREG-1520, Rev. 0, March, 2002.

(NRC, 2001) U.S. Nuclear Regulatory Commission, "Integrated Safety Analysis Guidance Document," NUREG-1513, May 2001.

(NRC, 1998) U.S. Nuclear Regulatory Commission, "Nuclear Fuel Cycle Facility Accident Analysis Handbook," NUREG/CR-6410, March 1998.

(NRC, 1997) U.S. Nuclear Regulatory Commission, "Chemical Process Safety at Fuel Cycle Facilities," NUREG-1601, August 1997.

(NRC, 1988) U.S. Nuclear Regulatory Commission, "A Regulatory Analysis on Emergency Preparedness for Fuel Cycle and Other Radioactive Material Licensees," NUREG-1140, January 1988.

7.0   FIRE SAFETY

The purpose of this review is to determine, with reasonable assurance, whether International Isotopes Fluorine Products, Inc. (IIFP or the applicant), has designed a facility with adequate protection against fires and explosions that could affect the safety of licensed materials and thus present an increased radiological risk. The review should also establish that the applicant has considered the radiological consequences of fires and will institute suitable safety controls to protect workers, the public, and the environment.

Details related to items relied on for safety (IROFS) have been marked by the applicant as "Security-Related Information," pursuant to Title 10 of the *Code of Federal Regulations* (10 CFR) 2.390, and the Commission has determined to include this information in the public version of this report.

7.1   Regulatory Requirements

The regulatory basis for the fire safety review should be the general and additional contents of application, as required by 10 CFR 40.32, entitled "General Requirements for Issuance of Specific Licenses." In addition, the fire safety review should focus on providing reasonable assurance of compliance with the following regulations:

- 10 CFR 70.61, "Performance Requirements"

- 10 CFR 70.62, "Safety Program and Integrated Safety Analysis"

- 10 CFR 70.64, "Requirements for New Facilities or New Processes at Existing Facilities"

- 10 CFR 70.65, "Additional Content of Applications."

7.2   Regulatory Acceptance Criteria

Chapter 7 of NUREG-1520, "Standard Review Plan for the Review of a License Application for a Fuel Cycle Facility" (NRC, 2002), contains the guidance applicable to the U.S. Nuclear Regulatory Commission's (NRC's) review of the fire safety description section of the license application (LA) (IIFP, 2012a). This chapter is applicable in its entirety. Sections 7.4.3.1 through 7.4.3.5 of NUREG-1520 (NRC, 2002) provide the acceptance criteria applicable to this review.

7.3   Staff Review and Analysis

This section addresses the NRC staff's review of facility fire protection, including fire safety management measures, fire hazards analysis, facility fire protection, process fire safety, and fire safety and emergency response, as presented in the LA (IIFP, 2012a); the Integrated Safety Analysis (ISA) Summary (IIFP, 2012b).

7.3.1 Fire Safety Management Measures

The applicant will implement fire safety management measures as described in Chapter 11 of the LA (IIFP, 2012a). Management measures applicable to fire safety include configuration management, maintenance, training and qualifications, procedures development and implementation, audits and assessments, incident investigations and corrective actions process, records management and document control, and quality assurance program elements. These measures will ensure that IROFS related to fire protection are available and reliable during normal operations, anticipated (off-normal) events, and accidents. The applicant will follow the codes and standards, as listed in Table 7-1 of the LA (IIFP, 2012a), which are applicable to the individual fire safety management measures. Chapter 11 of this Safety Evaluation Report (SER) evaluates management measures.

7.3.1.1   Management Policy and Direction

The Regulatory Affairs and Quality Assurance Director is responsible for fire protection and is assisted by the Environmental Safety and Health Manager, who is responsible for the day-to-day safe operation of the facility, including fire safety. These individuals are part of the Facility Safety Review Committee which, with managers from other disciplines, integrates facility modifications. Fire safety personnel who are trained in fire protection and have nuclear fire safety experience assist the Environmental Safety and Health Manager. The fire protection staff is responsible for the following:

- fire protection program and procedural requirements

- fire prevention activities (i.e., administrative controls and training)

- maintenance, surveillance, and quality of the facility fire protection features

- control of design changes, as related to fire protection

- documentation and recordkeeping, as related to fire protection

- organization and training of the fire brigade

- pre-fire planning.

Fire prevention at the facility consists of administrative controls to (1) govern the handling of transient combustibles, (2) control ignition sources, (3) ensure that open flames or combustion generated smoke is not used for leak testing, (4) conduct periodic fire prevention inspections, (5) perform periodic housekeeping inspections, and (6) implement a system to control the disarming of the various types of fire detection or fire suppression systems. The inspection, testing, and maintenance of fire protection systems will comply with nationally recognized industry standards. Section 7.3.3 of this chapter provides further information concerning the fire detection and suppression systems.

7.3.1.2   Fire Safety Management Measures Conclusions

Based on its review, the NRC staff concludes that the applicant's fire safety management measures are acceptable for the following reasons:

- Consistent with the acceptance criteria in Section 7.4.3.1 of NUREG-1520 (NRC, 2002), the applicant's fire safety management measures identify a senior level manager who has the authority and staff to ensure that fire safety receives appropriate priority.

- Consistent with the acceptance criteria in Section 7.4.3.1 of NUREG-1520 (NRC, 2002), the applicant's fire safety management measures identify a facility safety committee staffed by managers of different disciplines to integrate facility modifications.

- Consistent with the acceptance criteria in Section 7.4.3.1 of NUREG-1520 (NRC, 2002), the applicant's fire safety management measures include fire prevention; inspection, testing, and maintenance of fire protection systems; fire brigade qualifications, drills, and training; and pre-fire plans as recommended by National Fire Protection Association Standard (NFPA) 801, "Standard for Fire Protection for Facilities Handling Radioactive Materials" (NFPA, 2008c).

- Consistent with the acceptance criteria in Section 7.4.3.1 of NUREG-1520 (NRC, 2002), the applicant's fire safety management measures are documented in sufficient detail to identify their relationship to, and functions for, normal operations; anticipated (off-normal) events; and accident safety (i.e., IROFS).

- Consistent with the acceptance criteria from Section 7.4.3.1 of NUREG-1520 (NRC, 2002), the applicant's fire safety management measures will ensure that the IROFS, as identified in the ISA Summary (IIFP, 2012b), are available and reliable and will ensure that the facility maintains fire safety awareness among employees, controls transient ignition sources and combustibles, and maintains a readiness to extinguish or limit the consequences of fire.

Based on the above findings, the NRC staff finds that the applicant's fire safety management measures meet the requirements of 10 CFR 40.32, 10 CFR 70.61, 10 CFR 70.64, and 10 CFR 70.65 as they pertain to the fire protection aspects of the facility.

7.3.2 Fire Hazards Analysis

As part of its ISA, the applicant performed a fire hazards analysis (FHA), consistent with the guidance in NUREG-1520 (NRC, 2002) and NFPA 801 (NFPA, 2008c), which considered each process area and described, by fire area, the fuel loading, fire scenarios, methods of consequence analysis, potential consequences, and mitigative controls.

Using the results from the FHA, the ISA Summary (IIFP, 2012b) describes, qualitatively, the potential credible fire accident scenarios and associated risks for the facility. The applicant postulated and evaluated the following key fire accident scenarios:

- fire involving a transport truck or external wild lands (including an evaluation of the cylinder staging and storage pads)

- fire involving a cylinder hauler

- fire or explosion involving underground natural gas and propane lines

- fire in various areas of the depleted uranium hexafluoride (DUF_6) conversion plant (depleted uranium tetrafluoride (DUF_4) and DUF_6 buildings)

- fire in various areas of the silicon tetrafluoride (SiF_4) plant (fluorine extraction process [FEP] building)

- fire in various areas of the boron trifluoride (BF_3) plant (FEP building).

7.3.2.1   Items Relied on for Safety Related to Fire Safety

The applicant identified a set of IROFS that would ensure that the likelihood of a fire causing high-consequence events is highly unlikely and the likelihood of a fire causing intermediate consequence events is unlikely. Table 6-1 of the ISA Summary (IIFP, 2012b) lists these IROFS.

The NRC staff considers the failure probability indices assigned to these IROFS to be achievable, with their respective bases, as described in Section 6.1 of the ISA Summary (IIFP, 2012b). Section 6.1 of the ISA Summary (IIFP, 2012b) provides proposed surveillance frequencies, safety margins, and other measures that will support the low failure probabilities assigned to these IROFS, which are in accordance with the NRC guidance provided in Table A-9 of NUREG-1520 (NRC, 2002). Additionally, the applicant has followed the requirements of 10 CFR 70.64(b) by incorporating the preference for engineered controls over administrative controls into the design. As described in Section 4.2 of the ISA Summary (IIFP, 2012b), general management measures will provide further support for all of the listed IROFS. Chapter 3 of this SER provides additional discussion of the ISA methodology.

The LA also describes fire protection measures that provide overall defense-in-depth protection of fire safety for operations. The following subsections, which discuss each fire-related accident scenario, evaluate all fire protection measures. The applicant's ISA Summary (IIFP, 2012b) adequately identified the IROFS and features of fire protection in accordance with the regulations and the guidance established in NUREG-1520 (NRC, 2002).

7.3.2.2   Fire Involving a Transport Truck or External Staging and Storage Pads

Concerning the cylinder staging and storage pads, the applicant will administratively limit transient combustible loading on the pads, to ensure cylinder integrity. Additionally, as a defense-in-depth measure, the cylinder pads are sloped to prevent the accumulation of flammable or combustible liquids.

The potential for a wild land fire to impact the site is not considered to be a credible event given the dry desert topography and lack of forestry and vegetation. Regardless, the applicant provides for vegetation control to ensure that brush is kept to a minimum.

The applicant determined that the likelihood of a fire being initiated and the IROFS failing, causing a release exceeding the consequence threshold of 10 CFR 70.61(b) or 10 CFR 70.61(c), is highly unlikely. The applicant based its analysis for the transport truck fire and fire on the cylinder staging and storage pads on the guidance provided in NUREG-1805, "Fire Dynamics Tools (FDT^s) Quantitative Fire Hazard Analysis Methods for the U.S. Nuclear Regulatory Commission Fire Protection Inspection Program," issued December 2004 (NRC, 2004), and NUREG-1520 (NRC, 2002). The NRC staff's evaluation determined that the applicant's analysis meets the acceptance criteria in Section 7.4.3.2 of NUREG-1520 (NRC, 2002) and is, therefore, acceptable.

The NRC staff finds that the applicant has demonstrated that the facility will comply with the performance requirements of 10 CFR 70.61 in the event of a fire from a transport truck or on external staging and storage pads.

7.3.2.3   Fire Involving a Cylinder Hauler

The applicant determined that the likelihood of a fire being initiated and IROFS failing, so that a release exceeding the consequence threshold of 10 CFR 70.61(b) or 10 CFR 70.61(c) occurs, is highly unlikely. The NRC staff evaluated the applicant's calculations and finds the results acceptable, given that the calculations were performed utilizing guidance previously approved by the NRC in "Confirmatory Calculations for Fire Protection Review of National Enrichment Facility Integrated Safety Analysis (ISA) Summary" (NRC, 2005). The NRC staff's evaluation determined that the applicant's analysis meets the acceptance criteria in Section 7.4.3.2 of NUREG-1520 (NRC, 2002) and is, therefore, acceptable.

The NRC staff finds that the applicant demonstrated that the facility will comply with the performance requirements of 10 CFR 70.61 in the event of a fire on a cylinder transporter used to transport DUF_6 cylinders to and from the DUF_6 autoclave building and the cylinder staging/storage pads.

7.3.2.4   Fire or Explosion Involving Underground Natural Gas and Propane Lines

Several underground natural gas pipelines are located in the vicinity of the proposed 40-acre site. The applicant performed an analysis based on the guidance provided in NRC Regulatory Guide 1.91, Revision 1, "Evaluations of Explosions Postulated to Occur on Transportation Routes Near Nuclear Power Plants," issued February 1978 (NRC, 1978). The leak or rupture of an underground natural gas pipeline, followed by detonation, would generate a blast pressure wave. The source magnitude of the blast would depend on several factors (pipe size, gas pressure, gas temperature, depth of the pipe beneath the ground surface). Atmospheric conditions (wind speed and stability class) would affect dispersion of the natural gas, which would also strongly influence the magnitude of the blast. The magnitude of the blast pressure wave generated by the blast would rapidly diminish with distance.
The annual probability that a nearby, fossil fuel pipeline could rupture, detonate, and cause a 1-psi blast pressure wave at a process building is less than 4×10^{-6}. The probability of this event qualifies as highly unlikely, in accordance with the guidance in NUREG-1520 (NRC, 2002); therefore, no further analysis of this event is required.

Based on the above, the applicant determined that the likelihood of a fire or explosion being initiated and causing a release exceeding the consequence threshold of 10 CFR 70.61(b) or 10 CFR 70.61 (c) is highly unlikely. The NRC staff conducted a detailed onsite review of the applicant's analysis and supporting documents (NRC, 2011). The applicant based its analysis on the guidance provided in Regulatory Guide 1.91 (NRC, 1978). The NRC staff's evaluation determined that the applicant's analysis meets the acceptance criteria in Section 7.4.3.2 of NUREG-1520 (NRC, 2002) and is, therefore, acceptable.

The NRC staff finds that the applicant has demonstrated that the facility will comply with the performance requirements of 10 CFR 70.61 in the event of a fire or explosion involving underground natural gas and propane lines.

7.3.2.5   Fire in Various Areas of the DUF₆ Conversion Plant

The applicant will receive DUF_6 from customers in solid form contained in 14-ton steel cylinders that are unloaded and temporarily stored in the full cylinder staging pad area until scheduled for feeding to the process. The DUF_6 cylinders are placed in a containment autoclave where the contents are vaporized. Steam-heated autoclaves are used to heat the DUF_6 cylinders to vaporize the material allowing it to enter the DUF_6 feed system to the reactor for processing. DUF_6 does not react with oxygen, nitrogen, carbon dioxide, or dry air; but it does react with water. For this reason, DUF_6 is handled in leak tight containers and processing equipment.

The DUF_6 vapor is fed to a specially designed reaction vessel where it undergoes an exothermic reaction to produce DUF_4 and anhydrous hydrogen fluoride (AHF). The DUF_4 solids are continuously withdrawn from the reactor bottom through a special material cooling screw mechanism. The DUF_4 is transferred to storage hoppers for use as raw material feed in the FEP building for producing SiF_4 and BF_3.

Offgases from the reaction vessel leave the cooling screw equipment and pass through sintered metal and carbon-bed trap systems in series to remove entrained particulates and residual traces of unreacted DUF_6. Particulate DUF_4 that is collected in the sintered metal filter system is recycled back to the DUF_4 powder system. The offgas flow, exiting the carbon-bed traps, is cooled by passing it through heat exchangers for collecting the byproduct AHF. Residual offgases exit the condenser equipment to a scrubbing system designed for removing trace quantities of fluorides.

The AHF that liquefies in the condenser equipment is drained to storage tanks that are located in the AHF staging containment building. The AHF is temporarily stored and then loaded into tank-truck trailers inside the fluoride truck loading building for shipment to customers.

A portion of the DUF_6 conversion plant (the DUF_6 autoclave building) is characterized as having ordinary or moderate fire hazards. Such a fire would dissipate shortly given the limited combustibles and sprinklers in the area.

A second fire scenario involves the cylinder hauler vehicle used to transport DUF_6 cylinders to and from the cylinder staging pads. The cylinder hauler remains external to the building and an overhead bridge crane moves the cylinder into or out of the autoclave. A fuel leak by the cylinder hauler or another vehicle adjacent to the autoclave area would have minimal impact as the vehicle would not enter the building and the fuels would be drained away from the building interior, given the slope of the vehicle pathways. The radiant heat from this limited fire would not affect the DUF_6 cylinders because of their robust construction. The likelihood of the radiant heat from this limited fire affecting the DUF_6 cylinders, especially given their robust construction, qualifies as highly unlikely in accordance with the guidance in NUREG-1520 (NRC, 2002); therefore, no further analysis of this event is required.

Steam is the heating medium used in the autoclaves to bring the DUF_6 cylinders to the proper temperature for release. A natural gas fired package boiler located in the utility building provides the steam. No open flames or steam are produced in the autoclave area.

Hydraulic fluid is a limited combustible material used in the autoclave door opening and closing mechanisms. A leak in the system might spray hydraulic fluid onto an ignition source that would ignite the fluid. The capacities of the hydraulic system, hydraulic fluid properties, and operating pressures are not yet determined but are expected to be small, which is supported by analysis of similar operations in the FHA. Because of limited combustibles in the area, a fire would be small and would be contained by the automatic sprinklers in the building.

The DUF$_4$ building contains only minor fire hazards, with the greatest single hazard being the hydrogen gas that is injected into the reaction vessel. The process hazards section of this SER (i.e., Section 7.3.4.1) discusses hydrogen.

The reaction vessel shell is enclosed in a series of four clam-shell-type electric heaters within an insulated jacket. Ducting for cooling air penetrates the insulation jacket in selected locations. The heaters are controlled through program logic controllers to provide heating for startup. After the reaction vessel has reached its operating temperature, the heaters are used in conjunction with a cooling air blower and ducting for fine control of reaction vessel temperatures for optimum performance. The reaction vessel shell heaters are not considered to be a major ignition source for the area. Combustibles are limited, and no hydraulic fluids are located in this area.

A vacuum system will transfer the finished product from the DUF$_4$ process area to the FEP process. A failure in the vacuum system and a system rupture could allow material to fall freely in the area. A fire in the area from other sources could possibly be increased as a result of the loose particles of DUF$_4$ (although DUF$_4$ is not combustible) in the area, but the automatic sprinklers in the building would contain such a fire. DUF$_4$, upon exposure to water, can very slowly release some hydrogen fluoride (HF) by hydrolysis. Post fire cleanup of the water minimizes the likelihood of any reactions, given the very slow reaction time, and no additional significant fire consequences are created.

Granular-sized carbon is used in the carbon-bed traps for adsorption of trace of amounts of DUF$_6$ from the reaction vessel offgas stream. The carbon is enclosed in six separate trap housings. Each trap contains about 0.34–0.42 cubic m (12–15 cubic ft) of the granular carbon material. The carbon trap housing (shell) is of robust mechanical integrity, meets the requirements of the American Society of Mechanical Engineers (ASME) Code, and operates at relatively low pressures (about 16–25 pounds-force per square inch absolute). The likelihood of rupturing the trap housing is low. When a trap requires replacement, it is isolated with valves and is moved to the decontamination building for replacement of the carbon material inside an enclosed hooded booth system. Carbon material is not changed inside the DUF$_4$ process building's process area. The potential of spillage of any particulate carbon inside a process building that would result in a significant source of combustible material is very low.

Based on the above, the applicant determined that the likelihood of a fire being initiated and the IROFS failing, so that a release exceeding the consequence threshold of 10 CFR 70.61(b) or 10 CFR 70.61(c) occurs, is highly unlikely. The applicant performed this analysis using the guidance provided in NUREG-1520 (NRC, 2002). In addition, the analysis meets the requirements of NFPA 801 (NFPA, 2008c). The NRC staff's evaluation determined that the applicant's analysis meets the acceptance criteria in Section 7.4.3.2 of NUREG-1520 (NRC, 2002) and is, therefore, acceptable.

The NRC staff finds that the applicant has demonstrated that the facility will comply with the performance requirements of 10 CFR 70.61 in the event of a fire in the DUF$_6$ conversion plant.

7.3.2.6   Fire in Various Areas of the SiF$_4$ Plant

Silicon dioxide (SiO$_2$) powder is mixed with the DUF$_4$ and continuously fed to an electrically heated rotary calciner in which the reaction temperature is about 480–650 degrees C (900–

1,200 degrees F). Solid particles of depleted uranium dioxide (DUO_2) discharge from the rotary calciner via a cooling screw mechanism and are cooled and sent to the storage hopper. Alarms and controls are in place to shut down the rotary calciner and cooling screw in the event of a loss of cooling on the transfer screw. The product (SiF_4) offgas stream is noncombustible. Stored and packaged product is also noncombustible.

DUO_2 is pyrophoric, but only at elevated temperatures, specific particle sizes, and with adequate oxygen to support combustion. Within the storage hopper, a nitrogen gas buffer is maintained as a defense-in-depth measure. A spill or release of depleted uranium oxide could potentially be an issue. Exposed DUO_2 does not spontaneously combust; however, if the material is exposed to an existing fire, it increases the amount of fuel loading in the area. The specific heat released is not anticipated to be of sufficient quantity to create structural damage. The automatic sprinklers in the area would prevent other combustibles from igniting; however, the reaction from the introduction of water could further distribute burning material in the building. Given the limited combustibles, any spread of the fire would be highly unlikely to create a scenario in which the consequence thresholds of 10 CFR 70.61 would be exceeded.

Based on the above, the applicant determined that the likelihood of a fire being initiated and of the IROFS failing, so that a release exceeding the consequence threshold of 10 CFR 70.61(b) or 10 CFR 70.61(c) occurs, is highly unlikely. The applicant performed this analysis using the guidance provided in NUREG-1520 (NRC, 2002). In addition, the analysis meets the requirements of NFPA 801 (NFPA, 2008c). The NRC staff's evaluation determined that the applicant's analysis meets the acceptance criteria in Section 7.4.3.2 of NUREG-1520 (NRC, 2002) and is, therefore, acceptable.

The NRC staff finds that the applicant has demonstrated that the facility will comply with the performance requirements of 10 CFR 70.61 in the event of a fire in the SiF_4 plant.

7.3.2.7   Fire in Various Areas of the BF_3 Plant

Diboron trioxide (B_2O_3) is mixed with DUF_4 and continuously fed to an electrically heated preheater followed by reaction of the mixture in an electrically heated rotary calciner in which the reaction temperature is about 480–650 degrees C (900-1,200 degrees F). Product (BF_3) offgas streams and stored product are noncombustible. Solid particles of DUO_2 discharge from the rotary calciner via a cooling screw mechanism and are cooled and sent to the storage hopper. Alarms and controls are in place to shut down the rotary calciner and cooling screw in the event of a loss of cooling on the screw.

DUO_2 is pyrophoric, but only at elevated temperatures, specific particle sizes, and with adequate oxygen to support combustion. Within the storage hopper, a nitrogen gas buffer is maintained as a defense-in-depth measure. A spill or release of depleted uranium oxide could potentially be an issue. Exposed DUO_2 does not spontaneously combust; however, if the material is exposed to an existing fire, it increases the amount of fuel loading in the area. The specific heat released is not anticipated to be of sufficient quantity to create structural damage. The automatic sprinklers in the area would prevent other combustibles from igniting; however, the reaction from the introduction of water could further distribute burning material in the building. Given the limited combustibles, any spread of the fire would be highly unlikely to create a scenario in which the consequence thresholds of 10 CFR 70.61 would be exceeded.

Based on the above, the applicant determined that the likelihood of a fire being initiated and of the IROFS failing, so that a release exceeding the consequence threshold of 10 CFR 70.61(b) or 10 CFR 70.61(c), occurs is highly unlikely. The applicant performed this analysis using the guidance provided in NUREG-1520 (NRC, 2002). In addition, the analysis meets the requirements of NFPA 801 (NFPA, 2008c). The NRC staff's evaluation determined that the applicant's analysis meets the acceptance criteria in Section 7.4.3.2 of NUREG-1520 (NRC, 2002) and is, therefore, acceptable.

The NRC staff finds that the applicant has demonstrated that the facility will comply with the performance requirements of 10 CFR 70.61 in the event of a fire in the BF_3 plant.

7.3.2.8   Fire Hazards Analysis Conclusions

The NRC staff has reasonable assurance that the applicant has identified and evaluated all fire-related accident scenarios credible for the proposed extraction and deconversion processes. The applicant has reasonably identified and evaluated possible fire initiators and consequences and has identified IROFS for preventing or mitigating fire accident scenarios that could result in intermediate or high consequences, in accordance with 10 CFR Part 70, "Domestic Licensing of Special Nuclear Material," and as described by the guidelines established in NUREG-1520 (NRC, 2002).

Consistent with the acceptance criteria in Section 7.4.3.2 of NUREG-1520 (NRC, 2002), the applicant performed an FHA for each process area. The FHA described, by fire area, the fuel loading, fire scenarios, methods of consequence analysis, potential consequences, and mitigative controls, which meets the requirements of NFPA 801 (NFPA 2008c) and is, therefore, acceptable. In addition, based on its review of the above information, the NRC staff concludes with reasonable assurance that the applicant's performance of its FHA meets the requirements of 10 CFR 40.32, 10 CFR 70.61, and 10 CFR 70.64, as they pertain to the fire protection aspects of the facility.

7.3.3   Facility Design

7.3.3.1   Facility Passive-Engineered Fire Protection Systems

Buildings containing uranium are within the DUF_6 conversion plant (DUF_6 autoclave building, DUF_4 process building, decontamination building, DUF_4 container storage building, and the DUF_4 container staging building) and within the SiF_4/BF_3 plants (FEP building and FEP oxide staging building), which have noncombustible structural steel beams and columns, on concrete slabs, with built-up composite roofing on metal deck. This construction is equivalent to Type II-000, in accordance with NFPA 220, "Standard for Types of Building Construction" (NFPA, 2006b). Thus, all structural members are of noncombustible or limited combustible construction, but no fire rating is required. To meet building code requirements and as a defense-in-depth measure, some buildings are separated into different fire areas by rated walls constructed in accordance with NFPA 221, "Standard for High Challenge Fire Walls, Fire Walls, and Fire Barrier Walls" (NFPA, 2006c).

To ensure life safety protection for the occupants, buildings are provided with means of egress, illumination, and protection in accordance with the New Mexico Commercial Building Code (NMCBC) (NMCBC, 2006). Barriers with fire resistance ratings consistent with the NMCBC and the FHA are provided to prevent fire propagation. All of the buildings are provided with emergency lighting for the illumination of the primary exit paths and the essential operations areas where personnel are required to operate valves, dampers, and other controls in an emergency. Emergency lighting is considered a critical load. Marking of means of egress, including illuminated exit signs with battery backup, are provided in accordance with the NMCBC and NFPA 101, "Life Safety Code" (NFPA, 2006a).

The potential for lightning strikes to the buildings is considered possible; however, the structural design with metal beam and columns and connections to the underground grounding loops surrounding the buildings is permissible in lieu of air terminals under NFPA 780, "Standard for the Installation of Lightning Protection Systems" (NFPA, 2008b), and is considered to be effective lightning protection.

The applicant's ISA Summary (IIFP, 2012b) has adequately addressed passive-engineered fire protection systems in accordance with the guidance established in NUREG-1520 (NRC, 2002) and meets the requirements of NFPA 801 (NFPA, 2008c). In addition, the applicant's passive-engineered fire protection systems meet the requirements of 10 CFR 40.32, 10 CFR 70.64, and 10 CFR 70.65 as they pertain to the fire protection aspects of the facility.

7.3.3.2   Facility Active Engineered Fire Protection Systems

7.3.3.2.1   Electrical Installation, Ventilation, and Fire Alarm System

All electrical systems at the facility are installed in accordance with NFPA 70, "National Electric Code" (NFPA, 2008a). Switchgear, motor control centers, panel boards, variable frequency drives, uninterruptible power supply (UPS) systems, and control panels are mounted in metallic enclosures and contain limited amounts of combustible material. Cable trays and conduits are metallic, and the cable in cable trays is flame retardant and tested in accordance with industry guidance found in Institute of Electrical and Electronics Engineers (IEEE) Standard 383, "Standard for Type Test of Class 1E Electric Cables, Field Splices, and Connections for Nuclear Power Generating Stations" (IEEE, 2003).

Roof-mounted exhaust fans and wall-mounted intake louvers provide ventilation for the DUF_4 and FEP process buildings. Steam is used as the main heat source for the process building environment. Process control room areas are heated, ventilated, and cooled by electrical heat pump units with electric auxiliary heat. The control room heating, ventilation, and air-conditioning (HVAC) units create positive pressure in each of the control rooms with alarms to indicate loss of pressure. Process equipment areas are open and of large volumes, so steam heating is practical. Cooling of other process and storage areas is provided with wall-mounted exhaust fans and intake louvers. The ventilation and HVAC systems meet NFPA 90A, "Installation of Air Conditioning and Ventilating Systems" (NFPA, 2009b), and NFPA 90B, "Installation of Warm Air Heating and Air Conditioning Systems" (NFPA, 2009c). The ventilation systems are not engineered for smoke control, but are designed to shut down in the event of a fire. The offsite fire department provides smoke control using portable smoke removal equipment.

7.3.3.2.2   Portable Fire Extinguishers

Portable fire extinguishers are installed and inspected throughout all buildings in accordance with NFPA 10, "Standard for Portable Fire Extinguishers" (NFPA, 2007a). Multipurpose fire extinguishers are provided generally for Class A (ordinary combustibles), Class B (flammable and combustible liquids), and Class C (electrical equipment) fires. The NRC staff finds acceptable the applicant's commitment to provide portable fire extinguishers appropriate for the type of fire in accordance with national fire standards.

7.3.3.2.3 Fire Water Supply and Drainage

The facility fire water supply consists of two 378,541-L (100,000-gal) water storage tanks designed and constructed in accordance with NFPA 22, "Standard for Water Tanks for Private Fire Protection" (NFPA, 2003). Separate storage tanks are used for the sanitary water supply. The site's well water supply provides fill and make up to the tanks, and the water supply is capable of filling fire protection water inventory in a single tank within an 8-hour period. The fire pumps consist of one electric-driven pump and one diesel-driven pump, both rated for 2,271 liters per minute (lpm) (600 gallons per minute [gpm]), at 6.89-bar (100-pounds-force per square inch absolute) pumps. Both pumps are horizontal centrifugal pumps designed and installed in accordance with NFPA 20, "Standard for Installation of Stationary Pumps for Fire Protection" (NFPA, 2007c). The combination of two water tanks and two fire pumps provides 100-percent redundancy for fire protection. The tanks are arranged such that one will be available for suction at all times. In addition to fixed standpipes and fire hose stations, the facility will be provided with fire hose on mobile apparatus or at strategic locations throughout the facility. The amount of hose provided will be sufficient to ensure that all points within the facility will be able to be reached by at least two backup hoses with a diameter of 64 millimeters (mm) (2.5 inches [in]), consistent with NFPA 1410, "Standard on Training for Initial Emergency Scene Operations" (NFPA, 2010b). These lines will have a minimum nozzle pressure of 4.5 bar (65 pounds-force per square inch gauge [psig]) for the fixed hose and 6.9 bar (100 psig) for the 64-mm (2.5-in) hose. The NRC finds acceptable the applicant's commitment to provide a fire water supply system consistent with national standards, which provides adequate redundancy for water supply to all points within the facility.

Buildings, building aprons, and process area outdoor pads, where chemicals or licensed materials are stored or processed, have curbs or dikes—or both—to prevent drainage of contaminated liquids outside the spill controlled areas. Water from activation of the sprinkler system or from firefighting activities could contain contaminated materials or flammable and combustible liquids. During the initial period of sprinkler activation or generation of fire water in an area, the water collects in the spill controlled area and is handled and treated as any other type of spillage or liquid. Areas that have dikes for nonuranium hazardous chemical or oil spill control have installed pumps that can either automatically or manually be activated to pump spilled liquids or water for treatment. Areas where licensed materials are processed or stored, and have curbing or dikes, are not automatically pumped. If fire water accumulates in those areas in excess of the holding capacity, the water may be pumped either to another licensed material curb or dike area, or to holding tanks in the decontamination building, or to the large HF recycle tank where it can be sampled and its disposition determined. Portable pumps are also available for emergency pumping of liquids to other holding areas, tanks, or treatment, if necessary. If the volume of the fire water reaches a level that exceeds the respective spill control area, it is pumped to other outside spill control areas not directly affected by the fire response. If the water drains and enters the plant storm sewer drain system, it then flows to the storm water retention basin where it is sampled and a decision is made about its disposal. The NRC finds acceptable the applicant's drainage system and back-up pumping capability to control contamination from fire suppression liquids.

7.3.3.2.4 Engineered Automatic Fire Suppression Systems

Automatic wet pipe sprinkler systems are provided in all buildings in the facility, with the exception of the lime silo storage shed, which is not normally occupied, has limited amounts of combustibles, and contains no licensed material.

These systems are designed and tested in accordance with NFPA 13, "Standard for the Installation of Sprinkler Systems" (NFPA, 2007b). Sprinkler system control valves are monitored under a periodic inspection program, and their proper positioning is supervised in accordance with NFPA 801 (NFPA, 2008c).

The applicant's ISA Summary (IIFP, 2012b) adequately addressed active-engineered fire protection systems in accordance with the guidance established in NUREG-1520 (NRC, 2002) and meets the requirements of NFPA 801 (NFPA, 2008c). In addition, the applicant's active-engineered fire protection system meets the requirements of 10 CFR 40.32, 10 CFR 70.64, and 10 CFR 70.65 as they pertain to the fire protection aspects of the facility.

7.3.3.3   Facility Design Conclusions

The NRC staff finds the applicant adequately addressed building construction, fire area determination, electrical installation, life safety, drainage, and lightning protection in the application. The FHA described ventilation characteristics as they relate to fire protection and fire hazards.

Consistent with the acceptance criteria in Section 7.4.3.3 of NUREG-1520 (NRC, 2002), the application documents the fire safety considerations used in the general design of the facilities containing licensed material or facilities that have the potential to impact the safety of radiological facilities and is, therefore, acceptable. In addition, the NRC staff's evaluation finds that the fire protection features of the applicant's proposed facility meet the requirements of 10 CFR 40.32, 10 CFR 70.61, 10 CFR 70.64, and 10 CFR 70.65 as they pertain to the fire protection aspects of the facility.

7.3.4   Process Fire Safety

The applicant plans to convert (deconversion) DUF_6 into uranium oxide for long-term, stable disposal. The company will also include a commercial plant to produce specialty fluoride gas products for sale. Using the fluorine derived from the deconversion process, the applicant will manufacture high-purity SiF_4 and BF_3 in the IIFP facility. The fluoride gas products are utilized in the electronic, solar, and semiconductor markets. In addition, AHF is a product of the deconversion process and is sold as a chemical for use in various industrial applications.

The applicant's FHA, which is discussed in Section 7.2.2 of this SER, analyzed many of the specific hazards within the process. The following subsection discusses the aspects of each process chemical that has fire safety implications.

7.3.4.1   Process Descriptions

Depleted Uranium Hexafluoride (DUF_6): DUF_6 is not flammable and does not disassociate to flammable constituents under conditions at which it will be handled at the facility. DUF_6 does not react with oxygen, nitrogen, carbon dioxide, or dry air; but does react with water or water vapor. Hydrocarbons can be explosively oxidized if they are mixed with DUF_6 in the liquid state or at elevated temperatures. For this reason, the DUF_6 processes at the facility do not use nonfluorinated hydrocarbon lubricants. DUF_6 pumps are lubricated using a perfluoropolyether (PFPE) oil that is noncombustible. The used PFPE oil will be collected, packaged, and shipped offsite for disposal at a licensed low-level radioactive waste facility.

Hydrogen Fluoride (HF): HF is a byproduct of the chemical reaction of DUF_6 with water vapor. HF is extremely reactive in both gaseous and aqueous form. HF alone is neither flammable nor combustible. It can, however, react exothermically with water to generate sufficient heat to ignite nearby combustibles.

Depleted Uranyl Fluoride (DUO_2F_2): DUO_2F_2 is also a byproduct of the chemical reaction of DUF_6 with water vapor. DUO_2F_2 is stable in air to 300 degrees C (572 degrees F). It is neither flammable nor combustible and will not decompose to combustible constituents under conditions that will exist at the facility.

Depleted Uranium Dioxide (DUO_2): DUO_2 is pyrophoric, but only at elevated temperatures, specific particle sizes, and with adequate oxygen to support combustion. Within the process, a nitrogen gas buffer is maintained as a defense-in-depth measure. A spill or release of DUO_2 could potentially be an issue. Exposed DUO_2 does not spontaneously combust; however, if the material is exposed to an existing fire, it increases the amount of fuel loading in the area. The specific heat released is not anticipated to be of sufficient quantity to create structural damage to any of the various process buildings. The automatic sprinklers in the process areas would prevent other combustibles from igniting; however, the reaction from the introduction of water could further distribute burning material in the building. Given the limited combustibles in the process areas, the spread of the fire would become highly unlikely.

Storage and Handling of DUF_6: DUF_6 cylinders are stored or handled in the cylinder staging/storage pads and in the DUF_6 autoclave building. On the staging/storage pads, fire concerns include the cylinder transport vehicle, a fire exposure from nearby vegetation, and fire exposure from a nearby vehicle accident. The applicant evaluated these various fire scenarios and either concluded that they did not pose a threat to the stored cylinders or that, with adequate controls, the threats could be mitigated by IROFS. Combustible loadings in the DUF_6 autoclave building are limited, and transient combustibles will be controlled. Therefore, any fire originating in these areas will be limited.

Hydrogen Control: Hydrogen is produced for use in the DUF_4 plant and may also be generated at battery charging stations in the facility (all forklifts are battery powered). The gaseous hydrogen supply for the various processes is generated onsite using a vendor-supplied, pre-engineered system which reforms natural gas followed by purification using pressure swing absorption.

Hydrogen control in battery-charging stations will be provided by measures identified in NFPA 70E, "Standard for Electrical Safety in the Workplace" (NFPA, 2009a), including hydrogen detection. Natural and mechanical ventilation will be provided to ensure that hydrogen concentrations do not exceed 25 percent of the lower explosive limit. Redundant ventilating fans with explosive proof motors are run continuously or are interlocked to start upon the detection of hydrogen in the area. Redundant airflow sensors will indicate fan run status.

Combustible Material Hazards: Materials of construction for the process buildings are predominantly noncombustible (e.g., steel, aluminum, and concrete floors). A minimum of fixed combustibles is expected to be present in the operations areas, and the applicant plans to control transient combustibles to minimize potential fire hazards, and general housekeeping procedures. Other quantities of combustible materials are as follows:

- Minor quantities of hydraulic fluids or lube oils are used in the overhead cranes and small forklifts.

- Combustible coke is used in the coke box scrubber; however, it is kept wet as part of the process and is highly unlikely to become sufficiently dry to become a com'bustible hazard.

- Natural gas is the primary fuel for the steam boilers.

- Diesel fuel supplies the backup generators and the diesel powered fire pump. In addition, quantities of diesel fuel are onsite at the remote, dedicated fueling station for the cylinder hauler.

- Granular-sized carbon is used in the carbon-bed traps in various process areas.

The NRC staff finds the applicant's list of combustible materials accurate for the described operations and finds acceptable the applicant's commitments discussed above to limit combustible loading.

7.3.4.2   *Process Fire Safety Conclusions*

No fire safety IROFS are directly inherent to the process design. Section 7.2.2 of this SER discusses the fire safety IROFS used to protect the licensed material within the processes. In its review, the NRC staff has taken into account the potential presence of the identified combustibles in the various accident scenarios. The applicant's ISA Summary (IIFP, 2012b), which provides the supporting safety basis for the LA (IIFP, 2012a), documents the identification of fire hazards and related analyses.

Consistent with the acceptance criteria in Section 7.4.3.4 of NUREG-1520 (NRC, 2002), the application identifies the hazardous chemicals, processes, and design standards used to ensure safety in areas that have fire hazards that may threaten licensed material and is, therefore, acceptable. In addition, the NRC staff's evaluation finds that the fire protection features of the applicant's facility meet the requirements of 10 CFR 40.32, 10 CFR 70.61, 10 CFR 70.64, and 10 CFR 70.65 as they pertain to the process fire safety aspects of the facility.

7.3.5   Fire and Emergency Response

The facility will maintain a fire brigade made up of employees trained in firefighting techniques, first-aid procedures, and emergency response. The fire brigade is organized, operated, trained, and equipped in accordance with NFPA 600, "Standard on Industrial Fire Brigades" (NFPA, 2010a) for incipient firefighting capability. The intent of the facility fire brigade is to be able to handle all minor fires and to provide a first response effort designed to supplement the local fire department for major fires at the plant. The field incident commander, working with the plant's emergency director, will coordinate offsite fire department activities. The fire brigade is staffed so as to ensure a minimum of five brigade members available per shift. Section 7.3.3.2 of this SER describes the onsite water supply available to the fire brigade and other responders.

The applicant provides periodic training on the facility's emergency training procedures to personnel of offsite organizations. Facility emergency response personnel meet at least biennially with each offsite assistance group to accomplish training and review items of mutual interest, including relevant changes to the program. The primary agencies that will be available for this response are the fire and rescue agency of the City of Eunice, NM, and fire department of the City of Hobbs, NM. These agencies are signatories to the Lea County, NM, mutual aid agreement and can request assistance from any of several adjacent municipal fire departments and the fire/emergency response services. The applicant has received letters from the referenced fire departments, which include a commitment to fire protection and emergency drills and define the fire protection and emergency response commitments between the organizations.

The Hobbs Fire Department is the primary response agency, and is comprised of a roster of approximately 70 paid personnel, staffing three fire stations in a three-shift rotation. The department has structural engines, ladder truck, heavy rescue truck, grass fire trucks, a water tanker, and several command vehicles and ambulances, each equipped to provide advanced level life support. Firefighters are trained to Firefighter Level I and Emergency Medical Technician (EMT) Basic as a minimum, in accordance with New Mexico standards. Shift-assigned ambulance personnel are EMT paramedics under New Mexico standards.

Eunice Fire and Rescue is the secondary response agency and comprises a roster of approximately 20 volunteers. Eunice has structural fire engines, grass fire trucks, a water tanker, command vehicles, and ambulances, each equipped to provide intermediate level life support. Firefighters are trained, as a minimum, to Firefighter Level I and ambulance personnel, as a minimum, to EMT Basic, in accordance with New Mexico standards. In the event of a fire, the IIFP fire brigade responds and the Hobbs and Eunice Fire and Rescue Departments are notified to respond. If the fire is incipient, the IIFP fire brigade fights the fire using hand portable/wheeled fire extinguishers and 38-mm (1.5 in) hose lines. The estimated response time to the facility is less than 15 minutes for the Hobbs Fire Department and between 20 and 30 minutes from Eunice Fire and Rescue.

The NRC staff finds that the onsite fire brigade, onsite water supply, onsite hose lines, and mutual aid from adequately equipped fire departments can provide defense-in-depth protection from releases from all identified and credible fire scenarios, satisfy the requirements of 10 CFR 70.64(b), and are in accordance with the guidance in NUREG-1520 (NRC, 2002).

Consistent with the acceptance criteria in Section 7.4.3.5 of NUREG-1520 (NRC, 2002), the LA (IIFP, 2012a) documents the available fire protection systems and fire emergency response organizations and is, therefore, acceptable. In addition, the applicant's emergency response capability meets the requirements of 10 CFR 40.32, 10 CFR 70.61, 10 CFR 70.64, and 10 CFR 70.65, as they pertain to the fire protection aspects of the facility.

7.3.6 Baseline Design Criteria

The NRC staff reviewed the Fluorine Extraction Process and Depleted Uranium Deconversion Plant and its fire safety program to determine applicability and level of compliance with NFPA 801 (NFPA, 2008c) and applicable standards referenced within. Table 7-1 of the LA (IIFP, 2012a) lists the fire codes and standards considered by the applicant to be applicable to the facility. The NRC staff finds the use of these consensus codes and standards to be in accordance with the guidance of Section 7.4.3 of NUREG-1520 (NRC, 2002) with regard to nationally recognized codes and standards that may be used to measure reasonable assurance of fire safety. The design provides for adequate protection against fires and explosions. Therefore, the NRC staff considers that the use of the above codes and standards satisfies the requirements of 10 CFR 70.64(a) and 10 CFR 70.64(a)(3).

Consistent with the acceptance criteria in Section 7.4.3 of NUREG-1520 (NRC, 2002), the LA (IIFP, 2012a) and ISA Summary (IIFP, 2012b) document the defense-in-depth provided to enhance safety by reducing the challenges to IROFS. This chapter of the SER documents the evaluation of the notable fire protection related defense-in-depth measures. Furthermore, the IROFS outlined in Section 7.2.2.1 of this SER show a clear preference for engineered controls over administrative controls. Administrative controls were typically used to supplement engineered controls or when the use of an engineered control was not practicable. Therefore, given the outlined methodology, the NRC staff finds that the facility's design satisfies the requirements of 10 CFR 70.64(b), as they pertain to the fire protection aspects of the facility.

7.4   Evaluation Findings

The applicant has established a fire protection function meeting the acceptance criteria from Chapter 7 of NUREG-1520 (NRC, 2002). The function includes a facility safety review committee responsible for integrating modifications to the facility and a fire safety manager responsible for day-to-day program implementation. Fire prevention, inspection, testing, and maintenance of fire protection systems and the qualification, drills, and training of facility personnel are in accordance with applicable NFPA codes and standards. (Note that SER Section 11.3.3 describes fire protection training requirements.)

The applicant has conducted risk analyses in accordance with NFPA 801, "Standard for Fire Protection for Facilities Handling Radioactive Material" (NFPA, 2008c). The FHA identified credible fire scenarios that bound the fire risk. The ISA used these scenarios and identified fire protection IROFS. A memorandum of understanding with the local fire department documents the required assistance and the annual exercises. Procedures are in place to allow the fire department efficient access to process areas during fire emergencies. Worker egress is designed and maintained in accordance with NMCBC (NMCBC, 2006) and NFPA 101 (NFPA, 2006a).

The applicant demonstrated that it incorporated appropriate fire safety considerations in the design of its facilities. The applicant also demonstrated that the facility has appropriate active fire protection systems.

The NRC staff concludes that the applicant's capabilities meet the criteria in Chapter 7 of the SRP. In addition, the NRC staff concludes that the applicant provided sufficient information in accordance with requirements of 10 CFR 40.32, and 10 CFR 70, Subpart H, regarding potential fire hazards, consequences, and required controls for the proposed processes. The NRC staff finds that the applicant's proposed equipment, facilities, and procedures provide assurance that adequate fire protection will be provided and maintained for those IROFS to meet the safety performance requirements and BDC of 10 CFR Part 70.

7.5   References

(IEEE, 2003) Institute of Electrical and Electronics Engineers, Inc., "Standard for Type Test of Class 1E Electric Cables, Field Splices, and Connections for Nuclear Power Generating Stations," IEEE Standard 383, December 12, 2003.

(IIFP, 2012a) International Isotopes Fluorine Products, Inc., "Fluorine Extraction Process and Depleted Uranium Deconversion Plant (FEP/DUP) License Application, Revision B," May 2012, Agencywide Documents Access and Management System (ADAMS) Accession No. ML12123A245.

(IIFP, 2012b) International Isotopes Fluorine Products, Inc., "ISA Summary Rev. B for IIFP," May 2012, Agencywide Documents Access and Management System (ADAMS) Accession No. ML12123A245.

(NFPA, 2010a) National Fire Protection Association, "Standard on Industrial Fire Brigades," NFPA 600, 2010.

(NFPA, 2010b) National Fire Protection Association, "Standard on Training for Initial Emergency Scene Operations," NFPA 1410, 2010.

(NFPA, 2009a) National Fire Protection Association, "Standard for Electrical Safety in the Workplace," NFPA 70E, 2009.

(NFPA, 2009b) National Fire Protection Association, "Installation of Air Conditioning and Ventilating Systems," NFPA 90A, 2009.

(NFPA, 2009c) National Fire Protection Association, "Installation of Warm Air Heating and Air Conditioning Systems," NFPA 90B, 2009.

(NFPA, 2008a) National Fire Protection Association, "National Electric Code," NFPA 70, 2008.

(NFPA, 2008b) National Fire Protection Association, "Standard for the Installation of Lightning Protection Systems," NFPA 780, 2008.

(NFPA, 2008c) National Fire Protection Association, "Standard for Fire Protection for Facilities Handling Radioactive Materials," NFPA 801, 2008.

(NFPA, 2007a) National Fire Protection Association, "Standard for Portable Fire Extinguishers," NFPA 10, 2007.

(NFPA, 2007b) National Fire Protection Association, "Standard for the Installation of Sprinkler Systems," NFPA 13, 2007.

(NFPA, 2007c) National Fire Protection Association, "Standard for Installation of Stationary Pumps for Fire Protection," NFPA 20, 2007.

(NFPA, 2007d) National Fire Protection Association, "National Fire Alarm Code," NFPA 72, 2007.

(NFPA, 2006a) National Fire Protection Association, "Life Safety Code," NFPA 101, 2006.

(NFPA, 2006b) National Fire Protection Association, "Standard on Types of Building Construction," NFPA 220, 2006.

(NFPA, 2006c) National Fire Protection Association, "Standard for High Challenge Fire Walls, Fire Walls, and Fire Barrier Walls," NFPA 221, 2006.

(NFPA, 2003) National Fire Protection Association "Standard for Water Tanks for Private Fire Protection," NFPA 22, 2003.

(NMCBC, 2006) New Mexico Commercial Building Code, 2006.

(NRC, 2011) U.S. Nuclear Regulatory Commission, "January 19, 2011, Meeting Summary, 'Overview of the Fossil Fuel Pipeline Explosion Evaluation for the International Isotopes, Inc., FEP & DUF_6 Deconversion Facility.'"

(NRC, 2005) U.S. Nuclear Regulatory Commission, "Confirmatory Calculations for Fire Protection Review of National Enrichment Facility Integrated Safety Analysis (ISA) Summary," March 2005. Agencywide Documents Access and Management System (ADAMS) Accession No. ML050910020

(NRC, 2004) U.S. Nuclear Regulatory Commission, "Fire Dynamics Tools (FDT^s) Quantitative Fire Hazard Analysis Methods for the U.S. Nuclear Regulatory Commission Fire Protection Inspection Program," NUREG-1805, December 2004.

(NRC, 2002) U.S. Nuclear Regulatory Commission, "Standard Review Plan for the Review of a License Application for a Fuel Cycle Facility," NUREG-1520, Rev. 0, March, 2002.

(NRC, 1978) U.S. Nuclear Regulatory Commission, "Evaluations of Explosions Postulated To Occur on Transportation Routes Near Nuclear Power Plants," Regulatory Guide 1.91, Rev. 1, February 1978.

8.0   EMERGENCY MANAGEMENT

The purpose of reviewing the International Isotopes Fluorine Products, Inc.'s (IIFP or the applicant), Emergency Plan (EP) (IIFP, 2012a) is to determine whether IIFP has established adequate emergency response facilities and procedures to protect workers, the public, and the environment.

8.1   Regulatory Requirements

The regulations in Title 10 of the *Code of Federal Regulations* (10 CFR) 40.31(j)(1) require applicants that are requesting authorization to possess uranium hexafluoride (DUF_6) in excess of 50 kilograms (kg) (110 pounds [lb]) in a single container or 1,000 kg (2,200 lb) total to submit either (1) an evaluation that shows that the maximum intake of uranium by a member of the public from a release would not exceed 2 milligrams (mg) or (2) an EP which addresses the radiological hazards of an accidental release of source material and any associated chemical hazards directly incident thereto. The regulations in 10 CFR 40.31(j)(3) also outline the regulatory requirements for the information that must be included in an EP. The regulations in 70.64(a)6) require emergency planning to be incorporated into the baseline design criteria.

8.2   Regulatory Acceptance Criteria

Title 10 CFR 40.31(j)(3) and Section 8.4.3 of NUREG-1520, "Standard Review Plan for the Review of a License Application for a Fuel Cycle Facility" (NRC, 2002), outline the acceptance criteria for the U.S. Nuclear Regulatory Commission's (NRC's) review of the EP.

8.3   Staff Review and Analysis

8.3.1   Facility Description

Section 2.0 of the EP (IIFP, 2012a) describes the licensed activity, the facility, the site, and the area near the site. The information provided includes the following:

- a brief description of the deconversion process

- a discussion of chemicals of concern—including form, physical state, location, and quantity

- a detailed description of the site location and layout

- a description of the major structures to be located at the site

- a description of the ventilation systems—including stack heights, maximum flow rates, and filter efficiencies

- a description of the area near the site, including area land use information

- demography of the area, water use, and climate

- detailed maps of the facility and surrounding area.

The NRC staff finds that the information provided by the applicant as discussed above, taken together with the information provided by the applicant as discussed in Section 8.3.2 of this SER, below, meets the acceptance criteria of 10 CFR 40.31(j)(3)(i) and Section 8.4.3.1.1 of NUREG-1520 (NRC, 2002). Therefore, the NRC staff finds the applicant's facility description adequate.

8.3.2 Onsite and Offsite Emergency Facilities

Section 9.0 of the EP (IIFP, 2012a) contains descriptive information regarding the emergency response equipment and facilities. The primary Emergency Operations Center (EOC) is the facility's control room. The IIFP's control room/EOC controls communications to all principal points within and outside the facility. The IIFP's control room/EOC contains procedures and operational engineering information to assist in routine operations and in emergency response.

The IIFP EOC and its personnel support the following functions:

- Assess abnormal conditions.

- Notify additional facility personnel, if needed.

- Make offsite notifications.

- Perform or direct accident mitigation.

- Direct facility operations.

- Implement onsite protective actions.

If the nature and location of the emergency situation requires it, or if the control room/EOC becomes uninhabitable, the Emergency Director may move the emergency response personnel to the designated backup (alternate) EOC. The backup EOC is located in a different and physically separate building (identified in the IIFP facility's EP Implementing Procedures [EPIPs]) from the primary control room/EOC. The same documentation and communications capabilities available in the primary control room/EOC will also be available in the alternate onsite EOC. In addition, the applicant will designate an offsite EOC location (identified in the facility EPIPs), arranged as part of the community offsite assistance agreements, for use in the event the entire facility requires evacuation. The offsite EOC will have appropriate documents and communications equipment.

Section 7.4 of the EP (IIFP, 2012a) describes offsite emergency support and equipment. This section provides information regarding fire, emergency medical services, and local law enforcement. The applicant will sign a memorandum of understanding (MOU) between IIFP and the Hobbs, NM, fire department for fire and medical emergency services 3 months before the startup of operations or receipt of licensed materials or hazardous process chemicals. IIFP will also sign MOUs with the Lea County Medical Center, Lea County Emergency Management, Lea County Sheriff's Department, and the New Mexico Department of Homeland Security and Emergency Management at least 3 months before the startup of operations or receipt of licensed materials or hazardous process chemicals. These MOUs are not part of the License Application, but will be mutually signed by IIFP and the appropriate agency prior to operations.

Sections 8.2.2, 8.5.1.3, and 10.6.1 of the EP (IIFP, 2012a) describe the emergency monitoring equipment that is available for assessing radioactive releases, including personnel monitoring, air sampling, gaseous effluent monitoring, equipment for sampling soil, groundwater and vegetation, and portable radiological monitors.

Sections 5.2 and 5.3.5 of the EP (IIFP, 2012a) describe the emergency monitoring equipment that is available for assessing process system upsets, including alarms in the control room, and fluoride detectors. The applicant will inventory and test emergency equipment on a quarterly or monthly basis, depending on the type of equipment, as discussed in Section 10.6 of the EP (IIFP, 2012a). The appropriate manager will be notified of any deficiencies that cannot be corrected so that repair or replacement can be expedited.

Section 9.2 of the EP (IIFP, 2012a) describes the facility's communications equipment, which includes facility telephones with facsimile capability, the public address system, alarms, and two- way radios. The communications systems are designed so that a failure in one system does not leave the facility without communications capability. For offsite communications, the facility's telephone system is the primary means of communication; the facility's radio system is the alternate means of emergency communication with offsite authorities. Onsite and offsite radios are tested quarterly.

The applicant identified and described the onsite and offsite emergency facilities. The NRC staff finds that the information provided by the applicant meets the acceptance criteria in Section 8.4.3.1.2 of NUREG-1520 (NRC, 2002) and, together with new information discussed in Section 8.3.3 of this SER, meets the requirements of 10 CFR 40.31(j)(3)(ii) for new facilities because the applicant identified and described the onsite and offsite facilities that could be relied on in an emergency. The applicant's onsite and offsite emergency organization and facilities have sufficient equipment and resources, and the NRC staff finds them acceptable.

8.3.3 Types of Accidents

Section 4.0 of the EP (IIFP, 2012a) identifies postulated events that have high and intermediate consequences.

Since the IIFP processes only depleted uranium, in the form of DUF_6, depleted uranium tetrafluoride (DUF_4), and a blend of depleted uranium oxides, the offsite radiological consequences associated with plant accidents is limited. The applicant stated that no nuclear criticality potential exists at the facility, and there is no material onsite that contains fission products or transuranic elements. A potential exists to release depleted uranyl fluoride (UO_2F_2) into the environment as a reaction product from a DUF_6 release. The applicant stated that no credible accident has been identified to pose intermediate- or high-radiological consequences to the public. Revision B of the IIFP Integrated Safety Analysis (ISA) Summary (IIFP, 2012b), Sections 4.1.2 and 7, contain figures projecting doses and toxic substance concentrations as a function of distance and time for various meteorological stability classes, including computer models and assumptions. The NRC staff finds these statements to be accurate due to the fact that the facility processes depleted uranium and does not possess special nuclear material.

The NRC staff finds that established controls are in place to prevent or mitigate significant chemical or radioactive release at the plant site. Credible accident initiators have been identified and subsequent limits and controls established to control the frequency and severity of credible accidents such that acceptable risk to the public is maintained.

As stated above, the applicant identified and described the types of accidents identified in the EP for which protective actions may be needed. The NRC staff finds that the information provided by the applicant meets the acceptance criteria in Section 8.4.3.1.3 of NUREG-1520, (NRC, 2002) and 10 CFR 40.31(j)(3)(ii) for new facilities because IIFP has identified and described the types of accidents that may require protective actions. Therefore, the NRC staff finds that applicant's description of the types of accidents is acceptable.

8.3.4   Classification of Accidents

IIFP has established emergency action levels consistent with Appendix A, "Examples of Initiating Conditions," to NRC Regulatory Guide 3.67, "Standard Format and Content for Emergency Plans for Fuel Cycle and Materials Facilities" (NRC, 1992). Section 6.1 of the EP (IIFP, 2012a) describes the criteria used to classify an emergency as either an alert or a site area emergency declaration and defines both types of incidents. The applicant established that the threshold for escalating an event from an alert to a site area emergency declaration is based on the significance of the release to the environment and the possibility that the event could require a response by an offsite response organization. The EPIPs provide the processes for making the appropriate classification. Sections 6.1.1 and 6.1.2 of the EP (IIFP, 2012a) provide examples of site-specific incidents and the emergency classification that will be declared for each event. The Shift Superintendent is responsible for accident classification and assumes the responsibility of the Emergency Director until relieved by the Chief Operations Officer/Plant Manager (COO/PM) or his or her designee.

The NRC staff finds that the applicant adequately identified and described the EP classification system. The information provided by the applicant meets the requirements of 10 CFR 40.31(j)(3)(iii) and the acceptance criteria in Section 8.4.3.1.4 of NUREG-1520 (NRC, 2002) for new facilities because IIFP has identified and described the two classifications of an alert and site area emergency, the classification that is expected for each accident identified in the EP, the emergency action levels at which an alert or site area emergency will be declared, and the personnel positions and alternates with responsibility for accident classifications. The applicant's process for classification of accidents is, therefore, acceptable.

8.3.5   Detection of Accidents

Sections 2.2 and 8.3.3 of the EP explain the methods and systems available to detect accidents at the facility, including the following:

- UF_6, anhydrous hydrogen fluoride, silicon tetrafluoride, boron trifluoride, or uranium powder releases

- natural phenomena

- fires propagating between areas

- fires involving excessive transient combustibles

- process system upsets.

The operating procedures contain directions for manual actions to be taken to bring the facility to a safe condition.

The NRC staff finds that the applicant has adequately identified and described the methods and systems for the detection of accidents. The information provided by the applicant meets the requirements of 40.31 (j)(3)(iv) and the acceptance criteria in Section 8.4.3.1.5 of NUREG-1520 (NRC, 2002) for new facilities because IIFP identified and described the means of detecting an accident, the means of detecting any release of radioactive material or hazardous chemicals, the means of alerting the operating staff, and the anticipated response of the operating staff. Therefore, the NRC staff finds that the applicant's program for detection of accidents is acceptable.

8.3.6   Mitigation of Consequences

Section 8.3 of the EP (IIFP, 2009) describes actions and equipment that the applicant will use to mitigate the consequences of accidents at the facility. The major hazard is a chemical spill or release. The main features used at the facility to mitigate the consequences of accidents associated with a chemical release, natural phenomena, or fire include automatic isolation, interruption or termination of specific operations, fire detection and suppression systems, operator response to abnormal conditions or alarms, and process area dikes as a means of secondary containment of spilled materials. Section 8.4 of the EP (IIFP, 2012a) describes personnel evacuation and accountability plans, as well as the staging, maintenance, and use of protective equipment and supplies.

As stated above, the applicant identified and described the mitigation of consequences. The information provided by the applicant meets the requirements of 10 CFR 40.31(i)(3)(v) and the acceptance criteria in Section 8.4.3.1.6 of NUREG-1520 (NRC, 2002) for new facilities because IIFP identified and described the measures and equipment to be used for safe shutdown and the mitigation of consequences to workers onsite, as well as the public offsite. Therefore, the NRC staff finds that the applicant's program for mitigation of consequences is acceptable.

8.3.7   Assessment of Releases

Section 8.2 of the EP (IIFP, 2012a) describes the actions that the applicant will take to assess the extent of an accident at the facility. In case of an alert declaration, the spill control team monitors releases to provide assessment of dose or intake projections of facility personnel. The Environmental and Safety Officer (ESO) will advise on any additional sampling or monitoring for assessing effects on the environment or public, if needed. The applicant stated that, for an alert classification, it is unlikely that the assessment of dose and intake effects would extend beyond personnel within the facility. In case of a site area emergency declaration, in addition to monitoring performed for an alert declaration, monitoring personnel will conduct radiation and chemical surveys of assembly areas, the EOC, and appropriate areas around the emergency scene. Environmental air sampling will be conducted at selected and predetermined locations for uranium and fluorides. The ESO may determine the need to conduct other environmental sampling depending on the potential impact to the plant and offsite area.

As stated above, the applicant identified and described the assessment of releases. The information provided by the applicant meets the requirements of 10 CFR 40.31(j)(3)(vi) and the acceptance criteria in Section 8.4.3.1.7 of NUREG-1520 (NRC, 2002) for new facilities because IIFP identified and described the procedures to be used to assess the release of radioactive material or hazardous chemicals incident to the processing of licensed material. The applicant's program for assessment of releases is, therefore, acceptable.

8.3.8 Responsibilities

Sections 7.0 and 7.3, respectively, of the EP (IIFP, 2012a) describe the responsibilities of facility personnel during normal operations and emergency situations. In case of an emergency, the Shift Superintendent assumes the duties of the Emergency Director until the COO/PM, or designee, arrives. As stated in Section 7.2 of the EP, the Emergency Director ensures the response effort. Specific responsibilities of the Emergency Director include the following:

- declaring or terminating emergency events, as stated in Section 7.3.1 of the EP

- notifying appropriate offsite organizations and the NRC, as stated in Sections 6.2.1 and 6.2.2 for alert and site area emergency declarations, respectively

- making recommendations to offsite authorities concerning protective actions for the public offsite, as stated in Section 8.4.2 of the EP

- determining onsite protective actions for facility personnel, as stated in Section 8.4.1 of the EP

- initiating special sampling and monitoring as required, as stated in Sections 6.2.1 and 6.2.2 of the EP

- requesting support from offsite organizations, as stated in Section 7.4 of the EP

- approving press statements before their release, as stated in Section 7.3.1 of the EP.

Section 7.3 of the EP (IIFP, 2012a) summarizes the responsibilities of the emergency response organization (ERO) positions. Section 7.4 of the EP (IIFP, 2012a) provides the local offsite assistance to the facility for fire, emergency medical services, and local law enforcement. IIFP will sign MOUs with local agencies 3 months before the startup of operations or receipt of licensed materials or hazardous process chemicals, which will be reviewed annually and renewed at least every 4 years, or more frequently if necessary. As stated in Section 10.1 of the EP (IIFP, 2012a), departmental administrative procedures are established which assign responsibility for the development, review, approval, and update of the EP and its supporting procedures. The EP is reviewed by emergency preparedness personnel for accuracy and updated annually, as needed.

As stated above, the applicant identified and provided a description of responsibilities. The information provided by the applicant meets the acceptance criteria in Section 8.4.3.1.8 of NUREG-1520 (NRC, 2002) for new facilities because IIFP identified and described the ERO and administration that ensure effective planning, implementation, and control of emergency preparedness activities. The applicant's description of the emergency preparedness responsibilities is, therefore, acceptable.

8.3.9 Notification and Coordination

As discussed in Section 8.3.4 of this Safety Evaluation Report (SER), the Emergency Director is responsible for the classification of emergencies, which is outlined in Section 6.1 of the EP (IIFP, 2012a). Section 6.2 of the EP (IIFP, 2012a) provides a clear commitment to promptly notify offsite EROs of an emergency, including the notification to the NRC Operations Center within 1 hour. Section 7.4 of the EP (IIFP, 2012a) adequately describes the provisions for assistance from offsite EROs.

In the NRC staff's opinion, these sections adequately describe the agreements held between IIFP and local offsite EROs and agencies, procedures for access to the site for these response organizations, and equipment and services available from these organizations.

As discussed in Section 8.3.8 of this SER, the Emergency Director is responsible for (1) declaring an alert or site area emergency, (2) activating the onsite ERO, (3) notifying offsite emergency response authorities of an emergency, (4) notifying the NRC Operations Center, (5) deciding which onsite protective actions to initiate, (6) deciding which offsite protective actions to recommend and providing that information to offsite authorities, (7) deciding to request support from offsite organizations, and (8) deciding to terminate the emergency or enter recovery mode.

As stated above, the applicant identified and described the provisions for notification and coordination. The information provided by the applicant meets the requirements of 10 CFR 40.31(j)(3)(viii) and the acceptance criteria in Section 8.4.3.1.9 of NUREG-1520 (NRC, 2002) for new facilities because IIFP identified and described the emergency notification procedures that enable the organization to correctly classify emergencies, notify emergency response personnel, and initiate or recommend appropriate actions in a timely manner. The applicant's program for notification and coordination is, therefore, acceptable.

8.3.10 Information To Be Communicated

Section 6.3 of the EP (IIFP, 2012a) adequately describes the type of information to be given to offsite EROs and the NRC during an emergency. IIFP will use an event notification form as a script for initial notification of an emergency at the facility to appropriate offsite emergency response authorities, which will then be transmitted electronically as a confirmation to the notification. The emergency notification form also serves as a communications update to offsite organizations during an emergency event. If a site area emergency is declared, recommendations will be made to offsite organizations for use in advising the public. Facility meteorological data provided in the EOC may be used to provide more specific information.

As stated above, the applicant identified and described the notification and coordination procedures. The information provided by the applicant meets the requirements of 10 CFR 40.31(j)(3)(ix) and the acceptance criteria in Section 8.4.3.1.10 of NUREG-1520 (NRC, 2002) for new facilities because IIFP identified and described the use of a standard reporting checklist, the types of information to be provided, the preplanned protective action recommendations, the offsite authorities to be notified, and the recommended actions to be implemented by offsite organizations for each accident discussed in the EP. The applicant's description of the information to be communicated is, therefore, acceptable.

8.3.11 Training

Section 10.2 of the EP (IIFP, 2012a) describes the training IIFP will provide to workers on how to respond to an emergency. All workers receive general employee training, which includes quality assurance, radiation protection, safety, emergency and administrative procedures, and applicable regulations. The applicant also provides training in radiation protection and emergency procedures specific to each type of job function. The facility emergency response personnel receive additional training to provide specific information about how the ERO responds during emergency conditions and includes EOC staffing, determining and estimating potential offsite releases of radiation and chemicals, and interface with offsite assistance organizations. This training is required before an individual is assigned to the emergency organization, and IIFP provides refresher training at least once every year.

The IIFP administrative procedures contain the specific topics, performance objectives, content, training schedules, and number of training hours required for each position. The Environmental Safety and Health (ESH) Manager, or designee, reviews the contents of the formal radiation protection and hazard communications training programs and updates the information as required at least every 2 years to ensure that the programs are current and adequate. Individuals requiring unescorted access to the plant controlled access area receive annual retraining.

Periodic training is offered to offsite assistance organization personnel in accordance with IIFP EPIPs. IIFP facility personnel meet at least every 2 years with each offsite assistance group to accomplish training and review items of mutual interest, including relevant changes to the program. This training includes facility tours, information concerning facility access control (normal and emergency), potential accident scenarios, emergency action levels, notification procedures, exposure guidelines, personnel monitoring devices, communications, contamination control, and the offsite ERO role in responding to an emergency at the IIFP facility, as appropriate.

As stated above, the applicant identified and described its emergency preparedness training program. The information provided by the applicant meets the requirements of 10 CFR 40.31(j)(3)(x) and the acceptance criteria in Section 8.4.3.1.11 of NUREG-1520 (NRC, 2002) for new facilities because IIFP identified and described the topics and general content of the training programs. The areas addressed include the administration of the training programs, the training provided for the use of protective equipment, the training program for onsite personnel who are not members of the EROs, and special instructions and orientation tours provided for offsite organizations. The applicant's proposed training program is, therefore, acceptable.

8.3.12 Recovery and Plant Restoration (Safe Shutdown)

Section 12.0 of the EP (IIFP, 2012a) describes key positions and responsibilities of the recovery organization and describes how reentry and restoration will be accomplished. Once significant offsite releases and the emergency condition are terminated, deliberate reentry and restoration activities may be applied. Visual, chemical, and radiological monitoring activities will be performed as necessary to identify any ongoing minor releases and to identify any areas with residual chemical or radioactive contamination. Reentry into areas affected by hazardous chemical or radioactive releases will be authorized in accordance with plant safety and health physics procedures.

Visual and engineering assessments of plant systems will be implemented to determine the need for repairs, modification, and functional testing. Recovery activities will be conducted in accordance with plant radiation protection procedures and all employee exposures will be controlled in accordance with the facility's program to keep exposures and doses as low as reasonably achievable. Before resuming normal operations, the COO/PM, or designee, will notify the appropriate local, State, and Federal authorities, informing them that the facility is in compliance with applicable environmental regulations. Within 15 days after any emergency event involving the offsite release of hazardous materials, the COO/PM, or designee, will submit a written report on the incident to the U.S. Environmental Protection Agency's Regional Administrator that characterizes the extent of the release and provides an assessment of hazards to human health.

As stated above, the applicant identified and described the safe shutdown process. The information provided by the applicant meets the requirements of 10 CFR 40.31(j)(3)(xi) and the acceptance criteria in Section 8.4.3.1.12 of NUREG-1520 (NRC, 2002) for new facilities because IIFP identified and described the methods and responsibilities for assessing damage, procedures for determining actions to reduce releases of radioactive material or hazardous chemicals, provisions for accomplishing required restoration actions, and key positions in the recovery organization. The applicant's program for safe shutdown, recovery, and facility restoration is, therefore, acceptable.

8.3.13 Exercises and Drills

Section 10.3 of the EP (IIFP, 2012a) identifies adequate provisions for periodic drills and biennial exercises to test the adequacy of EPIPs and emergency equipment and instrumentation and to ensure that all ERO personnel are familiar with and proficient in their duties.

Section 10.4 of the EP (IIFP, 2012a) provides that post exercise evaluations will be conducted by those involved and appropriate improvements will be implemented. Areas evaluated include the adequacy of the EP, procedures, equipment, facilities, personnel training, and overall response effectiveness. The ESH Manager is responsible for the planning, scheduling, and conducting of emergency response drills and exercises for the facility. Offsite EROs are invited to participate in the biennial exercise, and the NRC is invited to participate or observe. IIFP will submit exercise objectives and scenarios to the NRC for review and comment at least 30 days before the exercise. Section 10.3.2 of the EP (IIFP, 2012a) includes an adequate provision for quarterly communications checks to verify the operability of equipment used by the ERO to communicate with offsite agencies and response organizations.

As stated above, the applicant identified and described its approach to exercises and drills. The information provided by the applicant meets the requirements of 10 CFR 40.31(j)(3)(xii) and the acceptance criteria in Section 8.4.3.1.13 of NUREG-1520 (NRC, 2002) for new facilities because IIFP identified and described the conduct of drills and exercises in a manner that demonstrates the capability of the organization to plan and perform an effective response to an emergency.

In Section 13 of the EP (IIFP, 2012a) the applicant stated that it will maintain compliance with the Emergency Planning and Community Right-to-Know Act of 1986 with respect to any hazardous materials possessed at the plant site. The information provided by the applicant meets the acceptance criteria in Section 8.4.3.1.1 of NUREG-1520 (NRC, 2002) and 10 CFR 40.31(j)(3)(xiii); therefore, the NRC staff finds the applicant's plan for the treatment of hazardous chemicals acceptable.

8.3.14 Responsibilities for Developing and Maintaining the Emergency Plan and its Procedures

Section 10.1 of the EP (IIFP, 2012a) describes the EPIPs that state the specific duties, responsibilities, emergency action levels, and actions to be taken by responders identified in Section 7.3 of the EP (IIFP, 2012a). The section also describes administrative procedures in place that ensure that all individuals and groups assigned responsibilities in an emergency have easy access to a current copy of each procedure that pertains to their functions. It also describes other administrative procedures that assign responsibility for the development, review, approval, and update of the EP and supporting procedures. The applicant reviews and updates the EP annually as needed and will not implement changes that would decrease the effectiveness of the EP without prior NRC approval. IIFP will provide any proposed change that affects an offsite organization to that organization for review and comment at least 60 days before the change being implemented, unless mutually agreed otherwise. The applicant provides that revisions to the EP and supporting procedures be distributed to all affected parties and submitted to the NRC within 3 months of the revision. Section 10.5 of the EP also describes the annual audit to be performed on the EP as part of the applicant's quality assurance program.

As stated above, the applicant identified and described the responsibilities for developing and maintaining the emergency program and its procedures. The information provided by the applicant meets the acceptance criteria in Section 8.4.3.1.24 of NUREG-1520 (NRC, 2002) for new facilities because IIFP identified and described the responsibilities for developing and maintaining the emergency program and its procedures.

8.4   Evaluation Findings

The NRC staff evaluated IIFP's EP (IIFP, 2012a) for the proposed facility. The applicant has established an EP for responding to the radiological hazards resulting from a release of radioactive material or hazardous chemicals relating to the processing of licensed material, in accordance with 10 CFR 40.31(j)(1)(ii). The NRC staff reviewed IIFP's EP with respect to the requirements of 10 CFR 40.31(j)(3), 70.64(a)(6) and the acceptance criteria in Section 8.4.3 of NUREG-1520 (NRC, 2002). The NRC staff concludes that the IIFP EP is adequate to demonstrate compliance with the regulatory requirements. In particular, the applicant has ensured that: (1) the facility is properly configured to limit releases of radioactive materials or hazardous chemicals in case of an accident; (2) a capability exists for measuring and assessing the significance of accidental releases of radioactive materials or hazardous chemicals; (3) appropriate emergency equipment and procedures are provided onsite to protect workers against radiation and other chemical hazards that might be encountered after an accident; (4) a system has been established to notify Federal, State, and local government agencies and to recommend appropriate protective actions to protect members of the public; and (5) the necessary recovery actions have been established to return the facility to a safe condition after an accident.

The requirements of the EP are implemented through approved written procedures. Changes that decrease the effectiveness of the EP may not be made without NRC approval. The NRC will be notified of other changes that do not decrease the effectiveness of the EP within 6 months of making the changes.

8.5   References

(IIFP, 2012a) International Isotope Fluorine Products, Inc., "Emergency Plan Rev B of IIFP License Application," May 2012, Agencywide Documents Access and Management System (ADAMS) Accession No. ML12123A245.

(IIFP, 2012b) International Isotopes Fluorine Products, Inc., "ISA Summary Rev. B for IIFP," May 2012, Agencywide Documents Access and Management System (ADAMS) Accession No. ML12123A245.

(NRC, 2002) U.S. Nuclear Regulatory Commission, "Standard Review Plan for the Review of a License Application for a Fuel Cycle Facility," NUREG-1520, Rev. 0, March, 2002.

(NRC, 1992) U.S. Nuclear Regulatory Commission, "Standard Format and Content for Emergency Plans for Fuel Cycle and Materials Facilities," Regulatory Guide 3.67, 1992.

9.0   ENVIRONMENTAL PROTECTION

The purpose of the U.S. Nuclear Regulatory Commission's (NRC's) review of the International Isotopes Fluorine Products, Inc.'s (IIFP or the applicant), Environmental Protection Plan for its proposed Fluorine Extraction Process and Uranium Deconversion Process (FEP/DUP) plant is to determine whether the applicant's proposed environmental protection measures are adequate to protect the environment and the health and safety of the public, as required by Title 10 of the *Code of Federal Regulations* (10 CFR) Part 20, "Standards for Protection against Radiation"; 10 CFR Part 40, "Domestic Licensing of Source Material"; 10 CFR Part 51, "Environmental Protection Regulations for Domestic Licensing and Related Regulatory Functions"; and 10 CFR Part 70, "Domestic Licensing of Special Nuclear Material." NUREG-1748, "Environmental Review Guidance for Licensing Actions Associated with NMSS Programs," issued August 2003, provides general procedures for the environmental review of licensing actions regulated by the Office of Nuclear Material Safety and Safeguards.

9.1   Regulatory Requirements

- 10 CFR Part 20 provides the effluent control and treatment measures necessary to meet the dose limits and dose constraints for members of the public specified in Subpart B, "Radiation Protection Programs"; Subpart D, "Radiation Dose Limits for Individual Members of the Public"; and Subpart F, "Surveys and Monitoring." Subpart F also specifies the survey requirements. Subpart K, "Waste Disposal," specifies the waste disposal requirements; Subpart L, "Records," specifies the records requirements; and Subpart M, "Reports," specifies the reporting requirements.

- 10 CFR Part 51 provides that the applicant must establish effluent and environmental monitoring systems to provide the information required by 10 CFR 51.60(a).

- 10 CFR 40.32(c) provides that the application will be approved if the proposed equipment, facilities, and procedures are adequate to protect health and minimize danger to life and property.

- 10 CFR 40.65, "Effluent Monitoring Reporting Requirements," provides the reporting requirements for radiological effluent monitoring for a 10 CFR Part 40 licensee.

- 10 CFR 70.65(b)(6) provides that an applicant for a facility must provide an Integrated Safety Analysis (ISA) Summary that includes a list of the items relied on for safety (IROFS).

9.2   Regulatory Acceptance Criteria

Section 9.4.3 of NUREG-1520, Revision 0, "Standard Review Plan for the Review of a License Application for a Fuel Cycle Facility" (NRC, 2002), outlines the acceptance criteria for the NRC's review of the IIFP Environmental Protection Program.

9.3 Staff Review and Analysis

The applicant has established and implemented personnel training and qualification requirements in accordance with approved written procedures. Chapter 11 of the license application (LA) (IIFP, 2012a) describes the applicant's training program; Chapter 2 of the LA (IIFP, 2012a) describes its qualification requirements for key management positions.

9.3.1 Personnel Training

As noted in Section 11.3 of the LA (IIFP, 2012a), the applicant's training programs are provided through shared responsibility of the environmental health and safety (EHS) disciplines and line management. The training programs are designed primarily to ensure that personnel who perform activities relied on for safety have the appropriate skills to design, operate, and maintain the facility in a safe manner. Training requirements are applicable primarily to those personnel who have a direct relationship to various aspects of IROFS. In addition to its safety focus, the training program is also designed to ensure that assigned personnel are trained and tested as necessary to perform the responsibilities important to the protection of the environment.

Technical training consists of initial, on-the-job, continuing, and special training, as applicable to assist personnel gain an understanding of specific, assigned technical duties. Professional development, which uses internal or external professionals via formal workshops, tutorials, and selected training programs, assists personnel in gaining additional understanding of technical practices common to their assigned job functions. Job-specific training is performance based. Lesson plans are based on job performance requirements and are reviewed by line management and by the responsible organization for the subject matter. Under its training program, the applicant also provides continuing or periodic retraining to ensure that personnel remain proficient and retain important knowledge and skills. Retraining may be required as a result of facility modifications or changes in procedures or in the Quality Assurance Program, which would result in new or changed information. The applicant evaluates trainee understanding and proficiency through observation, demonstration, or examinations, and documents training results.

The applicant also performs periodic evaluations of its training program to assess program effectiveness. These evaluations identify program strengths and weaknesses. The evaluations are also used to help determine whether training content matches current job needs and whether corrective actions are needed to improve training program effectiveness. In addition, the applicant may also perform independent audits to evaluate the overall training program effectiveness.

9.3.2 Personnel Qualifications

Chapter 2, "Organization and Administration" of the LA (IIFP, 2012a), identifies key management and supervisory positions, hierarchy, and functions, including personnel qualifications for each key position. Figure 2-2, "Plant Operation Organization," of the LA (IIFP, 2012a), provides an overview of those management positions and functions; Section 2.2 of the LA (IIFP, 2012a) describes those key management positions and qualifications that relate to, among other things, environmental protection.

Chapter 2 of this Safety Evaluation Report (SER) describes the overall personnel qualifications. This section highlights the qualifications of key environmental protection personnel.

The project environmental assessment lead (EAL) is a professional staff contractor working under the project oversight and support contract that is approved by the President/Chief Executive Officer (CEO) of International Isotopes Inc. (INIS)/IIFP. During the design and construction stages, the EAL ensures that environmental technical and licensing support, as requested by the Chief Operations Officer (COO)/Commercial Facility Project Director or IIFP's Regulatory/Quality Assurance Director, is provided for evaluation and assessment of design or engineering modifications or construction activities. The EAL also provides technical support for the Federal, State, and local environmental-related permit applications. A primary responsibility of the EAL during the design/construction stage of the project is to prepare responses for and interact with the NRC about requests for additional information related to the licensing review of the IIFP's Environmental Report (ER) (IIFP, 2011). The above responsibilities transfer to the IIFP's ESH Manager and designated environment staff as those positions are filled and the transition to operations is completed. The EAL shall have, as a minimum, a bachelor's degree in engineering or a scientific field and a minimum of 5 years in a chemical, radiological, or nuclear facility; with at least 3 of those years in responsible environmental assignments. The EAL must have experience in interacting with regulatory agencies and a working knowledge of relevant regulatory requirements.

The COO/Plant Manager (PM) appoints the ESH Manager at the facility, with the concurrence of the INIS Regulatory Affairs and Quality Assurance Director (RAQD). The ESH Manager reports to the IIFP COO/PM and also has a reporting and interacting relationship with the RAQD on matters of ESH policies, regulatory requirements, plant safety, and environmental compliance. In addition, the ESH Manager has the authority and responsibility to elevate any ESH concerns to corporate management and the INIS President/CEO. The IIFP ESH Manager establishes and oversees the Radiation Protection (RP), Licensing, Integrated Safety Analysis, Industrial Safety, Environmental Protection, Fire Protection, and Emergency Preparedness/Security Programs to ensure compliance with applicable Federal, State, and local regulations and laws. Those programs are designed to ensure the health and safety of employees and the public, as well as the protection of the environment. The ESH Manager shall have, as a minimum, a bachelor's degree in engineering, science, or a related field and 5 years of responsible assignments of ESH activities at chemical, radiological, or nuclear facilities.

The environmental lead reports to the IIFP ESH Manager and has responsibilities in support of the ESH Manager, which include, but are not limited to: (1) developing environmental programs and procedures, (2) leading monitoring and measuring activities, (3) developing and maintaining environmental-related permits, (4) assisting in training of employees in environmental matters, (5) conducting audits and inspections, (6) preparing and providing environmental data and reports, and (7) interacting with Federal, State, and local representatives in ensuring compliance with permit requirements and conditions. The environmental lead shall have a bachelor's degree in engineering or a scientific field and at least 2 years of environmental-related experience in a chemical, radiological, or nuclear facility.

The NRC staff reviewed the applicant's training program identified in Section 11 of its LA (IIFP, 2012a) and the personnel qualifications identified in Section 2 of its LA (IIFP, 2012a). This review included the qualification and training of managers, supervisors, and technical staff who are associated with environmental protection. The NRC staff finds that the applicant has established that the training, testing, and qualification of these personnel meets the requirements found in 10 CFR 40.32(b) and with guidance found in Section 9.4.3.2 of NUREG-1520, Revision 0 (NRC, 2002), which in turn incorporates, by reference, the training and qualifications information in Section 11.4.3.3 of NUREG-1520, Revision 0 (NRC, 2002). Section 11.4.3.3 of NUREG-1520 (NRC, 2002) identifies training areas and provides that managers and staff are to possess levels of education and experience commensurate with the responsibilities of their positions. The NRC staff finds the personnel qualifications of the proposed facility's staff adequate because the minimum qualifications of personnel are commensurate with the assigned responsibility and the training of personnel who perform regulated activities and they also meet the acceptance criteria in Section 11.4.3.3 of NUREG-1520 (NRC, 2002)

9.3.3 Radiation Safety

9.3.3.1   As Low As Reasonably Achievable Goals for Air and Liquid Effluent Control

Title 10 CFR 20.1101, "Radiation Protection Programs," requires each licensee to develop, document and implement an RP program. This environmental review of the IIFP RP Program focuses on the applicant's proposed methods to maintain public doses as low as reasonably achievable (ALARA), in accordance with 10 CFR 20.1101. Acceptance criteria found in Section 9.4.3.2.1 of NUREG-1520, Revision 0 (NRC, 2002), state that the applicant's proposed ALARA Program is acceptable if the program complies with 10 CFR 20.1101 according to Regulatory Guide 8.37, "ALARA Levels for Effluents from Materials Facilities," issued July 1993 (NRC, 1993). In addition, Section 9.4.3.2.1 of NUREG-1520 (NRC 2002) states the applicant should establish its ALARA goals at a modest fraction (10 to 20 percent) of the values in Appendix B, "Annual Limits on Intake (ALIs) and Derived Air Concentrations (DACs) of Radionuclides for Occupational Exposure; Effluent Concentrations; Concentration for Release to Sewerage" to 10 CFR Part 20, Columns 1 and 2 and Table 3, and the external exposure limit in 10 CFR 20.1302(b)(2)(ii); or, if the applicant proposes to demonstrate compliance with 10 CFR 20.1301, the dose limit for members of the public. The applicant's constraint approach is acceptable if it is consistent with the guidance in Regulatory Guide 4.20, "Constraint on Releases of Airborne Radioactive Material to the Environment for Licensees Other Than Power Reactors," issued December 1996 (NRC, 1996), and provides sufficient detail to demonstrate specific application of the guidance to proposed routine and nonroutine operation, including anticipated events. The applicant should also describe its proposed effluent controls to maintain public doses ALARA and demonstrate a commitment to reduce unnecessary exposure to members of the public and releases to the environment. According to Section 9.4.3.2.1 in NUREG-1520, Revision 0 (NRC, 2002), the applicant also needs to commit to annual reviews of the content and implementation of the RP Program, which includes the ALARA Program. These reviews should consider analysis of trends in release concentrations, environmental monitoring data, and radionuclide usage; determinations of whether operational changes are needed; and evaluations of designs for system installations or modifications. In addition, the results of the annual reviews should be reported to senior management, along with recommendations for changes to facilities or procedures necessary to achieve ALARA goals.

Sections 4.2 and 9.2 of the applicant's LA (IIFP, 2012a) and Section 4.12.2 of its ER (IIFP, 2011) describe the applicant's ALARA and Radiation Protection Programs. The applicant stated that it maintains and uses treatment systems, as appropriate, to ensure that releases of radioactive material to unrestricted areas remain below the limits specified in 10 CFR 20.1301 and in accordance with ALARA policy. The applicant has an Environmental Protection Program for the proposed IIFP facility. The applicant stated that the primary purpose of this program is to ensure that exposure of the workers, public, and the environment to radioactive materials used in facility operations is kept ALARA. As discussed in Chapter 6 of the applicant's ER (IIFP, 2011), compliance with the ALARA concept is a part of the applicant's Environmental Protection Program. The applicant relies on air and liquid effluent controls to maintain public doses ALARA. The applicant's ALARA goal is 20 percent of the values found in Appendix B to 10 CFR Part 20.

The NRC staff finds that the applicant's approach is sufficiently detailed to demonstrate that IIFP is in compliance with the regulatory dose limits found in 10 CFR 20.1301; that air and liquid dose constraints meet the acceptance criteria in Section 9.4.3.2.1 of NUREG-1520, Revision 0 (NRC, 2002); and that the applicant's ALARA Program for controlling gaseous and liquid effluents is consistent with the guidance found in Regulatory Guide 8.37 (NRC, 1993). The NRC staff determines that the ALARA Goals for Air and Liquid Effluent Control are adequate because the applicant committed to use effluent controls to maintain public doses ALARA

9.3.3.2   Air Effluent As Low As Reasonably Achievable Goal

The applicant proposed an ALARA goal for radiological effluents to air of less than 0.1 millisievert per year (mSv/yr) [10 millirem per year (mrem/yr)] total effective dose equivalent (TEDE), which is within the 10 CFR 20.1101 constraint of 0.1 mSv/yr (10 mrem/yr) TEDE for the maximally exposed member of the public. The NRC staff determined that the applicant's proposal satisfies the ALARA goal of 0.1 mSv/yr (10 mrem/yr) recommended in Regulatory Guide 8.37, Regulatory Position C.1.2, "ALARA Goals" (NRC, 1993), and meets the applicable acceptance criterion found in Section 9.4.3.2.1 of NUREG-1520, Revision 0 (NRC, 2002). Therefore, the NRC staff finds the applicant's proposal to be acceptable because ALARA goals are set to a fraction of the values in Table 2, Columns 1 and 2, and Table 3 of Appendix B to 10 CFR Part 20.

9.3.3.3   Liquid Effluent As Low As Reasonably Achievable Goal

As noted in Sections 4.2, and 9.2.1 of the applicant's LA (IIFP, 2012a) and Section 4.12.2 of its ER (IIFP, 2011), no liquid effluent discharges are expected because of the anticipated low volume of contaminated liquid waste and the effectiveness of the treatment process. Thus NRC staff finds that liquid effluent does not provide a significant pathway for radiological exposure to the general public. The applicant stated that liquid plant effluents are maintained on the IIFP site, and there is no discharge of process wastewater radioactive wastes. Therefore, staff concluded, average annual release concentrations of liquid effluents meet the limits in 10 CFR 20.1302, "Compliance with Dose Limits for Individual Members of the Public," and will not exceed the values in Appendix B to 10 CFR Part 20. Liquid effluents result in negligible increases to the environmental or public radiological exposures. NRC staff determined that ALARA levels for liquid effluents are consistent with the guidance given in Regulatory Guide 8.37 (NRC, 1993). The RP procedures incorporate the ALARA philosophy into facility operations and ensure that exposures are kept below the 10 CFR 20.1101(d) limits. The level of radioactivity from liquid effluents to the public and the environment must be below the regulatory goal of 0.1 mSv/yr (10 mrem/yr) recommended in Regulatory Guide 8.37 (NRC, 1993) and meet the applicable acceptance criteria found in Section 9.4.3.2.1 of NUREG-1520, Revision 0 (NRC, 2002), and is, therefore, acceptable to the NRC staff.

The NRC staff finds the ALARA goal with respect to liquid effluents acceptable because it complies with the acceptance criteria in NUREG-1520 Section 9.4.3.2.1, Revision 0 (NRC, 2002), and 10 CFR 20.1101(b) and (d).

9.3.3.4   Air Effluent Controls To Maintain Public Doses As Low As Reasonably Achievable

Sections 4.6.1 and 9.2.1 of the applicant's LA (IIFP, 2012a) and Section 4.12.2 of its ER (IIFP, 2011) discuss the air effluent controls for the proposed IIFP facility. Most of the airborne depleted uranium is removed through filtration by prefilters, high-efficiency filters, and carbon-bed filters before entering the three-stage scrubber system. After scrubbing, the gaseous effluent is discharged into the atmosphere. In Section 2.1.3 of its ER (IIFP, 2011), the applicant provided a detailed description of the filtration and scrubber systems, which overall achieve greater than 99 percent efficiency. Table 2-1 of the ER (IIFP, 2011) provides design efficiencies for process vent offgas treatment equipment. The stacks are sampled continuously and routinely analyzed to measure the radioactivity of the discharged gases.

As noted in Section 4.12.2.2 of the ER (IIFP, 2011), the applicant estimated offsite radiological impacts to key receptors (critical populations) from routine effluent releases using the GENII model (Version 2.06). The key receptor locations for determining dose impacts included the resident nearest to the IIFP facility and the maximally exposed individual (MEI) (at the northwest boundary). The MEI is a hypothetical person living at the point of highest projected total uranium concentrations near the site boundary. The applicant evaluated the impact from inhalation of gaseous effluents, immersion in a passing effluent plume, exposure to direct radiation from deposited radioactivity on the ground surface (ground plane exposure), and ingestion of contaminated food products. Because the applicant does not anticipate contamination of drinking water, its analysis did not consider radiological contamination of drinking water. The applicant's analysis included dose equivalent assessments for four age groups—adults, teens, children, and infants—for these pathways.

9.3.3.5   Staff Evaluation of Air Effluent Controls

The NRC staff evaluated the air effluent controls and effects described above. The applicant calculated the dose equivalents associated with gaseous effluents for the MEI and the nearest resident, by pathway, for the total body in adults, teens, children, and infants. Tables 4-25 and 4-26, respectively, of the ER (IIFP, 2011) present these results. The calculated committed effective dose equivalent (CEDE) for the adult MEI from the proposed IIFP facility emissions is 8.4×10^{-6} mSv/yr (8.4×10^{-4} mrem/yr). For the adult, full-time resident nearest the facility, the calculated CEDE from the IIFP facility is 6.4×10^{-5} mSv/yr (6.4×10^{-3} mrem/yr). Doses for public receptors at other sites of interest (e.g., schools and hospitals) would be lower than the MEI because the airborne concentrations of uranium are lower at these more distant locations. The doses for the MEI and nearest resident are well below (by orders of magnitude) the U.S. Environmental Protection Agency's (EPA's) 0.1 mSv/yr (10 mrem/yr) standard in 40 CFR Part 190, "Environmental Protection Standard for Nuclear Power Operations." This estimated maximum public dose is also well below the 0.1 mSv (10 mrem) ALARA constraint on air emissions described in 10 CFR 20.1101. Because public receptors at other sites of interest are more distant than the MEI, their doses would be even lower because of uranium dispersion at the more distant locations.

The NRC staff finds that the applicant's air effluent controls ensure that radiation levels to the public remain well below regulatory limits and ALARA air effluent goals, and that the applicant's approach to effluent control meets the applicable acceptance criteria found in Section 9.4.3.2.1 of NUREG-1520, Revision 0 (NRC, 2002). Therefore, the NRC staff finds these controls acceptable. The NRC staff also finds that the applicant has demonstrated that its air effluent controls reduce releases to provide adequate protection of the environment and of the health and safety of the public.

9.3.3.6   Liquid Effluent Controls to Maintain Public Doses As Low As Reasonably Achievable

As noted in Section 4.12.2 of the applicant's ER (IIFP, 2011), the general public may be impacted by radioactive material from liquid and gaseous effluent discharges associated with controlled releases from the uranium process lines during routine operations of the proposed facility. However, as noted in Section 9.3.1.3 of this SER, no liquid discharges from the proposed facility are expected because of the anticipated low volume of contaminated liquid waste and the effectiveness of treatment processes. Also, as noted in Section 4.13.3.2 of the applicant's ER (IIFP, 2011), the proposed facility does not discharge any process effluents to natural surface waters or grounds; and there is no tie into a publicly owned treatment works. All process wastes are recycled. The applicant does not expect rainfall runoff from the cylinder storage pads to be a significant exposure pathway. Runoff water is directed to an onsite retention basin for evaporation of the collected water. No liquids from retention basins are discharged offsite. Therefore, the applicant does not anticipate any radiological contamination of drinking water. Thus, the applicant stated that liquid effluents do not provide a significant pathway for radiological exposure to the public, and is within the effluent release requirements in Appendix B to 10 CFR Part 20.

9.3.3.7   Staff Evaluation of Liquid Effluent Controls

The applicant stated, as noted above and in Section 9.2.1.2 of the LA (IIFP, 2012a), process water discharges are treated and are contained onsite either by recycling and reusing in the process or by evaporating. Cooling water is recycled. The proposed facility liquid effluent collection and recycle systems provide a means to control liquid waste and maintain a process water practical mass balance using flow-surge tanks, scrubber solution regeneration/recycle, and evaporation equipment. Uranium is precipitated from wastewater streams and disposed as low-level radioactive waste. In addition, storm water runoff is collected and transported to a retention basin via the plant storm water sewer system. The storm water is stored temporarily in the retention basin until it is sampled and then evaporated or discharged. Any discharged storm water contains only trace levels of radiological contamination, and the NRC staff determined that these releases are within the effluent release requirements in Appendix B to 10 CFR Part 20.

The NRC staff finds that the applicant's liquid effluent controls ensure that radiation levels to the public remain well below regulatory limits and ALARA liquid effluent goals and that the applicant's approach to effluent control meets the applicable acceptance criteria found in Section 9.4.3.2.1 of NUREG-1520, Revision 0 (NRC, 2002). Therefore, the NRC staff finds these controls to be acceptable. The NRC staff also finds that the applicant has demonstrated that its liquid effluent controls reduce releases to provide adequate protection of the environment and of the health and safety of the public.

9.3.3.8   As Low As Reasonably Achievable Program Reviews and Reports to Management

In Section 4.2 of its LA (IIFP, 2012a), the applicant described its ALARA Program for the proposed facility. Section 4.3.2 of this SER also describes the ALARA Program. The ALARA Program provides for an annual review of the content and implementation of the RP Program, including the Effluent Control Program. The ALARA Committee, which is part of the Facility Safety Review Committee (FSRC), is responsible for conducting annual ALARA reviews. The FSRC is an independent advisory committee that reports directly to the COO/PM.

Membership of the ALARA Committee includes the following positions:

- COO/PM

- RP Manager

- selected department managers

- ESH Manager

- selected supervisors and hourly personnel.

The scope of the ALARA Committee's activities include, at a minimum, annual review of the following:

- site radiological operating performance; including trends in airborne concentrations, personnel exposures, and environmental monitoring results

- operations and exposure records to identify opportunities to reduce exposures

- employee training and methods for using information on the job to keep exposure ALARA

- potential modifications of procedures and equipment when changes reduce exposures at reasonable cost.

In addition, the ALARA Committee reviews major changes in authorized activities affecting RP practices and evaluates contamination minimization and removal activities.

The proceedings, findings, and recommendations of the ALARA Committee are reported in writing to the COO/PM and appropriate line management and area managers responsible for operations reviewed by the Committee. Such reports are retained for a minimum of 3 years. Based on expected improvements, updated performance data, economics, and consideration of other site priorities, management decides which of the ALARA Committee recommendations should be pursued. If a specific recommendation is pursued, a task owner is assigned, and the action is tracked to completion.

As discussed in Sections 4.1.1.3 and 4.3 of the LA (IIFP, 2012a), the RP Manager is responsible for the overall implementation of the RP Program and has direct access to the COO/PM who, in turn, has the overall responsibility for safety and activities conducted at the proposed facility.

As described in Section 4.3.1 of this SER, the applicant implements the ALARA Program in accordance with 10 CFR 20.1101 through approved written procedures and policies. The approach described meets the above-referenced requirements in 10 CFR Part 20 and the applicable acceptance criteria found in Section 9.4.3.2.1 of NUREG-1520, Revision 0 (NRC, 2002), and is, therefore, acceptable to the NRC staff because the applicant describes how facility design procedures for operation will, to the extent practicable, minimize contamination of the facility and environment and minimize the generation of radioactive waste.

9.3.3.9   Waste Minimization

As required by 10 CFR 20.1406, "Minimization of Contamination," an applicant for a license must describe how the facility design and procedures for operation minimize, to the extent practicable, contamination of the facility and the environment; facilitate eventual decommissioning; and minimize, to the extent practicable, the generation of radioactive waste. Section 9.4.3.2.1 of NUREG-1520, Revision 0 (NRC, 2002), addresses applicable acceptance criteria. The applicant's program for waste minimization is acceptable if the applicant describes how the facility's design procedures for operation minimize, to the extent practicable, contamination of the facility and the environment; facilitate decommissioning; and minimize, to the extent practicable, the generation of radioactive waste. In addition, the acceptance criteria in NUREG-1520 Section 9.4.3.2.1 requires for the program to have senior management support, provide methods to characterize waste generation and waste management costs, provide for periodic waste minimization assessments, provide for technology transfer to seek and exchange technical information on waste minimization, and provide methods to implement and evaluate waste minimization recommendations.

In Section 4.7.8 of its LA (IIFP, 2012a) and in Section 4.13.4 of its ER (IIFP, 2011), the applicant committed to design and operate its facility to meet the requirements in 10 CFR 20.1406 to minimize contamination; facilitate eventual decommissioning; and minimize, to the extent practicable, the generation of radioactive waste. The applicant has assigned high priority to minimizing the generation of waste through reduction, reuse, or recycling. For example, the applicant incorporated closed-loop cooling systems in the design to reduce water use. In addition, the applicant maintains ALARA controls during facility operation to account for standard waste minimization practices, as directed in 10 CFR Part 20. As part of its waste management program described in Section 3.12 of the applicant's ER (IIFP, 2011), minimization practices include reclamation, recycle, reuse, compaction, and design features or procedures to avoid or reduce the generation of wastes.

In Section 1.1.6 of its LA (IIFP, 2012a), the applicant described its solid waste management program at the proposed facility for industrial (nonhazardous), radioactive, and hazardous wastes. Solid waste is grouped into one of these waste categories. The applicant may send wastes that are candidates for volume reduction, recycling, or treatment to licensed treatment facilities that have the ability to reduce the volume of most Class A low-level radioactive waste. The applicant also described its program to reuse or recycle process effluents.

The NRC staff evaluated and determined that the applicant meets the requirements specified in 10 CFR 20.1406, which provides that IIFP adopt procedures to minimize the generation of radioactive waste. Furthermore, the applicant's waste minimization practices are consistent with the guidance provided in Regulatory Guide 4.21, "Minimization of Contamination and Radioactive Waste Generation: Life-Cycle Planning," issued June 2008 (NRC, 2008), and meet applicable acceptance criteria in Section 9.4.3.2.1of NUREG-1520, Revision 0 (NRC, 2002). Therefore, the NRC staff finds the applicant's proposed waste minimization program acceptable. The NRC staff finds that the applicant's implementation of its program for managing solid radiological and nonradiological wastes related to facility operation, including its volume reduction and recycling programs, reduce unnecessary exposures to these wastes and ensure adequate protection of public health and safety and the environment.

9.3.4 Effluent and Environmental Monitoring

The regulations in 10 CFR 20.1302 require a licensee to make or cause to be made, as appropriate, surveys of radiation levels in unrestricted and controlled areas and radioactive material in effluents released to unrestricted and controlled areas to demonstrate compliance with the dose limits for individual members of the public in 10 CFR 20.1301.

Section 9.4.3.2.2 of NUREG-1520, Revision 0 (NRC, 2002), addresses the acceptance criteria for effluent monitoring. NUREG-1520 Revision 0 (NRC, 2002) states that an applicant's effluent monitoring program is acceptable if radioactive materials concentrations in airborne and liquid effluents are ALARA and are below the limits specified in Table 2 of Appendix B to 10 CFR Part 20, or if site-specific limits are established in accordance with 10 CFR 20.1302(c). In addition, if the applicant proposes to demonstrate compliance with 10 CFR 20.1301 by using calculations of the TEDE, the applicant needs to perform pathway analyses using appropriate models, codes, and assumptions according to the acceptance criteria in NUREG-1520, Revision 0 (NRC, 2002), Section 9.4.3.2.2.

Consistent with NUREG-1520, Revision 0 (NRC, 2002), each of the following criteria should to be addressed. The applicant should identify and monitor all liquid and airborne effluent discharge locations and continuously sample airborne effluents from all routine and nonroutine operations, unless periodic sampling or other means has been justified. Sample collection and analysis methods and frequencies need to be appropriate for the effluent medium and the radionuclides being sampled. Radionuclide specific analyses need to be performed on appropriate samples using justified methods. In addition, the minimum detectable concentration for sample analysis needs to be adequate (sufficiently sensitive) for comparison to the concentration limits in Appendix B to 10 CFR Part 20, and laboratory quality control procedures need to be adequate to validate the analytical results. The applicant also needs to establish action levels and proposed action if action levels are exceeded. In addition, the applicant needs to completely and accurately describe all applicable Federal or State discharge limits for gaseous and liquid effluents applicable to the proposed facility. Leakage detection systems also need to be in place to detect leaks from tanks, ponds, or lagoons that could affect groundwater, surface water, and soils. The applicant needs to control and maintain releases to sewer systems to meet the requirements of 10 CFR 20.2003, "Disposal by Release into Sanitary Sewerage." The applicant also needs to have reporting procedures that meet 10 CFR 40.65, "Effluent Monitoring Reporting Requirements." In addition, the applicant's procedures and facilities for solid and liquid waste handling, storage, and monitoring need to result in safe storage and timely disposition of the material.

Section 9.4.3.2.2 in NUREG-1520, Revision 0 (NRC, 2002), contains the acceptance criteria for environmental monitoring. An applicant's environmental monitoring program should include the establishment of background and baseline radionuclide concentrations in environmental media. Monitoring should include sampling and analyses for air, surface water, groundwater, soil, sediments, and vegetation, as appropriate; and identify adequate and appropriate sampling locations and frequencies for each environmental medium and the analyses to be performed for each medium. Monitoring procedures should employ acceptable analytical methods and instrumentation, and instrumentation should be appropriately maintained and calibrated. If the applicant proposes to use its own laboratory for environmental sample analyses, the applicant should commit to providing third-party verification of its methods. In addition, the applicant should identify appropriate action levels and actions to be taken if action levels are exceeded for each environmental medium and radionuclide. The applicant should select action levels based on pathway analyses that demonstrate that, below those concentrations, doses meet the

ALARA and 10 CFR Part 20, Subpart B, limits. The applicant also should specify minimum detectable concentrations for sample analyses at least as low as those selected for effluent monitoring in air and water consistent with the selected action levels. Data analysis methods and criteria should be provided for evaluating and reporting the results of environmental sampling and indicate when an action level is being approached in time to take corrective actions. The applicant also should provide a description of the status of all licenses, permits, and other approvals for facility operation that is complete and accurate. In addition, the program should be adequate to assess environmental impacts from potential radioactive and nonradioactive material releases as identified in high- and intermediate-consequence events in the ISA Summary (IIFP, 2012b).

The IIFP Radiation Safety Program incorporates the applicant's effluent monitoring program. The applicant described its effluent and environmental monitoring programs for radiological and nonradiological effluents released from the proposed facility in Section 9.2.2 of its LA (IIFP, 2012a) and in Chapter 6 of its ER (IIFP, 2011). In ER Chapter 6 (IIFP, 2011), the applicant indicated that it performs measurements and monitoring necessary to demonstrate that the amount of radioactive material present in effluent from the proposed facility is kept ALARA, in compliance with 10 CFR Part 20. The applicant also identified guidance in Regulatory Guide 4.16, Revision 2, "Monitoring and Reporting Radioactivity in Releases of Radioactive Materials in Liquid and Gaseous Effluents from Nuclear Fuel Processing and Fabrication Plants and Uranium Hexafluoride Production Plants," issued December 2010 (NRC, 2010), to ensure that it adheres to the ALARA principle such that there is no undue risk to the public health or safety at or beyond the IIFP site boundary.

9.3.4.1   Air Effluent Monitoring

The IIFP Radiation Safety Program includes the air effluent monitoring program. Radioactive air effluent from the facility is discharged only through monitored pathways (release points). The NRC staff stated that uranium isotopes and daughter products are expected to be the prominent radionuclides in the gaseous effluent.

9.3.4.2   Expected Concentrations

Expected concentrations of radioactive materials in airborne and liquid effluents are addressed in Section 9.2.2 of the applicant's LA (IIFP, 2012a); and Section 4.12.2 of its ER (IIFP, 2011) address expected concentrations of radioactive materials in airborne and liquid effluents, which were estimated using conservative assumptions. As described in Section 9.3.3.2 of this SER, the NRC staff finds that the expected concentrations of radioactive materials in airborne effluents is well below the regulatory limits specified in 10 CFR 20.1302(c). The applicant demonstrated compliance with air effluent limits by calculating the TEDE to the individual who is likely to receive the highest dose, in accordance with 10 CFR 20.1302(b)(1).

The NRC staff finds that the applicant's control of these concentrations is ALARA and below limits specified in Table 2 of Appendix B to 10 CFR Part 20. In addition, the applicant's approach to controlling these concentrations meets the applicable acceptance criteria found in Section 9.4.3.2.2 of NUREG-1520, Revision 0 (NRC, 2002), and is, therefore, acceptable to the NRC staff.

9.3.4.3 Total Effective Dose Equivalent

The applicant established constraints on atmospheric releases for the proposed facility such that no member of the public is expected to receive a TEDE in excess of 0.1 mSv/yr (10 mrem/yr) from these releases. Written procedures approved by IIFP management dictate that atmospheric releases be monitored and measured. In Section 4.12.2 of the ER (IIFP, 2011), the applicant provided a detailed description of the calculation of the TEDE to the individual who is likely to receive the highest dose using GENII (version 2.08), which implements dosimetry models recommended by the International Commission on Radiological Protection (ICRP). For both the inhalation and ingestion exposure pathways, the applicant used the exposure-to-dose conversion factors in Federal Guidance Report 11 (EPA, 1988). For direct dose from material deposited on the ground plane or from the passing plume, the applicant used the conversion factors in Federal Guidance Report 12 (EPA, 1993). The ingestion pathway models performed by the applicant were taken from Regulatory Guide 1.109, Revision 1, "Calculation of Annual Doses to Man from Routine Releases of Reactor Effluents for the Purpose of Evaluating Compliance with 10 CFR Part 50, Appendix I," issued October 1977 (NRC, 1977), and then multiplied by the relative age-dependent dose factor found in ICRP Publication 72, "Age dependent Doses to the Members of the Public from Intake of Radionuclides—Part 5, Compilation of Ingestions and Inhalation Coefficients" (ICRP, 1996). The applicant calculated the direct dose equivalent from the 30-year life expectation of the proposed facility by using the MCNP4C2 computer code (ORNL, 2000). In Tables 4-25 and 4-26 of its ER (IIFP, 2011), the applicant provided calculated dose equivalents for the MEI and nearest resident (adult, teen, child, and infant) to the proposed facility, which the NRC staff noted are well below (by orders of magnitude) the regulatory limit in 10 CFR Part 20 and well within the EPA regulatory standard in 40 CFR Part 190. The NRC staff determined that the TEDE calculation to the individuals likely to receive the highest dose in accordance with 10 CFR 20.1302(b)(1) was appropriate because it accurately represent the facility, site and surrounding area. The NRC staff finds that the applicant's approach acceptable because it meets the requirements in 10 CFR 20.1302(b)(1) and meets the applicable acceptance criteria found in Section 9.4.3.2.2 of NUREG-1520, Revision 0 (NRC, 2002).

9.3.4.4 Locations, Continuous Sampling, and Analysis Methods of Airborne Effluents

Sections 9.2.2 and 1.1.6.3 of its LA (IIFP, 2012a) and Section 6.1.1 of its ER (IIFP, 2011) describe the applicant's Gaseous Effluent Monitoring Program. Discharges of gaseous effluents have the highest potential of introducing uranium into the environment. Public exposure from routine operations at the proposed facility may occur as a result of gaseous effluents, including controlled releases from the uranium deconversion process lines during decontamination and maintenance of equipment. The applicant's Gaseous Effluent Monitoring Program is designed to determine the quantities and concentrations of gaseous discharges to the environment. The sources of air emissions from the proposed facility are the process building stacks and dust collector stacks, which are described in Chapter 9 of the applicant's LA (IIFP, 2012a) and in Chapter 6 of the applicant's ER (IIFP, 2011). Table 6-1 of the applicant's ER (IIFP, 2011) identifies the type and frequency of sample analyses for the Gaseous Effluent Sampling Program.

As noted above in Section 9.3.3.4 of this SER, the ventilation system air stream passes through a series of high efficiency particulate air and high efficiency gas absorption filters before being vented to the atmosphere through the process building stacks and monitored for uranium and hydrogen fluoride (HF). Continuous air sampler filters are analyzed each week for gross alpha and beta radiation.

The applicant stated that it annually reviews the trends in gaseous emissions and liquid effluent monitoring data to determine whether changes are needed in systems or practices to achieve the ALARA effluent goals. The applicant stated it will revise the program as appropriate to maintain its effectiveness as changes are noted, such as those related to operations or other factors identified in Chapter 6 of the ER (IIFP, 2011).

In Section 6.2 of its ER (IIFP, 2011), the applicant stated that it collects HF continuously on particulate filters that are in vent stacks and analyze these samples weekly. As noted in Section 6.1.2 of its ER (IIFP, 2011) and in Section 9.2.2.1 of the LA (IIFP, 2012a), the applicant also collects uranium isotopes continuously on particulate filters in air samplers located around the proposed facility and initially analyze these samples daily.

The NRC staff finds that the applicant has identified and monitors all airborne effluent discharge locations to determine contributions to dose limits, in accordance with requirements in 10 CFR Part 20. The NRC staff also finds that the applicant's effluent monitoring meets the acceptance criteria found in Section 9.4.3.2.2 of NUREG-1520, Revision 0 (NRC, 2002), which is applicable to both airborne and liquid effluents. The NRC staff further finds that the applicant continuously monitors samples of airborne effluents from all routine and nonroutine operations in a manner that meets the acceptance criteria found in Section 9.4.3.2.2 of NUREG-1520, Revision 0 (NRC, 2002); and ensures that the air sample collection and analysis methods and frequencies described in the IIFP Radiological Monitoring Program are appropriate to meet the acceptance criteria found in Section 9.4.3.2.2 of NUREG-1520, Revision 0 (NRC, 2002), which is applicable to both airborne and liquid effluent discharges. Therefore, the NRC staff finds the applicant's airborne effluent monitoring and analysis program to be acceptable.

9.3.4.5 Radionuclide-Specific Analyses

In Section 9.2.2.1 of its LA (IIFP, 2012a), the applicant indicates they perform radionuclide specific analyses on selected composite samples, which Section 6.1.1 of its ER (IIFP, 2011) identifies. Because uranium in gaseous effluent may exist in a variety of compounds (e.g., depleted uranium hexafluoride, uranium oxide, depleted uranium tetrafluoride, and depleted uranyl fluoride [UO_2F_2]), the radiation program staff maintain, review, and assess the effluent data to ensure that the gaseous effluents meet with regulatory release criteria for uranium. Monitoring reports, in which the quantities of individual radionuclides are estimated on the basis of methods other than direct measurement, include an explanation and justification of how the results were obtained. In Section 9.2.2.1 of its LA (IIFP, 2012a), the applicant committed to follow guidance specified in Regulatory Guide 4.16, Revision 2 (NRC, 2010). The applicant will submit to the NRC a semiannual report that includes the concentrations of the principal radionuclides released in the unrestricted area, as well as other information necessary to evaluate the radiation doses from effluent releases to the public in liquid and gaseous effluents. The report also includes the minimum detectable concentration (MDC) for analysis and the error for each data point.

The NRC staff finds that the applicant's proposed radionuclide-specific analyses and reporting meet the requirements of 10 CFR 40.65 and the guidance specified in Regulatory Guide 4.16, Revision 2 (NRC, 2010) because the applicant has committed to submit a report which specifies the quantity of each radionuclide released to unrestricted areas. The proposed radionuclide-specific analyses and reporting also meet the applicable acceptance criteria found in Section 9.4.3.2.2 of NUREG-1520, Revision 0 (NRC, 2002) because the effluents are below the limits specified in Table 2 of Appendix B to 10 CFR Part 20; and they are sampled according to procedures. The radionuclide specific analysis is acceptable to the NRC staff.

9.3.4.6   Minimum Detectable Concentrations

The applicant's Gaseous Effluent Monitoring Program is designed to determine the quantities and concentrations of gaseous effluents discharged into the environment. Uranium isotopes and "daughter" decay products are expected to be the prominent radionuclides in the gaseous effluent. In Section 6.1.1 of its ER (IIFP, 2011), the applicant indicated that the MDC of 3.7×10^{-11} becquerels per milliliter (Bq/ml) (1.0×10^{-15} microcuries per milliliter (uCi/ml) is a detection requirement for the Gaseous Effluent Monitoring Program for all gross alpha analyses performed on gaseous effluent samples. The Gaseous Effluent Monitoring Program is not required to detect concentrations in samples that are below the MDC value. The NRC staff finds that the MDC value for effluents in air is below the concentration limits found in Table 2 of Appendix B to 10 CFR Part 20.

The NRC staff finds that the MDCs for various listed media are sufficiently low to meet action level, regulatory, and permit requirements and are consistent with applicable acceptance criteria found in Section 9.4.3.2.2 of NUREG-1520, Revision 0 (NRC, 2002), and are, therefore, acceptable to the NRC staff.

9.3.4.7   Laboratory Quality Control

Section 9.2.2.1 of the applicant's LA (IIFP, 2012a) and Section 6.1.2.2 of its ER (IIFP, 2011) address laboratory quality control (QC). The applicant's laboratory QC procedures conform to the guidance in Regulatory Guide 4.15 "Quality Assurance for Radiological Monitoring Programs (Inception through Normal Operations to License Termination)—Effluent Streams and the Environment" (NRC, 2007). Laboratory QC procedures include the use of established standards, such as those used by the National Institute of Standards and Technology (NIST), as well as standard analytical procedures, such as those established by the National Environmental Laboratory Accreditation Conference (NELAC).

The applicant committed to use written procedures to ensure that sampling and monitoring equipment is properly maintained, calibrated, and in good working condition. Onsite and contractor laboratories participate in third-party intercomparison programs to validate the laboratories's performance. Examples of such programs include the Mixed Analyte Performance Evaluation Program and the U.S. Department of Energy's Quality Assurance Program. The applicant also stated that it requires that all radiological and nonradiological laboratory vendors be certified by the NELAC or an equivalent State laboratory accreditation agency.

The NRC staff finds that the applicant's laboratory QC procedures are adequate to validate the analytical results to ensure compliance with the monitoring requirements in 10 CFR Part 20 and meet applicable acceptance criteria in Section 9.4.3.2.2 of NUREG-1520, Revision 0 (NRC, 2002). The laboratory QC procedures use recognized established standards and standard analytical procedures. Therefore, the NRC staff finds these procedures acceptable.

9.3.4.8   Action Levels

An action level for environmental measurements is the concentration (or mass) of an analyte that indicates that some action needs to be taken, such as initiating an investigation or, if the level is sufficiently high, shutting down operations. Action levels provide guidance for ensuring that concentrations of radioactivity are within the limits found in Appendix B to 10 CFR Part 20. The applicant has implemented a program of corrective actions to ensure that the cause for exceeding the action level can be identified and corrected; applicable regulatory agencies are notified, if required; lessons learned are communicated to appropriate personnel; and applicable operational procedures are revised accordingly, if needed. As noted in Section 9.2.2.1 of its LA (IIFP, 2012a), the applicant set administrative action levels at 50 percent of values found in Table 2 of Appendix B to 10 CFR Part 20, which is sufficiently below compliance levels needed to permit implementation of corrective actions before regulatory limits are exceeded. Approved written procedures specify these action levels according to the type of sample and the specific analysis.

The applicant's proposed development of action levels and related actions meets the regulatory requirements of Appendix B to 10 CFR Part 20 and the applicable acceptance criteria found in Section 9.4.3.2.2 of NUREG-1520, Revision 0 (NRC, 2002), because action levels are set to 50 percent of values found in Table 2 of Appendix B to 10 CFR 20. This allows implementation actions to be taken when the limits are exceeded. The NRC staff finds the action levels acceptable.

9.3.4.9   Federal and State Permits

Section 1.4 of the ER (IIFP, 2011) identifies and describes statutory and regulatory require-ments of the NRC and other Federal agencies and departments, as well as the State of New Mexico, which are applicable to the construction and operation of the proposed facility. In addition, Section 1.5, Table 1-4, lists required Federal and State permits, as well as their status. The applicant did not identify any applicable local government requirements. Section 1.3.2.2 of the ER (IIFP, 2011) lists the preliminary Federal and State codes and standards for the design and construction of the proposed facility.

The NRC staff finds that the applicant accurately described applicable Federal and State standards for discharges—as well as any permits issued by Federal, and State for gaseous and liquid effluents, as documented in Sections 1.4 and 1.5 of the ER (IIFP, 2011). This meets the applicable acceptance criteria found in Section 9.4.3.2.2 of NUREG-1520, Revision 0 (NRC, 2002), and is, therefore, acceptable to the NRC staff.

9.3.4.10   Air Effluent Monitoring Summary Evaluation

Based on the NRC staff's review of the applicant's LA (IIFP, 2012a) and ER (IIFP, 2011), described in Section 9.3.3.1 of this SER, the NRC staff finds that the air effluent monitoring program maintains air effluent concentrations below the regulatory limits in Appendix B to 10 CFR Part 20 by detecting and measuring concentrations of radioactivity in air effluents during operation of the proposed facility. Furthermore, the NRC staff finds that the applicant's air effluent monitoring program meets the applicable acceptance criteria found in Section 9.4.3.2.2 of NUREG-1520, Revision 0 (NRC, 2002), and is, therefore, acceptable.

9.3.4.11   Liquid Effluent Monitoring

The IIFP Radiation Safety Program encompasses the liquid effluent monitoring program. As noted in Section 9.3.1.1 of this SER and in Sections 4.2 and 9.2.1 of the applicant's LA (IIFP, 2012a) and Chapter 4.12.2 of the ER (IIFP, 2011), no liquid effluent discharges are expected by the applicant from the facility because of the anticipated low volume of contaminated liquid waste and the effectiveness of the treatment process. Liquid plant effluents are maintained on the IIFP site, and there is no discharge of process wastewater radioactive wastes. There is no offsite release of liquid effluents to unrestricted areas. Thus, the NRC staff determined that liquid effluents result in negligible increases to the environmental or public radiological exposures.

9.3.4.12   Detection of Leaks to Groundwater, Surface Water, or Soil

As noted in Section 9.2.2.1 of its LA (IIFP, 2012a), the applicant's conceptual design includes appropriate spill and leak control pads and containment dikes. The applicant stated that liquid is monitored to ensure compliance with the limits in Appendix B to 10 CFR Part 20, and with the National Pollutant Discharge Elimination System permit levels for fluoride and other constituents specified in the permit. The applicant stated that compliance is demonstrated through effluent and environmental sampling data. The applicant conducts sampling at the site's storm water retention (evaporation) basin and at the DUF_6 cylinder storage pad. Groundwater sampling is conducted at four locations onsite. As described in Section 10.3.2.3 of this SER, the NRC staff expects any discharges of liquid effluents are to be negligible and within the regulatory limits of 10 CFR Part 20, Appendix B. Therefore, the NRC staff finds the applicant's detection of leaks to groundwater, surface water or soil acceptable.

9.3.4.13   Releases to Sewer Systems

As noted by the applicant in Section 9.2.2.1 of its LA (IIFP, 2012a) and in Section 6.1.1.2 of its ER (IIFP, 2011), all liquid effluents are to be maintained on the facility's site. The proposed facility does not discharge any process effluents into natural surface waters or grounds. No liquid effluents are released to the sewer system, and there is no tie into the publicly owned treatment works. Sanitary waste is treated onsite and rendered suitable for horticultural purposes.

Because there are no radiological liquid effluents releases to the sewer system and no radiological releases from the sewer system to the environment, the NRC staff finds that the releases to the sewer systems are controlled and maintained in a manner sufficient to meet the regulatory requirements of 10 CFR 20.2003 and to meet the acceptance criteria found in Section 9.4.3.2.2 of NUREG-1520, Revision 0 (NRC, 2002). Therefore, the NRC staff finds the applicant's method for controlling liquid effluent releases to be acceptable.

9.3.4.14   Reporting Procedures

In Section 9.2.2.1 of its LA (IIFP, 2012a), the applicant committed to implementing the recording guidance specified in Regulatory Guide 4.16, Revision 2 (NRC, 2010). As noted in the discussion of radionuclide-specific analyses in Section 10.3.3.1 of this SER, the NRC expects the applicant to submit a single semiannual effluent report that includes the concentrations of principal radionuclides released in the unrestricted area, as well as other information necessary to evaluate the radiation doses to the public from liquid and gaseous effluent releases. The report also would include the MDC for analysis and the error for each data point.

The NRC staff finds that the applicant has committed to reporting procedures that meet the requirements of 10 CFR 40.65 and are consistent with the guidance specified in Regulatory Guide 4.16, Revision 2 (NRC, 2010). The procedures also meet the applicable acceptance criteria found in Section 9.4.3.2.2 of NUREG-1520, Revision 0 (NRC, 2002), and are, therefore, acceptable to the NRC staff.

9.3.4.15 Liquid and Solid Waste Handling

The applicant addressed liquid and solid waste management in Section 3.12 of its ER (IIFP, 2011) and Section 9.2.2.1of its LA (IIFP, 2012a). As discussed in Sections 10.3.2.3 and 10.3.2.5 of this SER, as well as in Section 9.2.2.1 of the applicant's LA (IIFP, 2012a), fluoride bearing waste liquors are treated and recycled in the plant's scrubber system. Uranium removed from liquid streams is collected and sent to a licensed, low-level radioactive waste disposal site, along with waste uranium oxides produced by the deconversion process. Nonprocess waste liquids that are determined to contain regulated or hazardous contaminants are collected and disposed at an offsite licensed facility.

Solid waste is grouped into three categories: industrial (nonhazardous), radioactive, and hazardous waste. The applicant may send wastes that are candidates for volume reduction, recycling, or treatment to licensed treatment facilities that have the ability to reduce the volume of most Class A low-level radioactive waste. (See the LA, Section 1.1.6, Table 1-3, "Estimated Annual Waste Generated at the IIFP Facility" [IIFP, 2012a]).

The NRC staff finds that the applicant's program for management of liquid and solid wastes reduces unnecessary exposures in accordance with the 10 CFR Part 20 ALARA requirements; meets the waste handling acceptance criteria found in Section 9.4.3.2.2 of NUREG-1520; Revision 0 (NRC, 2002), and is, therefore, acceptable.

9.3.4.16 Liquid Effluent Monitoring Summary Evaluation

Based on the NRC staff's review of the applicant's LA (IIFP, 2012a) and ER (IIFP, 2011) described in Section 9.3.2.2 of this SER, the NRC staff finds that the Liquid Effluent Monitoring Program during operation of the proposed facility detects and measures concentrations of radioactivity in liquid effluent in a manner sufficient to demonstrate that liquid effluent concentrations remain below the regulatory limits in 10 CFR Part 20 and meet the applicable acceptance criteria found in Section 9.4.3.2.2 of NUREG-1520, Revision 0 (NRC, 2002). Therefore, the NRC staff finds the IIFP Liquid Effluent Monitoring Program acceptable.

9.3.4.17 Environmental Monitoring

Section 9.2.2.2 of the applicant's LA (IIFP, 2012a) and Chapter 6 of its ER (IIFP, 2011) describe the applicant's environmental monitoring program. Table 6.2 of the ER (IIFP, 2011) summarizes the program, including the medium (e.g., air, groundwater, soil, vegetation), sample locations, sample type, and parameter (e.g., gross alpha, gamma, beta) and frequency. Section 4.12.2 of the ER (IIFP, 2011) also discusses environmental monitoring during facility operation.

9.3.4.18 Background and Baseline Radionuclide Concentrations

The primary objective of the applicant's Radiological Environmental Monitoring Program (REMP) is to provide verification that the operations at the facility do not result in detrimental radiological impacts on the environment. To meet this objective, the applicant collects and analyzes representative samples from various environmental media for the presence of facility-related radioactivity. The REMP includes the collection of data during preoperational years to establish baseline radiological information against which operational radiological information can be compared. Such comparisons provide a means of assessing the magnitude of potential radiological impacts on members of the public and of demonstrating compliance with applicable RP regulatory requirements.

The NRC staff finds that the applicant has established an acceptable program to collect background and baseline concentrations of radionuclides in environmental media through sampling and analyses under the REMP for the proposed facility. These data are used as part of its RP Program and trending analyses to ensure that doses to individual members of the public are within the dose limits identified in 10 CFR Part 20, Subpart D. In addition, the REMP meets the applicable acceptance criteria in Section 9.4.3.2.2 of NUREG-1520, Revision 0 (NRC, 2002), and is, therefore, acceptable to the NRC staff.

9.3.4.19   *Sampling and Analyses for Monitoring*

Table 6.2 of the applicant's ER (IIFP, 2011) summarizes the REMP, including the medium sampled, sampling locations, sample type (e.g., thermal luminescent dosimeters, grab samples), and analyte/parameter frequency (e.g., total uranium sampled quarterly).

As noted in Section 9.3.3.1 of this SER, the applicant stated that air monitoring includes continuous sampling of the facility release points, as well as weekly composite samples for gross alpha activity and concentrations of uranium isotopes from active air monitors placed around the restricted area fence line. The applicant conducts periodic surveys in and around outdoor storage areas and use dosimeters at the fence line to ensure that direct radiation doses are maintained ALARA.

The groundwater monitoring program includes analysis of four onsite wells. Well samples are collected semiannually. Sampling begins before commencement of operations to establish baseline groundwater conditions and continue throughout operations and decommissioning.

The soil and vegetation monitoring includes onsite and offsite sample locations. Samples are collected quarterly and analyzed for uranium concentrations. Sampling continues throughout operations and decommissioning. Sediment samples are collected semiannually from the storm water runoff retention basins onsite and analyzed for uranium concentrations.

The NRC staff finds that the applicant's monitoring program, which includes sampling and analyses for monitoring air, surface water, groundwater, soil, sediments, and vegetation, meets the applicable acceptance criteria in Section 9.4.3.2.2 of NUREG-1520, Revision 0 (NRC, 2002), and is, therefore, acceptable to the NRC staff. The NRC staff also finds that the applicant's descriptions in Section 9.2.2 of its LA (IIFP, 2012a) and Section 6.1 of its ER (IIFP, 2011) adequately identify sample media, locations, types, and sampling frequencies and are consistent with applicable guidance found in Section 9.4.3.2.2 of NUREG-1520, Revision 0 (NRC, 2002), and are, therefore, acceptable to the NRC staff.

9.3.4.20   *Monitoring Procedures*

As noted in the NRC staff's discussion in Section 9.3.3.1 of this SER and Section 6.1.2.2 of the applicant's ER (IIFP,2011), IIFP uses written procedures to ensure that sampling and monitoring equipment is properly maintained; calibrated at regular intervals, including functional testing and routine checks; and is in good working condition. These procedures conform to guidance in Regulatory Guide 4.15 (NRC, 2007) and key parts of the REMP. The QC procedures include the use of established standards, such as those provided by NIST, and standard analytical procedures, such as those established by the NELAC. Samples are analyzed onsite for facility related radiological constituents and may be shipped to a qualified independent laboratory for analyses. Laboratories will participate in third-party comparison studies to validate their performance. Examples of third-party programs include the Mixed Analyte Performance Evaluation Program and the DOE Quality Assurance Program. Section 11.8 of the applicant's LA (IIFP, 2012a) also addresses, among other things, aspects of the Quality Assurance Program related to maintaining monitoring related records, as well as to developing and implementing monitoring-related procedures.

The NRC staff finds that the applicant's monitoring procedures conform to guidance found in Regulatory Guide 4.15 (NRC, 2007), because it employs nationally recognized analytical methods, instrumentation, laboratory validation methods, and laboratory procedures that are adequate to validate analytical results; and meet the applicable acceptance criteria in Section 9.4.3.2.2 of NUREG-1520, Revision 0 (NRC, 2002). Therefore, the NRC staff finds these procedures to be acceptable.

9.3.4.21   Action Levels

An action level is the concentration (or mass) of an analyte that indicates that some action needs to be taken, such as an investigation or, if the level is high enough, shutting down operations. Procedures specify action levels according to the type of samples and the specific analysis. The applicant stated that documented procedures include action levels for monitored environmental parameters, as appropriate, to provide guidance to ensure compliance with appropriate regulatory limits specified in 10 CFR Part 20, Subpart B. Documented procedures set response actions for elevated measurements at increasing levels of priority, ranging from (1) increasing monitoring frequency, (2) adjusting operations, and (3) performing corrective actions to prevent regulatory compliance levels from being exceeded. The applicant has designed administrative action levels to include, when practical, provision for automatic shutdown of the facility in the event action levels are exceeded.

Administrative action levels for physiochemical monitoring, sampling protocols, and emissions and effluent monitoring are performed for routine operations with provisions for additional evaluation in response to potential accidental release. The applicant will implement action levels before facility operation to ensure that chemical discharges remain below the limits specified in the facility discharge permits. The applicant's ESH organization staff selected the sampling and monitoring locations in accordance with facility permits and good sampling practices.

The NRC staff finds that the applicant's use of parameter action levels ensure that concentrations of radioactivity remain below the regulatory limits in 10 CFR Part 20 and meet the applicable acceptance criteria in Section 9.4.3.2.2 of NUREG-1520, Revision 0 (NRC, 2002). Therefore, the NRC staff finds the use of these actions levels to be acceptable.

9.3.4.22   Minimum Detectable Concentrations

Table 6-5 of the applicant's ER (IIFP, 2011) lists the MDCs for both effluent and environmental samples and identifies the medium (e.g., groundwater), activity (e.g., gross alpha, isotopic uranium), and typical MDCs (e.g., µCi/ml). The applicant will conduct physiochemical monitoring for required MDCs for fluoride and metals, among other things, via sampling of storm water, soil, sediment, vegetation, and groundwater, as noted in Table 6-3 of the applicant's ER (IIFP, 2011).

The NRC staff finds that the MDCs are sufficient to meet action level, regulatory, and permit requirements, as well as the requirements of environmental media monitoring programs. The MDCs also meet the applicable acceptance criteria in Section 9.4.3.2.2 of NUREG-1520, Revision 0 (NRC, 2002), and are, therefore, acceptable to the NRC staff.

9.3.4.23   Data Analysis

Field and laboratory analytical procedures, discussed above, address the collection of representative radiologic and physiochemical samples, use of appropriate sampling methods and equipment, proper sampling locations, and proper handling and analysis of samples.

The NRC staff finds that the data analysis methods and criteria that the applicant uses to evaluate and report environmental sampling results indicate when an action level is being approached in time to take corrective actions to ensure that concentrations remain within the regulatory limits in 10 CFR Part 20, Appendix B. The NRC staff finds that the applicant's data analysis methods and criteria meet the applicable acceptance criteria in Section 9.4.3.2.2 of NUREG-1520, Revision 0 (NRC, 2002), and are, therefore, acceptable.

9.3.4.24   Federal, State, and Local Requirements

Section 1.4 of the applicant's ER (IIFP, 2011) contains a complete description of required licenses, permits, and other approvals that are required by the Federal Government, as well as agencies of the State of New Mexico; Table 1-4 provides the status of each of these requirements. The applicant committed to follow applicable requirements for effluent monitoring activities described in the LA, Section 9.2.2 (IIFP, 2012a).

Based on the information in Table 1-4 of the ER (IIFP, 2011), the NRC staff finds that the information provided satisfies the applicable acceptance criteria in Section 9.4.3.2.2 of NUREG-1520, Revision 0 (NRC, 2002), and is, therefore, acceptable.

Measurement of Accidental Radioactive and Nonradioactive Releases to the Environment

In Chapter 3 of its LA (IIFP, 2012a), the applicant described its ISA, which evaluates potential risks and radiological and nonradiological (e.g., chemical) hazards from postulated unmitigated accident scenarios that could result in injuries to workers and the public or in significant environmental impacts. The applicant's ISA Summary (IIFP, 2012b) documents assessments of these accidents. The ISA Summary (IIFP, 2012b) also identifies active- and passive-engineered IROFS or administrative IROFS that prevent or mitigate the likelihood and consequences of those accident scenarios to acceptable levels. The applicant also addresses environmental effects of accidents in Section 4.12.3 of its ER (IIFP, 2011).

Because the IIFP facility only processes depleted uranium, the applicant mentioned that no nuclear criticality potential exists, and no materials that contain fission products or transuranic elements will be located onsite. Thus, the applicant stated that offsite radiological consequences associated with the facility are limited. Large inventories of depleted uranium material are present onsite. Despite these large inventories, the applicant has identified no credible accident to pose intermediate- or high-radiological consequences to the public. However, the applicant recognizes credible, intermediate-chemical consequence events could result from potential uranium oxide releases because of the acute chemical, as opposed to radiological, exposure of the uranium material. In addition, the applicant identified credible release scenarios that could result in both intermediate- and high-offsite consequences associated with acute chemical exposure. Thus, the only high- or intermediate consequence events postulated by the applicant are chemical in nature and involve the release of HF, either as a direct release of HF or as a release of a fluoride-bearing compound and subsequent release of HF as a reaction byproduct. However, these intermediate- or high-consequence events have a low probability of occurring.

The postulated credible accidents identified in the ISA Summary result from process upset conditions (e.g., loss of process and safety controls), natural phenomena (e.g., seismic events and tornados), and fire (IIFP, 2012b). Because of a combination of safety prevention limits and controls and mitigation measures, a significant process upset condition is not expected to result in the mitigated release of radiological or hazardous chemical material that would result in intermediate or high consequences to the public. Also, given the bounding of seismic events and tornados at the facility, as well as the preventive and mitigation methods employed if a significant fire occurs at the facility, a mitigated release of radiological or hazardous chemical material is not expected to result in intermediate or high consequences to the public.

Table 3-1 of the applicant's LA (IIFP, 2012a) identifies, among other things, consequence descriptions to the offsite public and the environment for high- and intermediate-severity accidents. Table 3-2 identifies EPA acute exposure guideline levels for uranium hexafluoride (UF_6), HF, and soluble uranium. The applicant also addressed radiological and non-radiological accident analyses, including mitigation measures to attenuate releases to the environment, in Section 4.12.3 of its ER (IIFP, 2011). As noted above, Chapter 3 of this SER presents the NRC staff's evaluation of these hazards.

As noted above in Sections 9.3.3.1, 9.3.3.2, and 9.3.3.3 of this SER, the applicant's monitoring program provides for effluent and environmental monitoring of airborne and liquid releases of uranium, UF_6, and HF. For example, air emissions are monitored continuously for uranium and HF. In addition, ambient air is monitored for activities from the UF_6 cylinder pads. Active air monitors (dosimeters) are also placed around the restricted area fence line for analysis of gross alpha activity and concentrations of uranium isotopes.

The NRC staff finds that the applicant's environmental monitoring program is adequate to assess impacts to the environment from potential radioactive and nonradioactive releases, as identified in high- and intermediate consequence accident sequences discussed in the ISA; adequately addresses related performance requirements in 10 CFR 70.61, "Performance Requirements," for individuals located outside the controlled area; and meets the applicable acceptance criteria in Section 9.4.3.2.2 of NUREG-1520, Revision 0 (NRC, 2002). Therefore, the NRC staff finds the applicant's approach to environmental monitoring to be acceptable.

9.3.5 Integrated Safety Analysis Summary

In Chapter 3 of this SER, the NRC staff provides its evaluation of the ISA Summary and documents its conclusion that the ISA Summary (IIFP, 2012b) is complete, provides reasonable estimates of the likelihood and consequences of each accident sequence, and provides sufficient information to determine whether adequate engineering or administrative controls are identified for each accident sequence. Chapter 11 of this SER contains the NRC staff's evaluation of management measures used to ensure that IROFS perform their intended safety functions. The NRC staff verified that environmental release limits would be met using existing IROFS. Therefore, no additional IROFS are identified for the proposed facility for reducing the environmental risks of natural phenomena and potential accidents.

Under 10 CFR Part 70, Subpart H, "Additional Requirements for Certain Licensees Authorized To Possess a Critical Mass of Special Nuclear Materials (10 CFR 70.60 through 10 CFR 70.76), an applicant is to ensure, among other things, compliance with various performance requirements. Title 10 CFR 70.61(c)(3) requires that the applicant apply controls such that a credible intermediate-consequence event is unlikely to occur or that the consequence of such an event does not exceed a 24-hour averaged release of radioactive material outside the restricted area in concentrations 5,000 times the values in Table 2 of Appendix B to 10 CFR Part 20.

In its ISA Summary (IIFP, 2012b), the applicant identified various sequences for radiological and non-radiological accidents which it evaluated to ensure adequate protection of worker health and safety. By ensuring that all credible, high-consequence events are rendered highly unlikely and that all intermediate-consequence events are unlikely, the applicant also ensured that the environmental performance requirements of 10 CFR 70.61(c)(3) are met. The NRC staff determined that offsite environmental consequences could occur only if uncontrolled, intermediate or high consequences to workers were also present. The NRC staff did not identify any accident sequence that would fail to meet the environmental performance requirements of 10 CFR 70.61.

9.3.6   10 CFR 70.61(c)(3)

The applicant stated that risk reduction is accomplished through a combination of preventive and mitigative measures, with an emphasis on preventive measures. Chapter 3 of this SER, which addresses accident sequences for intermediate- and high-consequence events, provides a more complete discussion. Chapter 3 also addresses preventive and mitigative measures.

The NRC staff finds that the applicant's ISA Summary complies with 10 CFR Part 70 because it meets the performance requirements of 10 CFR 70.61. The applicant's ISA Summary meets the applicable acceptance criteria in Section 9.4.3.2.3 of NUREG-1520, Revision 0 (NRC, 2002), as well and is, therefore, acceptable.

9.4   Evaluation Findings

The applicant has developed a program to implement adequate environmental protection measures during operation. These measures include (1) environmental and effluent monitoring and (2) effluent controls to maintain public doses ALARA as part of the radiation protection program. The NRC staff concludes that the applicant's program, as described in its application, is adequate to protect the environment and the health and safety of the public and complies with regulatory requirements imposed by the Commission in 10 CFR Part 20, 10 CFR Part 40, 10 CFR Part 51, and 10 CFR Part 70.

The NRC staff will issue a final Environmental Impact Statement (EIS) as part of this licensing action, as required by 10 CFR 51.20, "Criteria for and Identification of Licensing and Regulatory Actions Requiring Environmental Impact Statements." The final EIS will consider the environmental impacts of the construction, operation, and decommissioning of the proposed facility and compare alternatives, which will inform the NRC staff's recommendation concerning the LA for the proposed facility.

9.5   References

(EPA, 1988) U.S. Environmental Protection Agency, "Limiting Values of Radionuclide Intake and Air Concentration and Dose Conversion Factors for Inhalation, Submersion, and Ingestion," Federal Guidance Report No. 11, EPA- 520/1-88-020, Washington, D.C., September 1988.

(EPA, 1993) U.S. Environmental Protection Agency, Federal Guidance Report No. 12, External Exposure to Radionuclides in Air, Water, and Soil, Washington, D.C., 1993.

(ICRP, 1996) International Commission on Radiological Protection, "Age-dependent Doses to the Members of the Public from Intake of Radionuclides - Part 5 Compilation of Ingestion and Inhalation Coefficients," Publication 72, 1996.

(IIFP, 2012a) International Isotopes Fluorine Products, Inc., "Fluorine Extraction Process and Depleted Uranium Deconversion Plant (FEP/DUP) License Application, Revision B," May 2012, Agencywide Documents Access and Management System (ADAMS) Accession No. ML12123A245.

(IIFP, 2012b) International Isotopes Fluorine Products, Inc., "ISA Summary Rev. B for IIFP," May 2012, Agencywide Documents Access and Management System (ADAMS) Accession No. ML12123A245.

(IIFP, 2011) International Isotopes Fluorine Products, Inc., "Fluorine Extraction Process and Depleted Uranium Deconversion Plant (FEP/DUP) Environmental Report," Rev. A, 2011, Agencywide Documents Access and Management System (ADAMS) Accession No. ML100120758.

(NRC, 2010) U.S. Nuclear Regulatory Commission, "Monitoring and Reporting Radioactivity in Releases of Radioactive Materials in Liquid and Gaseous Effluents from Nuclear Fuel Processing and Fabrication Plants and Uranium Hexafluoride Production Plants," Regulatory Guide 4.16, Revision 2, December 2010.

(NRC, 2008) U.S. Nuclear Regulatory Commission, "Minimization of Contamination and Radioactive Waste Generation: Life-Cycle Planning," Regulatory Guide 4.21, June 2008.

(NRC, 2007) U.S. Nuclear Regulatory Commission, "Quality Assurance for Radiological Monitoring Programs (Inception through Normal Operations to License Termination) - Effluent Streams and the Environment," Regulatory Guide 4.15, 2007.

(NRC, 2002) U.S. Nuclear Regulatory Commission, "Standard Review Plan for the Review of a License Application for a Fuel Cycle Facility," NUREG-1520, Rev. 0, March, 2002.

(NRC, 1996) U.S. Nuclear Regulatory Commission, "Constraint on Releases of Airborne Radioactive Material to the Environment for Licensees Other Than Power Reactors," Regulatory Guide 4.20, December 1996.

(NRC, 1993) U.S. Nuclear Regulatory Commission, "ALARA Levels for Effluents from Materials Facilities," Regulatory Guide 8.37, July 1993.

(NRC, 1977) U.S. Nuclear Regulatory Commission, "Calculation of Annual Doses to Man from Routine Releases of Reactor Effluents for the Purpose of Evaluating Compliance with 10 CFR Part 50, Appendix I," Regulatory Guide 1.109, Revision 1, October 1977.

(ORNL, 2000) Oak Ridge National Laboratory, "MCNP4C Monte Carlo N-Particle Transport Code System," CCC-700 MCNP4C2, RSICC Computer Code Collection, Oak Ridge, TN, 2000. http://www.nea.fr/abs/html/ccc-0740.html

10.0 DECOMMISSIONING

The purpose of the U.S. Nuclear Regulatory Commission's (NRC's) review of the International Isotopes Fluorine Products, Inc. (IIFP or the applicant), Decommissioning Plan (DP) is to evaluate whether the application provides for decommissioning the facility safely and in accordance with NRC requirements.

At the time of the initial license application (LA) for a uranium hexafluoride deconversion facility, the applicant is required to submit a Decommissioning Funding Plan (DFP). The purpose of NRC's review of the DFP is to determine whether the applicant has considered decommissioning activities that may be needed in the future, has performed a credible site-specific cost estimate for those activities, and has presented the NRC with financial assurance to cover the cost of those activities in the future. The DFP, therefore, should contain an overview of the applicant's proposed decommissioning activities, the methods used to determine the cost estimate, and the financial assurance mechanism. This overview should contain sufficient detail to enable the reviewer to determine whether the Decommissioning Cost Estimate (DCE) is reasonably accurate.

10.1   Regulatory Requirements

The following NRC regulations require planning, financial assurance, and recordkeeping for decommissioning, as well as procedures and activities to minimize waste and contamination:

- Title 10 of the *Code of Federal Regulations* (10 CFR) Part 20, "Standards for Protection against Radiation," Subpart E, "Radiological Criteria for License Termination"

- 10 CFR 40.36, "Financial Assurance and Recordkeeping for Decommissioning"

- 10 CFR 40.42, "Expiration and Termination of Licenses and Decommissioning of Sites and Separate Buildings or Outdoor Areas"

10.2   Regulatory Guidance and Acceptance Criteria

Volume 3, "Financial Assurance, Recordkeeping, and Timeliness" of NUREG-1757, "Consolidated NMSS Decommissioning Guidance," issued September 2003 (NRC, 2003), contains the guidance applicable to NRC's review of the decommissioning section of the LA (IIFP, 2012a); and provides guidance for developing final DPs required under 10 CFR Part 20, Subpart E, and 10 CFR 40.42(g). The applicant will submit a final DP to the NRC at the time of decommissioning. At the time of initial licensing, as well as for license renewals, only an overview of the proposed decommissioning activities is needed to develop the DFP. This overview is a more generalized discussion of the detailed information that would be needed for the final DP described in NUREG-1757, Volume 3 (NRC, 2003).

Consistent with the guidance in NUREG-1757, Volume 3 (NRC, 2003), the application should include the following information:

- an overview of the proposed decommissioning activities that contains sufficient detail to enable the reviewer to determine whether the DCE is reasonably accurate

- proposed plans for minimizing contamination

- proposed plans for meeting the recordkeeping requirements of 10 CFR 40.36(f)

- proposed procedures to protect the health and safety of workers, the public, and the environment during decommissioning

- proposed environmental protection measures, specifically, the commitments to minimize waste associated with decommissioning

- proposed radiation protection program as it applies to radiological decontamination and the management of radiological effluents.

The overview of the decommissioning activities should cover the conceptual decontamination and DP, including the decommissioning program and steps, management and organization, health and safety, radiological decommissioning criteria, waste management, security and nuclear material control, recordkeeping, the decontamination process, and the minimization of contamination.

The following sections discuss these criteria and the applicant's information that addresses them.

10.3   Staff Review and Analysis

The NRC staff's review of the DFP focused on the applicant's conceptual decontamination and DP for the deconversion facility, the DCEs, and the financial assurance for decommissioning activities. The applicant identified the decommissioning activities that may be needed in the future for decommissioning and presented site-specific estimates of decommissioning costs for those activities. Using the cost data as a basis, the applicant stated that it has presented financial assurance to cover the costs required to release the deconversion facility site for unrestricted use. The following subsections discuss these decommissioning aspects, as described by the applicant, and the NRC staff's assessment of the applicant's proposed DP, cost estimates, and funding plan. Before license termination, the applicant will provide a detailed DP that will include specific activities which it will use to protect workers, the public, and the environment.

Chapter 10 of the LA presents the IIFP DFP for the proposed facility. The applicant developed the DFP following the guidance provided in NUREG-1757, Volume 1, Revision 2, "Decommissioning Process for Materials Licensees," issued September 2006 (NRC, 2006). IIFP committed to decontaminate and decommission the facility at the end of its operation so that the facility and grounds can be released for unrestricted use. In accordance with 10 CFR 40.42(d), the IIFP will review and update the DFP as necessary every year for the first 3 years of operation and at least once every 3 years thereafter. Before facility decommissioning, the applicant will prepare a DP in accordance with 10 CFR 40.42, "Expiration and Termination of Licenses and Decommissioning of Sites and Separate Buildings or Outdoor Areas," and submit it to the NRC for approval.

10.3.1 Conceptual Decontamination and Decommissioning Plan

The DFP contains an overview of the proposed decommissioning activities. The applicant will provide a detailed DP to the NRC at the time of decommissioning, in accordance with 10 CFR Part 20, Subpart E; 10 CFR 40.36, "Financial Assurance and Recordkeeping for Decommissioning"; and 10 CFR 40.42. This section also addresses the recordkeeping requirements in 10 CFR 40.36(f).

The NRC staff reviewed the proposed conceptual decontamination and DP and determined that it contains commitments to meet the radiological criteria for unrestricted use contained in 10 CFR 20.1402, "Radiological Criteria for Unrestricted Use," and to meet the timeliness in decommissioning requirements in 10 CFR 40.42(h). Therefore, as discussed below and based upon the applicant's commitments, the NRC staff finds the applicant's conceptual decontamination and DP acceptable.

10.3.2 Decommissioning Strategy

The overall strategy for decommissioning is to decontaminate or remove all materials from the site so that the facility can be released and the site made available for unrestricted use. At the end of useful plant life, the Fluorine Extraction Process and Depleted Uranium Deconversion Plant (FEP/DUP) will be decommissioned such that the site and remaining facilities may be released for unrestricted use, as defined in 10 CFR 20.1402. The NRC staff finds that IIFP committed to the use of guidance provided in NUREG-1505, Revision 1, "A Nonparametric Statistical Methodology for the Design and Analysis of Final Status Decommissioning Surveys - Interim Draft Report for Comment and Use," issued in 1998 (NRC, 1998); NUREG-1575, "Multi-Agency Radiation Survey and Site Investigation Manual (MARSIMM)," issued August 2000 (NRC, 2000); and NUREG-1757, Revision 2 (NRC, 2006), in developing initial and final site survey plans. These three guidance documents describe methodologies that NRC staff have found acceptable for implementing the Commission's decommissioning regulations in 10 CFR 20 Subpart E. The NRC staff finds that IIFP's commitment to follow them is sufficient to provide background and post decontaminationsite conditions to enable the free release of the site.

The applicant stated that it will decontaminate all remaining facilities where needed to acceptable levels for unrestricted use. Hazardous wastes will be treated or disposed of in licensed hazardous waste facilities. Disposal of radioactive or hazardous material will not occur at the plant site, but at licensed facilities located elsewhere. Following decommissioning, the facilities and site will be available for reuse.

Financial arrangements are made to cover costs required for returning the IIFP facility to unrestricted use. The applicant stated it will provide updates on cost and funding to the NRC every year for the first 3 years of operations, and then every 3 years thereafter, as described above. In addition, IIFP will submit a detailed, updated DP at a date near the end of plant life, in accordance with 10 CFR 40.42.

10.3.3 IIFP Facility Description

Chapter 1 of the LA (IIFP, 2012a) describes the IIFP FEP/DUP and site. The Integrated Safety Analysis (ISA) Summary (IIFP, 2012b) presents a detailed description of the plant site and facility and the safety aspects of the plant processes. Chapter 1 of the LA also describes the specific quantities and types of licensed materials used at the facility (see section entitled, "Institutional Information"). Also in Chapter 1 of the LA is a general description of how licensed materials are used at the facility (see section entitled, "General Information").

10.3.4 Design Features

Section 10.1.2 of the LA describes design features incorporated into the plant's initial design that will simplify eventual dismantling and decontamination. The plans are implemented through proper management and health and safety programs. Decommissioning policies address radioactive waste management, radioactive contamination control, physical security, and material control.

Section 10.1.2.2 of the LA describes major features incorporated into the facility design to facilitate decontamination and decommissioning. These features include the following:

- Building areas where uranium is processed and handled are separated physically from other building rooms and areas in which there is no need to have uranium present. These areas have separate ventilation and filtration systems to preclude contamination spread. Boundary control stations and hand/foot and portable monitors are used at applicable locations to verify that personnel and items exiting uranium process areas are not spreading radiological materials into nonuranium areas. The depleted uranium tetrafluoride (DUF_4) process building, fluoride extraction process (FEP) oxide staging building, plant operations decontamination building, DUF_4 container storage building, DUF_4 container staging building, and the FEP process building (in areas where licensed material is processed) meet these specific design features.

- All areas of the plant are sectioned into unrestricted and restricted areas. All procedures for these areas fall under the radiation protection program, and serve to minimize the spread of contamination and simplify the eventual decommissioning.

- The applicant will conduct routine radiological surveys throughout the operating lifetime of the facility to minimize the likelihood that radioactive contamination goes undetected and to provide a historical record which will simplify the site characterization process. The historical data collected by the Radiological Environmental Monitoring Program (REMP), described in Section 6.1.2 of the Environmental Report (IIFP, 2011d), will also be used to provide guidance for the final site survey. Samples will be collected to verify the historical data obtained via the REMP. Additional samples will be collected if areas of contamination are found.

- Non-radioactive process equipment and systems are minimized in locations subject to potential contamination. This limits the size of the restricted areas and limits the activities occurring inside these areas.

- Local air filtration is provided for areas with potential airborne contamination to preclude its spread. Containment equipment with hoods that exhaust through dust collectors, which are designed with high-removal efficiencies, are used where uranium materials are being packaged or withdrawn from process systems.

- The hazardous material processes include designs for purge and evacuation (P&E) systems and dust collection equipment as a means to provide effective cleanout of residual chemicals or dust from equipment or piping before opening systems for maintenance. The P&E and dust collector systems have multiple collection equipment in series (defense-in-depth) to ensure removal and treatment efficiency, redundancy, effectiveness, and reliability.

- Storm water runoff via the plant storm sewer system flows to a retention basin for evaporation, landscape watering, or discharge. Before discharging, collected storm water can be sampled if needed. The applicant stated domestic sanitary waste water is tertiary treated to meet all discharge standards and is either evaporated or used as harvested water for facility trees, grass, and shrubs. The facility is designed for no liquid process water discharges. Engineered systems are used to provide for regeneration of scrubbing solutions and recycle within the process systems.

10.3.5 Worker Exposure and Waste Volume Control

In Sections 9.2.1.4 and 10.1.2.3 of the LA (IIFP, 2012a), the applicant described features that serve to minimize worker exposure to radiation and minimize radioactive waste volumes during decontamination activities. As a result, the spread of contamination is minimized as well. These features include the following:

- During construction, the applicant will apply a washable coating to designated floors and walls in the restricted areas that have the higher potential to become radioactively contaminated during operation. The coating will serve to lower waste volumes during decontamination and simplify the decontamination process.

- Sealed, nonporous pipe insulation will be used in areas with higher potential to become contaminated. This will facilitate cleaning in the event of a spill and will reduce waste volume during decommissioning.

- Ample access will be provided for efficient equipment dismantling and removal of equipment that may be contaminated. This minimizes the time of worker exposure.

- Tanks will have access for entry and decontamination. Design provisions will also be made to allow complete draining of the wastes contained in the tanks.

- Connections in the process systems, provided for required operation and maintenance, will allow for thorough purging at plant shutdown. This will remove a significant portion of radioactive contamination before disassembly.

- Design drawings, produced for all areas of the plant, will simplify planning and implementing decontamination procedures. This in turn will shorten the durations that workers are exposed to radiation.

- Worker access to contaminated areas will be controlled to ensure that workers wear proper protective equipment and limit their time in the areas.

- Radioactive and hazardous wastes produced during decommissioning will be collected, handled, and disposed of in accordance with regulations applicable to the facility at the time of decommissioning. Generally, procedures will be similar to those described for wastes produced during normal operation. These wastes will ultimately be disposed in licensed radioactive or hazardous waste disposal facilities located elsewhere. Nonhazardous and nonradioactive wastes will be disposed in a manner consistent with good industrial practice and in accordance with applicable regulations.

- To facilitate decommissioning, the information relating to the facility design, facility construction, design modifications, site conditions before and after construction, onsite contamination, and results of monitoring and radiological surveys will be readily recoverable through the IIFP document control and management process.

10.3.6 Management Organization

Section 10.1.2.4 of the LA (IIFP, 2012a) describes an appropriate organizational structure that the applicant will develop to support the decommissioning strategy. The applicant stated that the organizational structure will ensure that adequate numbers of experienced and knowledgeable personnel are available to perform the technical and administrative tasks required to decommission the facility.

10.3.7 Health and Safety

In Section 10.1.2.5 of the LA (IIFP, 2012a), the applicant stated that the policy during decommissioning will be to keep individual and collective occupational radiation exposure as low as reasonably achievable (ALARA). A health physics program will identify and control sources of radiation, establish worker protection requirements, and direct the use of survey and monitoring instruments.

10.3.8 Waste Management

In Section 10.1.2.6 of the LA (IIFP, 2012a), the applicant stated that radioactive and hazardous wastes produced during decommissioning will be collected, handled, and disposed of in accordance with all regulations applicable to the facility at the time of decommissioning. Generally, procedures will be similar to those described for wastes produced during normal operation. These wastes will ultimately be disposed of in licensed radioactive or hazardous waste disposal facilities located elsewhere.

10.3.9 Physical Security and Material Control and Accounting

In Section 10.1.2.7 of the LA (IIFP, 2012a), the applicant stated that it will maintain physical security and material control and accounting as required during decommissioning in a manner similar to the programs in force during operation. The IIFP plan for completion of decommissioning, submitted near the end of plant life, will describe any necessary revisions to these programs.

10.3.10 Recordkeeping

In Section 10.1.2.8 of the LA (IIFP, 2012a), the applicant stated that records important for safe and effective decommissioning of the facility will be stored in the FEP/DUP records management system until the site is released for unrestricted use. Information maintained in these records includes the following:

(1)   Records of spills or other unusual occurrences involving the spread of contamination and cleanup around the facility, equipment, or site will be maintained. These records will include any known information identifying the involved nuclides; their quantities, forms, and concentrations; and survey results after cleanup of any spill area.

(2)   Routine radiological survey records of restricted and unrestricted areas will be retained indefinitely to support historical site assessment and facility characterization at the time of decommissioning.

(3)   As-built drawings and modifications of structures and equipment in restricted areas will be maintained where radioactive materials are used or stored. Required drawings will be referenced as necessary, although each relevant document will not be indexed individually. If drawings are not available, appropriate records of available information concerning these areas and locations will be substituted.

(4)   The following will be contained in a single records document, updated every 2 years, except for areas containing only sealed sources:

- all areas designated and formerly designated as restricted areas as defined under 10 CFR 20.1003, "Definitions"

- all areas outside of restricted areas that require documentation specified in item (1) above

- all areas outside of restricted areas where current and previous wastes have been buried will be documented under 10 CFR 20.2108, "Records of Waste Disposal"

- all areas outside of restricted areas that contain material such that, if the license expired, the licensee would be required to either decontaminate the area to meet the criteria for decommissioning in 10 CFR Part 20, Subpart E, or apply for approval for disposal under 10 CFR 20.2002, "Method for Obtaining Approval of Proposed Disposal Procedures."

(5)   Records of the cost estimate performed for the DFP or of the amount certified for decommissioning and records of the funding method used for assuring funds, if either a funding plan or certification is used, will be maintained.

10.3.11 Decommissioning Process

Section 10.1.3 of the LA (IIFP, 2012a) describes the IIFP's decommissioning process. Preparation for decommissioning is expected to begin for the facility upon a decision to cease operations permanently; and this preparation step is expected to be completed in approximately

1 year, including NRC's review and approval of the final plan. Actual decontamination and decommissioning would follow shortly after approval of the plan and the award of any subcontracts. Figure 10-1 of the LA (IIFP, 2012a) illustrates the DP schedule for the IIFP facility.

Before completely shutting down all processes, the bulk work-in-process (WIP) inventory of uranium materials would be processed as much as practical into depleted uranium oxide and the fluoride gas products, similar to normal operations. This activity would render the bulk materials into products for shipment to customers and into depleted uranium oxide approved for disposal, similar to normal plant operations. Based on the estimated maximum-average WIP inventories, the amount of time required to orderly process this material into its final form is between 12 and 15 days. After processing the bulk WIP, any residual inventory of uranium or contaminated materials would be included in the decommissioning steps that follow the decommissioning preparation and NRC approvals to proceed. As shown in Table 10-1 of the LA (IIFP, 2012a), the estimated residual amounts of uranium chemicals or uranium-contaminated chemicals expected to be disposed as low-level radioactive waste (LLW) are approximately 48 cubic m (1,700 cubic ft).

Before the start of decommissioning operations, the applicant will perform a radiological survey of the facility in conjunction with a historical site assessment. The final DP submitted to the NRC will present the findings of the radiological survey and historical site assessment. The applicant will prepare this DP in accordance with 10 CFR 40.42 and the applicable guidance provided in NUREG-1757, Volume 3 (NRC, 2003).

Decommissioning activities will include (1) outfitting of size reduction and packaging areas, (2) purging of process systems, (3) dismantling and removal of equipment, (4) sales of salvaged materials, (5) packaging and disposal of wastes, and (6) completion of a final radiation survey. Credit is not taken for any salvage value that might be realized from the sale of potential assets during or after decommissioning.

Using the IIFP approach, 10 CFR 40.42(k) and 10 CFR 20.1402 requires residual radioactivity to be reduced below specified levels before the facility may be released for unrestricted use. Current NRC guidelines for release serve as the basis for the decontamination costs estimated in Chapter 10 of the LA (IIFP, 2012a). Portions of the facility that do not exceed contamination limits may remain as is without further decontamination measures applied. The intent of decommissioning the facility is to remove all uranium process-related equipment from the buildings, such that only the building shells and site infrastructure remain. The removed equipment includes all piping and components from systems providing, for example, uranium hexafluoride (UF_6) or uranium tetrafluoride (UF_4) containment; uranium oxide containment; systems in direct support of uranium processing (such as refrigerant and chilled water); radioactive and hazardous waste-handling systems; and contaminated heating, ventilation, and air-conditioning (HVAC) filtration systems. The site infrastructure remaining after decommissioning is complete will include the steam facilities, electrical power facilities, water supply systems, sanitary water treatment systems, fire protection systems, HVAC systems, cooling water systems, and communication systems.

The applicant will outfit existing plant buildings, such as the decontamination building and material warehouse, to accommodate handling and packaging of components and materials for disposal. These areas will be the primary location for size reduction and packaging activities during the decommissioning process. Limited capabilities for decontamination will exist for mildly contaminated items that may be decontaminated to free release criteria in a cost-effective manner.

The applicant will decontaminate contaminated portions of the buildings as required. Potential contamination is limited to the structures in the restricted areas. Good housekeeping practices during normal operation will keep the other areas of the site clean, and routine radiological contamination surveys will ensure that radioactive contamination will not go undetected or be allowed to buildup to levels difficult to control. When decontamination is complete, the applicant will survey all areas and facilities on the site to verify that further decontamination is not required. Decontamination activities will continue until the entire site is demonstrated to be suitable for unrestricted use.

10.3.12   Size Reduction and Packaging Facility Outfitting

The NRC staff confirmed that the applicant's facility description that IIFP's proposed facilities can be adapted to accommodate the size reduction and packaging activities associated with decommissioning. IIFP identified the decontamination building and material warehouse as suitable for these purposes. IIFP estimated time for equipment installation as approximately 2 months. This timeframe supports the dismantling of the equipment. Section 10.1.4.3 of the LA (IIFP, 2012a) describes these facilities.

10.3.13   System Preparation

At the end of the useful life of each process line, the uranium process will be shut down and UF_6, UF_4, and uranium oxides will be removed to the extent practicable by normal process operation. This will be followed by evacuation and purging with nitrogen and the application of a fixative, where applicable. The shutdown and preparation of the decommissioning process is estimated to take approximately 3 months.

10.3.14   Dismantling

Dismantling requires cutting and disconnecting all components requiring removal. Dismantling operations are labor intensive and generally require the use of protective clothing. The work process will be optimized, considering the following:

- minimizing the spread of contamination and the level of protective clothing necessary using fixative coatings

- balancing the number of cutting and removal operations with the resultant size reduction and disposal requirements

- optimizing the rate of dismantling with the rate of size reduction of facility throughput

- providing storage and laydown space required, as impacted by retrieving, security, and other activities

- balancing the cost of salvage with the cost of disposal.

The applicant will decide the details of the complex optimization process near the end of plant life, taking into account specific contamination levels, market conditions, and available waste disposal sites. The DFP assumes that most items that were continuously in contact with UF_6, UF_4, or uranium oxide will be disposed of at a low-level radioactive waste disposal facility rather than employing rigorous decontamination techniques. Large contaminated components may be disassembled to separate contaminated and uncontaminated portions of the component. To avoid laydown space and contamination problems, dismantling will generally not be allowed to proceed faster than the downstream size reduction and packaging process.

The timeframe to accomplish both dismantling and size reduction at FEP/DUP is estimated to be approximately 18 months.

10.3.15   Decontamination and Size Reduction

Section 10.1.4 of the LA (IIFP, 2012a) describes in detail the decontamination and size reduction process. The description encompasses the decontamination and size reduction methodology, size reduction and packaging facilities, decontamination and size reduction procedures, and results.

10.3.16   Salvage of Equipment and Materials

Items to be removed from the facilities can be categorized as potentially reusable equipment, recoverable scrap, and wastes. However, based on a 40-year license, or beyond, operating equipment is assumed to have no reuse value. Wastes will also have no salvage value. With respect to scrap, some amounts of uncontaminated metal (steel, copper, Monel) may be recovered and sold. Contaminated materials will be disposed of as low-level radioactive waste. No credit is taken for any salvage value that might be realized from the sale of potential assets during or after decommissioning.

10.3.17   Disposal

All wastes produced during decommissioning will be collected, handled, and disposed of in a manner similar to that described for wastes produced during normal operation. Wastes will consist of normal industrial trash, nonhazardous chemicals and fluids, small amounts of hazardous materials, and radioactive wastes. The radioactive waste will consist primarily of piping, tanks, hoppers, and compactable trash generated during the dismantling process. Radioactive wastes will ultimately be disposed of in licensed low-level radioactive waste disposal facilities. Hazardous wastes will be disposed of in hazardous waste disposal facilities. Nonhazardous and nonradioactive wastes will be disposed of in a manner consistent with good industrial practice and in accordance with all applicable regulations. The DP, which the applicant will submit to the NRC before initiating the decommissioning of the plant will provide a complete estimate of the wastes and effluent to be produced during decommissioning.

10.3.18   Final Radiation Survey

At the end of useful plant life, the FEP/DUP will be decommissioned such that the site and remaining facilities may be released for unrestricted use as defined in 10 CFR 20.1402. IIFP stated that it will use guidance provided in NUREG-1505 (NRC, 1998), NUREG-1575 (NRC, 2000), and NUREG-1757, Volume 3 (NRC, 2003), in developing initial and final site survey plans capable of providing sufficient data on the background and post decontaminationsite conditions to enable the free release of the site.

A final radiation survey will be performed to verify proper decontamination to allow the site to be released for unrestricted use. The evaluation of the final radiation survey is based in part on an initial radiation survey performed before initial operation. Since the IIFP facility will only process depleted uranium, the initial and final site surveys required as part of the decommissioning process will include isotopic analysis for the uranium-238, uranium-235, and uranium-234 isotopes of uranium. The final survey will systematically and representatively measure radioactivity for the entire site and will be designed to detect any unreported spills and any generalized contamination that might accumulate over the period of operation of the facility. The intensity of the survey will vary depending on the location (i.e., the buildings, their immediate areas, and the remainder of the site).

Throughout the operating life of the facility, the applicant will conduct routine surveys of licensed material areas and maintain records of such surveys. These survey records will be used in conjunction with the REMP as part of the final survey evaluation and may reduce the amount of sampling in some areas for which the survey history indicates that no contamination has occurred. The applicant will document the survey procedures and results in a report which will include, among other things, a map of the survey site, measurement results, and the site's relationship to the surrounding area. The results will be analyzed and shown to be below allowable residual radioactivity limits; otherwise, IIFP will perform further decontamination.

For decommissioning funding purposes, the final site survey will consist of samples taken within the 40-acre IIFP restricted area, as well as at other locations outside the restricted area but within the 640-acre site boundary. Inside the restricted area, samples will be taken based on a sampling grid pattern of approximately 91 m by 91 m (100 yards by 100 yards). Additional samples will be collected within an area extending 3 m (10 ft) from process building walls on the basis of one sample per 9.3 square m (100 square ft) (i.e., one sample for every 3 m [10 ft] of building perimeter). It is unlikely that the area outside the restricted area, but within the site boundary, will be contaminated. Outside of the restricted area, but within the site boundary, the likelihood for contamination is extremely remote. Therefore, the grid will be expanded for this area such that samples will be taken on a grid pattern of approximately 610 m by 610 m (1,017 ft by 1,017 ft). A third part will provide the analysis of the samples since, at the time of performance of the final radiation survey, no analysis facilities will be available onsite. As part of the REMP, the applicant will perform a similar collection of samples for the initial site survey to provide a background value against which to compare the final site survey.

10.3.19   Decommissioning Impact on Integrated Safety Analysis

Although decommissioning steps are planned to be underway while some activities considered in the ISA continue to occur in the other areas of the plant, the current ISA does not fully evaluate these decommissioning risks. The applicant will perform an updated ISA before decommissioning activities begin to evaluate the risks from decommissioning operations on concurrent operations.

10.3.20   NRC Staff Determination

The NRC staff has reviewed the applicant's conceptual decontamination and DP and determined that IIFP has considered site-specific activities necessary to decontaminate and decommission the FEP/DUP. Therefore, as discussed in Sections 10.3.3 through 10.3.19 above, the NRC staff finds that the applicant's proposed process provides reasonable assurance that decommissioning can be performed in accordance with the guidance in NUREG-1757, Volume 3 (NRC, 2003), and the requirements in 10 CFR Part 20, Subpart E, and 10 CFR 40.42(g).

10.4 Site-Specific Cost Estimate

10.4.1 Regulatory Requirements

IIFP is required to submit a DFP under 10 CFR 40.36(a), which states that an "applicant for a specific license authorizing the possession and use of more than 100 mCi of source material in a readily dispersible form shall submit a decommissioning funding plan...."

Among other things, 10 CFR 40.36(d) states that the DFP must "contain a cost estimate for decommissioning and a description of the method of assuring funds for decommissioning ... including means for adjusting cost estimates and associated funding levels periodically over the life of the facility...." As part of its LA (IIFP, 2012a), IIFP submitted a DFP and cost estimate.

10.4.2 Summary of Staff Review

The NRC staff reviewed the information provided in the LA (IIFP, 2012a) and the responses to NRC's requests for additional information (RAIs) dated April 29, 2011 (IIFP, 2011a); June 29, 2011 (IIFP, 2011b); and August 12, 2011 (IIFP, 2011c) in order to determine whether the applicant's DFP and cost estimate are acceptable.

10.4.3 Staff Review of Decommissioning Funding Plan and Cost Estimate

Applicants for licenses under 10 CFR Part 40, "Domestic Licensing of Source Material," are subject to financial assurance requirements for decommissioning, decontamination, and reclamation in accordance with 10 CFR 40.36, "Financial Assurance and Recordkeeping for Decommissioning." Chapter 10 of the LA contains a DFP and cost estimate, as required by 10 CFR 40.36(a) and 10 CFR 40.36 (d), respectively. The NRC staff used NUREG-1757, Volume 3 (NRC, 2003), to guide its review.

The NRC staff reviewed the DFP, cost estimate, and responses to NRC's RAIs. In conjunction with IIFP's responses to the RAIs, the NRC staff finds the cost estimate acceptable because it (1) is based on reasonable and documented assumptions, (2) incorporates the costs of a third-party contractor, (3) does not take credit for any salvage value that might be realized from the sale of potential assets during or after decommissioning, (4) includes the estimated costs of radioactive waste disposal, and (5) includes a 25-percent contingency factor. The cost estimate also includes the estimated cost of deconverting the estimated amount of depleted uranium remaining at the site at the time of decommissioning. IIFP estimates that the cost to decommission the site and facility in a manner that is consistent with the unrestricted release criteria set forth in 10 CFR 20.1402 would be $14,204,535.

As set forth in 10 CFR 40.36(d), each DFP must also contain a cost estimate submitted concurrently with the LA, for decommissioning and a description of the method of assuring funds. In its August 12, 2011, submittal (IIFP, 2011c), IIFP requested an exemption to the requirement in 10 CFR 40.36(d) to submit the certificate of financial assurance along with the DFP. The certification of financial assurance requires a licensee to certify that financial assurance has been provided in the amount of the cost estimate for decommissioning. Specifically, the exemption would allow IIFP to provide the certification of financial assurance separately from the DFP. The NRC approves this exemption, but imposes two license conditions, as described below. These license conditions require the certification of financial assurance to be provided to the NRC with the financial instrument it intends to execute 6 months before commencing operations. The license conditions also require IIFP to provide the NRC with executed financial instruments at least 21 days before commencing operations. As described in detail in Section 1.2.3.5.2 of this SER, the NRC staff found IIFP's exemption request acceptable. [1]

As stated in Section 1.2.3.5 of this SER, the NRC has reviewed and approved this exemption request consistent with 40.36(d) and the categorical exclusion requirements in 51.22(c)(25).

As part of its RAI response, IIFP provided a draft payment surety bond (PSB), a draft standby trust agreement (STA), and a draft certification of financial assurance (certification) for review. The NRC staff reviewed the draft PSB, STA, and certification and finds that the language of the documents are consistent with the guidance in NUREG-1757, Volume 3, Appendix A.9, Appendix A.17, and Appendix A.2.4, respectively (NRC, 2003).

Consistent with the financial assurance exemption request described in Section 1.2.3.5.2 of this Safety Evaluation Report, the applicant will provide the NRC with its proposed final draft financial instruments for review in advance of IIFP obtaining NRC-licensed material at the site in quantities and form requiring decommissioning financial assurance (e.g., in quantities and form that exceed the regulatory thresholds of 10 CFR 40.22 and 10 CFR 40.36). In addition, IIFP will provide executed versions of the reviewed financial instruments before the receipt of NRC-licensed material; and, if they are acceptable, the NRC staff will provide its approval at that time. The NRC staff will impose the following license condition to which the applicant has agreed:

> Consistent with 10 CFR 40.36(a), prior to obtaining radioactive material under the NRC license, the licensee shall provide final copies of proposed financial assurance instruments and certification of financial assurance to the NRC for review at least 6 months before the anticipated date for obtaining radioactive material under the NRC license and shall provide final, executed copies of the NRC-reviewed financial assurance instruments and certification of financial assurance to NRC at least 21 days before the anticipated date for obtaining the radioactive material at the site. The amount of financial assurance provided shall be at least as great as the NRC-approved cost estimate.

> The licensee shall not obtain radioactive material under the NRC license, in any form, in an amount that would require financial assurance as set forth in 10 CFR 40.36, until the NRC reviews and approves the executed financial assurance instrument and certification of financial assurance.

---

1   As described in Section 1.2.3.5 of this SER, IIFP stated that it may use equipment obtained from the Sequoyah Fuels facility to build its facility. IIFP stated that if any of these components cannot be decontaminated, these components would be appropriately shipped to the site. To provide assurance that decommissioning funding is sufficient during the construction phase, the staff will impose another license condition that would require financial assurance if the total amount of contamination on equipment exceeds the regulatory limit set forth in 10 CFR 40.36.

As required by 10 CFR 40.36(d), licensees must update their cost estimates for decommissioning at least triennially. In its application, IIFP committed to updating its cost estimate "approximately every three years." The NRC staff finds that requiring IIFP to update its estimate more frequently than triennially would be prudent because the facility will not commence operations until sometime after license issuance and, consequently, the costs of decommissioning may change. Therefore, the NRC staff will impose the following license condition to which the applicant has agreed:

> The licensee shall provide an updated DFP and updated facility DCE to the NRC for review at least 6 months before the planned date for obtaining licensed material in any form, including contamination, in an amount that would require financial assurance as set forth in 10 CFR 40.36.

> On an annual basis, starting on the anniversary date of obtaining licensed material, and for 2 subsequent years the licensee shall provide an updated DFP and updated facility DCE to the NRC for review. With each annual update, if the cost estimate exceeds the amount of financial assurance provided, the licensee shall provide financial assurance in the amount of the updated cost estimate and an updated certification of financial assurance to the NRC for review and approval.

> All updates to the DFP and cost estimate for facility decommissioning and financial assurance instruments shall reflect current year U.S. dollars and shall encompass all current cost data—including but not limited to taking into account changes in inflation, possession limits, licensed material, labor rates, disposal and shipping rates, and site and facility factors. All costs shall be supported by a detailed basis, be based on the costs of a third-party contractor, and shall not take credit for any salvage value that might be realized from the sale of potential assets during or after decommissioning. The total cost estimate shall include a contingency factor of at least 25 percent.

10.4.4 Evaluation Findings

The NRC staff has evaluated the applicant's plans and financial assurance for decommissioning in accordance with the "Consolidated NMSS Decommissioning Guidance," NUREG-1757, (NRC, 2003). Relying on the information provided in the LA (IIFP, 2012a) and the responses to the NRC's RAIs dated April 29, 2011 (IIFP, 2011a); June 29, 2011 (IIFP, 2011b); and August 12, 2011 (IIFP, 2011c), the NRC staff finds that (1) the DFP and cost estimate for decommissioning and decontaminating the site and facility to unrestricted release criteria is consistent with NUREG-1757, Volume 3 (NRC, 2003); (2) the DFP and cost estimate is based on reasonable and documented assumptions; (3) the cost estimate is based on the costs of a third-party contractor; (4) the cost estimate includes a 25-percent contingency factor; (5) the cost estimate does not take credit for any salvage value that might be realized from the sale of potential assets during or after decommissioning; and (6) the PSB, STA, and certification are consistent with the applicable sections of NUREG-1757, Volume 3 (NRC, 2003). On the basis of this evaluation, the NRC staff has determined that the applicant's plans and financial assurance for decommissioning comply with 10 CFR 40.42 and 10 CFR 40.36 and are acceptable.

10.5 References

(IIFP, 2012a) International Isotopes Fluorine Products, Inc., "Fluorine Extraction Process and Depleted Uranium Deconversion Plant (FEP/DUP) License Application, Revision B," May 2012, Agencywide Documents Access and Management System (ADAMS) Accession No. ML12123A245.

(IIFP, 2012b) International Isotopes Fluorine Products, Inc., "ISA Summary Rev. B for IIFP," May 2012, Agencywide Documents Access and Management System (ADAMS) Accession No. ML12123A245.

(IIFP, 2011a) International Isotopes Fluorine Products, Inc., "Official Response to Financial Assurance RAIs," April 29, 2011, Agencywide Documents Access and Management System (ADAMS) Accession No. ML11129A083.

(IIFP, 2011b) International Isotopes Fluorine Products, Inc., "Letter from INIS containing the Public Revision B of the Financial Assurance RAI Responses and Follow-up RAI Responses," June 29, 2011, Agencywide Documents Access and Management System (ADAMS) Accession No. (Public) ML11195A176.

(IIFP, 2011c) International Isotopes Fluorine Products, Inc., "Letter from INIS re: Affidavit and Financial Assurance Responses to the Second Follow-up RAIs for the IIFP License Application," August 12, 2011, Agencywide Documents Access and Management System (ADAMS) Accession No. ML11234A074

(IIFP, 2011d) International Isotopes Fluorine Products, Inc., "Fluorine Extraction Process and Depleted Uranium Deconversion Plant (FEP/DUP) Environmental Report," Rev. A, 2011, Agencywide Documents Access and Management System (ADAMS) Accession No. ML100120758.

(NRC, 2006) U.S. Nuclear Regulatory Commission, "Consolidated NMSS Decommissioning Guidance," NUREG-1757, Vol. 1, Rev. 2, "Decommissioning Process for Materials Licensees," 2006.

(NRC, 2003) U.S. Nuclear Regulatory Commission, "Consolidated NMSS Decommissioning Guidance," NUREG-1757, Vol. 3, "Financial Assurance, Recordkeeping, and Timeliness," 2003.

(NRC, 2000) U.S. Nuclear Regulatory Commission, "Multi-Agency Radiation Survey and Site Investigation Manual," NUREG-1575, August 2000.

(NRC, 1998) U.S. Nuclear Regulatory Commission, "A Nonparametric Statistical Methodology for the Design and Analysis of Final Status Decommissioning Surveys - Interim Draft Report for Comment and Use," NUREG-1505, Rev. 1, 1998.

11.0 MANAGEMENT MEASURES

Management measures are functions that International Isotopes Fluorine Products, Inc. (IIFP or the applicant), will perform, generally on a continuing basis, which are applied to items relied on for safety (IROFS) to provide reasonable assurance that the IROFS are available and reliable to perform their function when needed. The applicant will implement management measures to ensure compliance with performance requirements, and the degree to which they will be applied will be a function of the item's importance in terms of meeting performance requirements, as evaluated in the Integrated Safety Analysis (ISA) (IIFP, 2012b). This chapter addresses each of the management measures included in the definition of management measures found in Title 10 of the *Code of Federal Regulations* (10 CFR) Part 70, "Domestic Licensing of Special Nuclear Material," including (1) configuration management (CM), (2) maintenance, (3) training and qualifications, (4) procedures, (5) audits and assessments, (6) incident investigations, (7) records management, and (8) other quality assurance (QA) elements.

The purpose of this review is to verify whether IIFP's license application (LA) for the Fluorine Extraction Process/Depleted Uranium Deconversion Process (FEP/DUP) plant provided conclusive information to ensure that the management measures applied to IROFS, as documented in the ISA Summary (IIFP, 2012b), offer adequate assurance that the IROFS will be available and reliable and consistent with the provisions of 10 CFR 70.61, "Performance Requirements." This review also determines whether the measures are applied to the IROFS in a graded manner commensurate with their importance to safety.

11.1   Regulatory Requirements

The following requirements for fuel cycle facility management measures, specified in 10 CFR Part 70, apply to the IIFP facility:

- 10 CFR 70.4, "Definitions," which states that management measures include (1) CM, (2) maintenance, (3) training and qualifications, (4) procedures, (5) audits and assessments, (6) incident investigations, (7) records management, and (8) other QA elements

- 10 CFR 70.62(a)(3), which states that records must be kept for all IROFS failures, describes required data to be reported, and sets time requirements for updating the records

- 10 CFR 70.62(d), which requires a licensee to establish management measures for application to engineered and administrative controls and control systems that are identified as IROFS, under 10 CFR 70.61(e), to ensure they are available and reliable

- 10 CFR 70.72, "Facility Changes and Change Process," which requires a licensee to establish a CM program to evaluate, implement, and track changes to the facility; structures, systems, and components (SSCs); processes; and activities of personnel.

11.2   Regulatory Acceptance Criteria

Section 11.4.3 of NUREG-1520, Revision 1, "Standard Review Plan for the Review of a License Application for a Fuel Cycle Facility," issued March 2002 (NRC, 2002), contains the acceptance criteria for the U.S. Nuclear Regulatory Commission's (NRC's) review of the FEP/DUP facility management measures program.

11.3   Staff Review and Analysis

As stated by the applicant, the IIFP project is currently in the development, conceptual design, and licensing phase. The applicant will apply management measures throughout all phases of construction, operations, and maintenance of the facility.

The applicant provided the IIFP design, construction, and operations organization description and organizational charts in the LA Chapter 2 (IIFP, 2012a). The IIFP President has overall responsibility for the establishment and implementation of the QA policies, goals, and objectives for IIFP. The application emphasizes that the responsibilities of line managers in the implementation and maintenance of the management measures policies and procedures are commensurate with delegated authority.

As described in the LA (IIFP, 2012a), the IIFP Chief Operations Officer (COO) will be responsible for assuring that management measures in accordance with graded QA are being implemented by the Design and Build (DB) contractor chosen by IIFP to perform detailed design and construction. Also, the application clearly delineates in the LA, Chapter 2 (IIFP, 2012a), that the IIFP QA Manager and Environmental Safety and Health (ESH) Manager have responsibilities and authorities independent of production, engineering, and maintenance organizational functions. The ESH Manager provides authorized oversight and technical direction to ensure that production never takes priority over the safety and protection of employees, the public, and the environment. The ESH Manager reports to the COO directly.

11.3.1 Configuration Management

The IIFP QA Program requirements and associated procedures implement the IIFP CM Program. The application describes the CM policy, design requirements, document control, change control, and assessments.

Configuration Management Policy

In Section 11.1.1, "Configuration Management Policy," of the LA (IIFP, 2012a), the applicant stated that CM will provide a means to establish and maintain a technical baseline for the facility based on clearly defined requirements. In addition, the applicant described design documents as those that provide design input, analysis, and results specifically for IROFS with the appropriate QA level. In Section 11.8.2.2 of the LA (IIFP, 2012a), the applicant defined the QA levels, which will be discussed later in this Safety Evaluation Report (SER). The CM Program will provide a formal review process, including interdisciplinary reviews, for design changes.

During design and construction, CM will be based on design control provisions and associated procedural control of the design documents. During construction, changes to drawings and specifications issued for construction, procurement, or fabrication will be systematically reviewed, verified, evaluated for impact, and approved before implementation. As the project progresses from design and construction to operation, CM responsibilities will be transferred from the Design Engineering Manager to the engineering organization.

The scope of the CM Program includes all IROFS identified in the ISA and any items which may affect the function of the IROFS. The scope of documents included in the CM program will expand throughout the design process. As described by the applicant, design documents that are subject to CM include calculations, safety analyses, design criteria, engineering drawings, system descriptions, technical documents, and specifications that establish design requirements for IROFS. During construction, the scope of CM includes documents such as vendor data, test data, inspection data, and applicable procedures. The CM Program will also address those documents generated through functional interfaces of the CM Program with other management measures, including QA, procedures, incident investigation audits and assessments, maintenance, records management, document control, training, and qualifications. The applicant will establish CM procedures that will provide for evaluation, implementation, and tracking of changes to IROFS, as well as processes, equipment, computer programs, and activities of personnel that impact IROFS.

The applicant stated that the CM objectives are to ensure design and operation within the design basis of IROFS. As described by the applicant in the LA Chapter 11 (IIFP, 2012a), configuration control will be implemented through controlling procedures which address the preparation, review (including interdisciplinary review), verification, approvals, and distribution for use of the detailed design. Engineering documents will be assessed for QA level classification. Changes to the approved design are subject to review to ensure consistency with the design basis of IROFS.

The applicant will perform periodic audits and assessments of the CM Program and of the design to ensure that the system meets its goals and that the design is consistent with the design basis. The result of audits and assessments, or incident investigations, may result in the development of prompt corrective actions. The applicant's corrective action process will be in accordance with the IIFP Quality Assurance Program Description (QAPD) (IIFP, 2012c) and associated procedures.

Activities included in CM are conducted during design control provisions, in accordance with a systematic process of preparation, review, and approval. The process will ensure consistency between the design and design bases of IROFS. In addition, CM includes the activities for IROFS during operations to ensure that the activities are within the limits and constraints established in the ISA. The CM Program also ensures that changes to the facility are controlled and are in accordance with the requirements of 10 CFR 70.72. Finally, CM includes the training records of personnel to ensure that only qualified personnel will be performing activities associated with the operation of IROFS.

The applicant adequately described how CM will be implemented for management measures throughout the organizational structure and staffing interfaces. The DB Engineering Manager will administer CM during design and construction, with input from the engineers responsible for each discipline. The Configuration Manager will have the responsibility to ensure the appropriate conduct of interdisciplinary reviews.

The applicant stated that the primary IIFP contractors will be responsible for development of their respective QA programs and CM elements. The contractor's programs must be consistent with the requirements of the IIFP QA Program. Section 11.1.1.4 of the LA (IIFP, 2012a) documents and describes the interfaces among IIFP and contractors performing quality-related activities. IIFP and contractors are responsible for identifying problems with quality and elevating their concerns to the appropriate management, if there is a potential problem with a system component.

<u>Design Requirements</u>

The Design Engineering organization establishes and maintains the design requirements during design and construction; the Engineering Manager establishes and maintains the design requirements during operations. The applicant stated that the CM Program would document and control design requirements.

The application stated that IROFS and any items that affect the function of the IROFS will be designated as QA Level 1 or QA Level 2. The ISA Summary lists the IROFS (IIFP, 2012b). The applicant committed in the LA, Section 11.1.2, "Design Requirements," to augment and maintain the IROFS list, as appropriate, so that it remains current during detailed design of the facility (IIFP, 2012a). The design documents associated with IROFS will be subject to interdisciplinary reviews and design verification. In addition, the applicant will evaluate changes to the design to ensure consistency with the design basis. Any computer codes used in the design of IROFS will be subject to the same design control measures, with additional requirements as appropriate for software control, verification, and validation.

The applicant stated that qualified individuals will prepare design documents—including the appropriate codes, standards, and licensing commitments—and identify any deviations or changes in the design documentation package. The application states that an additional qualified individual will review the design documentation package for concept and conformity to design inputs. The manager having overall responsibility for the design function will approve the document, while the Configuration Manager will ensure that the designated engineering organization documents the entire review process in accordance with approved procedures. This will ensure that the design documents specify appropriate quality standards, including quantitative or qualitative acceptance criteria. The QA Manager will audit the design control process through augmented audit teams.

Qualified individuals, other than those who performed the design but who may be from the same organization, will perform the design verification. Any verification performed by supervisors of the individuals doing the design will be documented and approved in advance by the supervisor's management. The applicant stated that any supervisor approving a design may not specify a singular design approach or rule out certain design considerations. The supervisor may not establish the inputs used in the design or be the only individual in the organization to perform the verification. The independent design verification will be accomplished before the design document is used by other organizations for design work or to support other activities, such as procurement, construction, or installation. However, when the independent verification is not practical because of time constraints, the applicant will identify and control the unverified portion of the document. The applicant stated that, in all cases, the design verification will be completed before relying on the item to perform its function or before the installation becomes irreversible. The applicant will review, check, and approve changes to the design and procurement documents commensurate with the original approval requirements.

The Corrective Action Program and its procedures will address design deficiencies affecting IROFS. The responsible manager will receive the nonconformance report in order to complete a review of the problem and assess the extent of the negative conditions. As required, the engineering organization will be informed for coordination of any necessary revisions.

The applicant described the design interface and stated that it will be maintained by communication among the principal users. During the operational phase, any design changes will be provided to pertinent personnel to ensure the correct performance of their duties.

<u>Document Control</u>

The applicant stated that it had established procedures to control the preparation and issuance of documents, such as the ISA, and all procedures that pertain to IROFS. IIFP will develop procedures for training, QA, maintenance, audits and assessments, emergency operating procedures, emergency operating plans, system modification documents, assessment reports, and others that seem applicable as part of CM. The applicant will put in place measures to ensure that documents—including revisions—are adequately reviewed, approved, and released for use by authorized personnel. The document control procedures establish the distribution requirements and controls to ensure that documents are transmitted and received in a timely manner at the appropriate locations. The controlled copies are distributed to the persons performing the activity for their use.

The applicant uses an electronic document management system to file project records and to make available the official (controlled) copy of the current documents. The system indexes the controlled documents through unique numbering of the documents, including revision numbers. If hard copies are needed, the applicant will provide them in accordance with the applicable procedures. Superseded or cancelled documents are appropriately labeled and maintained as records for the life of the project or the termination of the license, whichever occurs later.

<u>Change Control</u>

The applicant stated that procedures will control changes to the technical baseline. During the design stage of the project, the applicant will control the changes through the IIFP QAPD design control process (IIFP, 2012c). The applicant committed to an appropriate level of technical, management, and safety review and approval before implementation. The process will include interdisciplinary reviews as the applicant's primary mechanism to ensure consistency of the design with the design basis. During construction and operations, the applicant will perform the appropriate reviews to ensure (1) consistency with the design basis of IROFS and (2) consistent with the ISA, the design is constructed and operated or modified within the limits of the design basis.

The applicant will control changes to the design to ensure consistency through a systematic review of the design basis for the ISA and any other affected documents. During the design stage, the applicant committed to a systematic interdisciplinary review to ensure consistency between documents, including the design changes and safety assessment. Interdisciplinary reviews will also ensure that the design changes either (1) do not impact the ISA, (2) are accounted for in subsequent changes to the ISA, or (3) are not approved or implemented.

During the construction stage, the applicant committed to document, review, approve, and post any changes to documents issued for construction, fabrication, and procurement against each affected design document. The changes will continue to be evaluated through interdisciplinary review to ensure compliance with procurement specifications and drawings. As required by 10 CFR 70.72, the applicant must implement a change control process that includes the reporting of changes to the ISA made without prior NRC approval and must submit a license amendment for changes that require Commission approval before implementation.

During the operations stage, the applicant committed to document, review, and approve any design changes before their implementation. For changes made to the ISA, in accordance with 10 CFR 70.72, the applicant must have measures in place to ensure that changes to the onsite documentation are made promptly to avoid inadvertent access by facility personnel to outdated information that may affect the performance of their duties. The applicant stated that it had established measures to ensure that the quality of the facility's SSCs is not compromised by planned changes and that the system will maintain its quality through the modification. Administrative procedures that are approved by the ESH Manager or designee describe the modification process. The modification procedures will include the technical and quality requirements that must be met in order to implement the modification, as well as the requirements for initiating, approving, monitoring, designing, verifying, and documenting the modification.

The applicant will evaluate the modifications to ensure consistency among the facility's procedure, personnel training, testing program, and regulatory documents. Other areas that the applicant will consider during the evaluation of changes or modifications may include, but are not limited to, radiation exposure, lessons learned, QA aspects, potential operability or maintainability concerns, constructability concerns, post modification testing, environmental considerations, human factors, and modification costs. After the completion of the modification or change, the system is tested and personnel are trained to ensure correct operations. When the system becomes operational, all of the required documentation will be distributed to operations and maintenance staff, including formal notice to all appropriate managers.

Assessments

The applicant committed in the LA, Section 11.1.6, "Assessments" (IIFP, 2012a), to conduct periodic assessments of the CM Program to determine the program's effectiveness and to correct deficiencies. Assessments and system walkdowns will be planned, conducted, and documented in accordance with procedures. Incident investigations are conducted in accordance with the QA Program and associated corrective action procedures. The applicant stated that it will develop prompt, corrective actions as a result of incident investigation or in response to adverse audit/assessment results.

11.3.2 Maintenance

The applicant described the maintenance and functional testing programs for the operations phase of the facility. In the LA, Section 11.2 (IIFP, 2012a), the applicant committed to conducting planned and scheduled maintenance to ensure that equipment and controls are maintained in a condition of readiness to perform as intended when required.

The applicant will implement measures to ensure that the quality of these IROFS is not compromised by planned changes or maintenance activities in order to provide for the continued safe and reliable operation of the facility IROFS. The applicant will use a systems based program for planning, scheduling, tracking, and maintaining records for maintenance activities that affect IROFS. The maintenance function will be implemented in accordance with approved procedures for IROFS. The applicant will require, when applicable, that contractors working on or near IROFS identified in the ISA Summary (IIFP, 2012b) follow the same maintenance guidelines described for maintenance function activities.

In terms of written procedures for maintenance involving IROFS, the applicant committed to (1) review the work performed for premaintenance activities; (2) notify affected parties before performing work and completion of maintenance work; and (3) ensure maintenance technicians follow comprehensive procedures, among others listed in Section 11.2.1 of the LA (IIFP, 2012a).

Surveillance and Monitoring

The applicant stated that it uses surveillance and monitoring to detect degradation and adverse trends of IROFS so that action may be taken before component failure. The parameters to be monitored are selected based upon the applicant's ability to detect the predominant failure modes of the critical components. The applicant will establish performance criteria to monitor plant operations, IROFS function, and component parameters. Performance criteria will also be used to demonstrate that IROFS are being effectively controlled through appropriate predictive and repetitive maintenance strategies.

IIFP will perform surveillance of IROFS at specified intervals. The surveillance frequency will be established in accordance with the IROFS degree of safety importance. The surveillance activity supports the determination of performance trends for IROFS, indicating when potential performance degradation exists, adjusting the preventive maintenance frequencies, or taking any other corrective action. Moreover, the applicant will evaluate surveillance and monitoring results to determine any impact on the ISA or any updates needed. For surveillance tests that can only be done while the equipment is out of service, the maintenance procedures prescribe the proper compensatory measures.

The applicant described how incident investigations may identify root causes of failures. The Surveillance/Monitoring and Preventive Maintenance Programs will use incident investigation results as lessons learned when appropriate. IIFP will maintain records showing the current surveillance schedule, performance criteria, and test results in accordance with the records management system.

Corrective Maintenance

The applicant's corrective maintenance function will involve repairs or replacement of equipment that has unexpectedly degraded or failed. The corrective maintenance function provides a planned, systematic, integrated, and controlled approach for the repair and replacement activities associated with IROFS. After conducting corrective maintenance and before returning IROFS to operational status, the applicant will conduct a functional test to ensure that an IROFS will perform its intended safety function. If the performance of a replaced or repaired component differs from the original, the CM Program will review the change and preoperationally test it to ensure that the item will perform its safety function. The applicant will evaluate corrective maintenance results to determine any impact on the ISA or any updates needed.

Preventive Maintenance

The applicant stated that preventive maintenance (PM) includes preplanned and scheduled periodic refurbishment; partial or complete overhaul; or replacement of IROFS, if necessary, to ensure their continued safety function. The applicant will use the results of surveillance and monitoring, including failure history and the review of the records of IROFS failures, to plan PM activities.

As described in the application, the determination of PM frequency will seek to balance the objective of preventing failures through maintenance against the objective of minimizing the availability of IROFS because of PM. Feedback will be collected from PM, corrective maintenance, results from incident investigation, and identified root causes. The applicant will use the feedback, as appropriate, to modify the frequency or scope of PM activities. IIFP will document the rationale for any PM deviations from industry standards or vendor recommendations.

The applicant addressed calibration and testing within the PM function. Facility personnel will use PM Program procedures and calibration standards to calibrate equipment and monitor devices to plant safety and safeguards. The applicant will provide compensatory measures during testing of nonredundant IROFS to ensure that their function is performed until the item is back in service.

After conducting PM and before returning QA Level 1 or QA Level 2 items to operational status, the applicant will conduct a functional test, if necessary, to ensure that the item will perform its intended safety function. The applicant will evaluate PM activity results to determine their impact on the ISA or if any updates needed. IIFP will maintain records pertaining to PM of IROFS, and items affecting IROFS, in accordance with the records management system.

Functional Testing

The applicant stated that it will perform functional testing of IROFS, as appropriate following initial installation, as part of periodic surveillance testing and after corrective or preventive maintenance or calibration. The purpose of this testing will be to ensure that the item is capable of performing its safety function when required. The applicant committed in the LA, Section 11.2.2.4 (IIFP, 2012a), to performing functional tests in accordance with appropriate administrative controls and approved written procedures. The procedures define the method for the test and the required acceptable results that will be recorded and maintained as quality records.

The applicant's overall testing program is divided into two major testing programs: preoperational testing and post maintenance testing. The applicant described the preoperational testing program as testing conducted to initially determine various facility parameters and to initially verify the capability of SSCs to meet performance requirements. As stated by the applicant, the major objective of preoperational testing is to verify that IROFS are capable of performing their intended function.

Initial startup testing is part of the preoperational program and will be performed beginning with the introduction of deplete uranium hexafluoride (DUF_6) and ending with the startup of depleted uranium tetrafluoride (DUF_4), silicon tetrafluoride (SiF_4), and boron trifluoride (BF_3) operations. The applicant stated that the purpose of initial startup testing is to ensure safe and orderly DUF_6 feeding and to verify parameters assumed in the ISA. The applicant will maintain records of the preoperational and startup tests required before operation.

The operational testing program will consist of periodic and special testing. The periodic testing will be conducted to monitor various facility parameters and verify the continuing integrity and capability of IROFS. IIFP may conduct special testing at the facility in a nonrecurring nature, which does not fall under any other category of testing program.

The applicant will also establish a post maintenance testing program to provide assurance that IROFS will perform their intended function following maintenance activities. These tests will confirm that the maintenance performed was satisfactory, the identified deficiency was corrected, and the maintenance activity did not adversely affect the reliability of the IROFS. Before returning the equipment to service, the applicant will perform post maintenance testing to ensure that the equipment is ready for operations. If no acceptable results are obtained, the applicant will conduct and document corrective actions through the nonconformance report process.

11.3.3 Training and Qualifications

Section 11.3 of the application describes the training program for operations of the facility, including training for the preoperational functional testing and initial startup testing. The applicant will provide training to each individual commensurate with his or her role and responsibilities during the design, construction, and operations phases. The QA training will be for personnel performing QA Level 1 and QA Level 2 work activities; personnel performing nondestructive examination, inspection, and testing; and QA auditors.

The applicant stated that the principal objective of the training program is to ensure job proficiency of all facility personnel through effective training and qualification. The training program will be designed to accommodate future growth and meet commitments with applicable established regulations and standards. In addition, employees will be provided with training and continuing training, as applicable, to establish and maintain knowledge foundations and develop work performance skills.

As described by IIFP, qualification will be indicated by successful completion of prescribed training, demonstration of the ability to perform assigned tasks, and maintenance of requirements established by regulation. The applicant will use a graded approach to systematic training incorporating methods to accomplish the analysis, design, development, implementation, and evaluation of training. Exceptions from training requirements might be granted when justified, documented and approved in accordance written procedures.

11.3.3.1   Organization and Management of the Training Function

The applicant stated that line managers are responsible and have commensurate authority to develop and effectively conduct training. The training function will support line managers by facilitating the planning, directing, analyzing, developing, conducting, evaluating, and controlling of the performance-based training process.

The applicant will have facility procedures which specify the requirements for training of personnel performing activities related to IROFS. Procedures will allow for exceptions from training only when properly justified, documented, and approved. The applicant will update any lesson plans affected through the change control process of the CM Program when design changes or facility modifications are implemented.

The applicant will maintain programmatic and individual training records to support management information needs associated with personnel training, job performance, and qualifications. Individual records will include general employee training, technical training, and employee development training. This training will be conducted at the facility and may include special company-sponsored training conducted by others. The applicant will maintain training records in an accurate and retrievable method; such records will be retained in accordance with the records management procedures.

11.3.3.2   Analysis and Identification of Functional Areas Requiring Training

The applicant will perform a needs/job analysis and will identify tasks to ensure that it will provide appropriate training to personnel working on tasks related to IROFS. For each task, the needs/job analysis will include job hazards that are referred to as precaution and limitations. The applicant will evaluate the effectiveness of the training and materials and will update the list of activities selected for training, as necessitated by changes in procedures, processes, plant systems, equipment, or job scope.

11.3.3.3   Position Training Requirements

The applicant stated that it will develop minimum training requirements for positions whose activities are related to IROFS. IIFP will provide training for personnel of various abilities and experience backgrounds. Continuing training courses will be established to ensure that the appropriate personnel remain proficient.

The objective of the training is to ensure safe and efficient operation of the facility and compliance with applicable established regulations and requirements. Training requirements, as described by the applicant, will be applicable to, but not necessarily restricted to, those personnel within the plant organization who have a direct relationship to the operation, maintenance, testing, or other technical aspect of the facility IROFS.

11.3.3.4   Training Basis and Objectives

The applicant will design its training program to prepare initial and replacement personnel for safe, reliable, and efficient operation of the IIFP facility. The applicant will emphasize the safety requirements where human actions are important to safety. Established learning objectives will identify the training content and define satisfactory trainee performance for the task, or a group of tasks, selected for training from the job analysis. Learning objectives will be sequenced within training materials based on the relationship to one another and documented in lesson plans and training guides.

11.3.3.5   Organization of Instruction, Using Lesson Plans, and Other Training Guides

The applicant will develop lesson plans or other approved process-controlling documents from the learning objectives based on job performance requirements and under the guidance of the training functional organization. These documents will be reviewed and approved before they are issued or used by the training function and, generally, by the organization cognizant of the subject matter. Lesson plans or other approved process-controlling documents will be used for classroom training and on-the-job training, as required, and will include standards for evaluating acceptable trainee performance.

11.3.3.6   Evaluation of Trainee Learning

The applicant will use observation, demonstration, or oral or written tests to evaluate the trainee's understanding and command of learning activities. Individuals qualified in the subject matter will perform the evaluations.

11.3.3.7   Categories of Required Training

The applicant described the different categories of required training in the LA, Section 11.3.7.1 through the LA, Section 11.3.7.5 (IIFP, 2012a), which include the following:

- General Employee Training (GET): GET includes QA, radiation protection, environment, health and safety, emergency response, and administrative policies established by facility management and applicable regulations, among others. Unescorted access to the controlled access area is permitted only after completion of the appropriate level of GET.

- Radiological Safety Training: The radiation safety training program includes topics such as radiation and radioactive materials; risks involved in receiving low-level radiation exposure in accordance with 10 CFR 19.12, "Instruction to Workers"; notices, reports, and instruction to workers; inspection and investigation; and the basic criteria and practices for radiation protection.

- Industrial Safety/Industrial Hygiene: The applicant will conduct industrial safety/industrial hygiene training (other than those general aspects that are addressed in GET), including various aspects of industrial and chemical safety protection, as necessary, to train new employees in specific job duties and to provide refresher training topics to workers depending on employee job responsibilities.

- Emergency Preparedness/Security and Fire Brigade: This category includes the details on the development, maintenance, and implementation of the IIFP Emergency Plan (EP) and EP implementing procedures. It also includes the initial training for all new fire brigade members, classroom refresher training and drills, annual practical training, and leadership training for fire brigade and incident commanders.

- Technical Training: The applicant will design, develop, and implement technical training to assist facility employees in gaining an understanding of applicable fundamentals, procedures, and practices related to IROFS and in developing skills necessary to perform assigned work related to IROFS. It consists of four segments: initial training, on-the-job training and qualifications, continuing training, and special training.

11.3.3.8   Evaluation of Training Effectiveness

The applicant will evaluate the training program periodically to measure program effectiveness. The evaluation will consider feedback provided from trainees after completion of the classroom training sessions for program improvement. It will determine strengths and weaknesses, whether the program content matches current job needs, and whether corrective actions are needed to improve the program's effectiveness.

The training function will be leading any training program evaluations and implementing the corrective actions. Evaluation results will be documented highlighting the program's strengths and weaknesses. The applicant will review identified weaknesses; recommend improvements; and make changes to the affected procedures, practices, or training materials.

The QA department will audit the facility system and training program. Trainees and vendors can provide input related to the facility training program through surveys, questionnaires;

performance appraisals, staff evaluations, and overall training program effectiveness evaluation techniques.

11.3.3.9   Personnel Qualification

The applicant will provide the training and qualification requirements for QA personnel and key management positions. The applicant will establish and implement qualification and training requirements for process operator candidates in plant procedures.

11.3.3.10   Provisions for Continuing Assurance

The applicant will establish applicable continuing education or periodic retraining to ensure personnel proficiency and retention of knowledge and skills which are important to Operations. Personnel who perform activities relied on for safety will be evaluated periodically by written test, oral test, or on-the-job performance to determine whether they are capable of continuing performing these activities. The applicant will provide retraining or other appropriate action when results of the evaluations dictate the need. In addition, the applicant will provide retraining to personnel if any new or revised information results in plant modifications, procedure changes, and QA program changes. Unacceptable, individual performance will be transmitted to the appropriate line manager.

11.3.4 Procedures Development and Implementation

The applicant will conduct all activities involving IROFS or QA Level 1 and QA Level 2 items in accordance with approved procedures. Compliance with procedures is mandatory. The applicant will use procedures to control IROFS activities.

11.3.4.1   Type of Procedure

As stated by the applicant, procedures will be categorized as management control procedures or operating procedures/instructions. Management control procedures are those that will describe administrative and general practices approved and issued by management at a level appropriate to the scope of the practice, directing and controlling activities across the various organizational functions. Operating procedures will provide specific direction for task-based work and will be used to directly control process operations at the work place.

The applicant described four types of plant procedures that it will use to control QA Level 1 and QA Level 2 plant and administrative activities: (1) operating procedures, (2) management control procedures, (3) maintenance procedures, and (4) emergency procedures. It is clearly stated in the application that if any aspect of a procedure is unclear or incorrect as written, personnel shall safely stop the operation and/or activity, contact management and shall not restart until corrective action has been taken.

11.3.4.1.1   Operating Procedures

As described by the applicant, operating procedures will be used to directly control process operations. The operating procedures will include, among other things, the (1) purpose of the activity; (2) governing regulations; polices; and guidelines; (3) type of procedure; (4) steps for each operating process stage; (5) hazards and safety considerations; (6) operating limits; (7) measures to prevent exposure; (8) associated IROFS and their functions; and (9) the timeframe for which the procedure is valid.

The applicant states that it will implement a methodology for identifying, developing, approving, implementing, and controlling operating procedures. The method will ensure, as a minimum, that, (1) ISA results are considered when identifying needed procedures; (2) operating limits and IROFS are specified in the procedure; (3) procedures include required actions for off-normal conditions of operation; as well as normal operations; (4) if needed; safety checkpoints are identified at appropriate steps in the procedure; (5) procedures are validated through field tests; (6) procedures are approved by management personnel responsible and accountable for the operation; (7) mechanisms are specified for revising and reissuing procedures in a controlled manner; (8) QA elements and CM Program at the facility provide reasonable assurance that current procedures are available and used at all work locations; and (9) the facility training program trains the required persons in the use of the latest procedures available.

11.3.4.1.2   Management Control (Administrative) Procedures

As described by the applicant, management control procedures will provide programmatic requirements for policy or programs and administrative systems. The applicant will use management control procedures to perform activities that support production and control processes with IROFS or hazardous chemicals, or both, incident to the processing of licensed material. Management control procedures will include management measures such as (1) CM; (2) industrial safety/industrial hygiene; radiation safety; chemical safety; and fire safety; (3) QA; (4) design control; (5) facility personnel training and qualification; (6) audits and assessments; (7) incident investigations; (8) recordkeeping and document control; (9) reporting; and (10) procurement.

The procedures for construction will be an integral part of the procedures of the DB contractor that IIFP will choose. In case the DB contractor uses its own QA program, instead of IIFP's QA Program, the applicant will review and approve the QA program that will be used by the DB contractor.

11.3.4.1.3   Maintenance Procedures

Maintenance procedures will address the preventive and corrective maintenance, surveillances, functional or post maintenance testing, and the requirements for premaintenance activities involving reviews of the work to be performed and reviews of procedures for accuracy and completeness. The LA, Section 11.4.1.3 (IIFP, 2012a), mentions testing, calibration, and selection and qualification of maintenance personnel as just some of the activities that will be performed in accordance with written procedures.

Various safety disciplines, including fire, radiation, and industrial and chemical process safety, will review maintenance procedures. These procedures will describe, at a minimum, the following:

- premaintenance activities

- steps that require notification of all affected parties before performing work on completion of maintenance work

- controls on and specification of any replacement components or materials to be used to ensure like-kind replacement

- functional testing to verify operability of the equipment

- tracking and records management of maintenance activities

- safe work practices (e.g., lockout/tagout, confined space entry, control of exclusion area, radiation or hot work permits, and fire, chemical, and environmental requirements).

11.3.4.1.4   Emergency Procedures

The applicant will develop and maintain a documented controlled set of EP Implementation Procedures (EPIPs) applicable to the IIFP facility. The EPIPs will include emergency instructions pertinent to specific accident scenarios and other categorized nonroutine operational events clearly stating the duties, responsibilities, action levels, and actions to be taken by responders.

The ESH Manager will review periodically the EP and will update it as needed. The applicant will not implement any changes to the plan that decrease its effectiveness without prior NRC approval. Changes that will not decrease the effectiveness of the EP may be implemented without NRC prior approval, provided the applicant submits the changes to the NRC and appropriate organizations within 3 months of making the changes.

11.3.4.2   Procedures Process

The applicant described the process of development and modification of procedures. The process will utilize the following nine basic elements to accomplish procedure development, review, approval, and control: (1) identification, (2) development, (3) verification, (4) review and comment resolution, (5) approval, (6) validation, (7) issuance, (8) change control, and (9) periodic review.

11.3.4.2.1   Identification

As described by the applicant, site managers are responsible for identifying the tasks to be included in procedures within their areas of control. Procedures are required when actions are taken necessary to prevent or mitigate the consequences of accidents described in the ISA. At a minimum, the applicant will require a procedure for any task or activity that affects QA Level 1 and QA Level 2 SSCs. The applicant will evaluate new or revised NRC certification requirements to determine their impact on existing implementing procedures or to identify the need for new implementing procedures.

11.3.4.2.2   Development

The applicant will categorize procedures in different designations depending on the administrative or nonadministrative use and the safety or financial consequences of failing to adhere to the requirements. The applicant will use the following designations: "In Hand," which means that the procedure is of continuous use; "General Intent," which means that the procedure is for reference use; or "Information Use," which means that the procedure is for information only.

The user organization will be responsible for procedure development, preparation, and quality. The applicant will require input and review by affected parties and other selected parties, such as safety and quality, to ensure that safety and QA requirements are identified and included when QA Level 1 or QA Level 2 SSCs are involved.

The applicant utilized interviews with procedure users and process walkdowns to ensure that procedures are usable, reflect as-built conditions, guide production operations, and maintain management controls for safety and quality. The applicant will ensure that each procedure identifies and incorporates regulatory commitments, ISA, and QA Program requirements.

11.3.4.2.3   Verification

The applicant will review two basic attributes during the verification process: (1) technical accuracy, including all technical information, and (2) administrative verification to verify that the procedure has the correct format and style. Procedure verification will consist of a walkdown of the procedure in the field or a tabletop walkthrough to ensure its performance as written.

11.3.4.2.4   Review and Comment Resolution

The applicant will distribute draft new procedures and procedure changes, as needed, to technical and cross-discipline individuals not having direct responsibility for processing the new procedure or procedure change for their review. The applicant also described the comments/questions process that will ensure that any comments will be resolved with the originating organizations. If the original draft changes extensively, the revised draft procedure will be reverified, and the validation will be checked. The QA function will review QA implementing procedures for compliance and consistency with the QA Program.

11.3.4.2.5   Approval

The responsible manager will approve the procedures following the resolution of any review comments. Managers will also be responsible for ensuring that necessary training or required reading is completed before procedure implementation.

11.3.4.2.6   Validation

The applicant described the purpose of procedure validation as a process to ensure that technical errors or human factor issues will not be inadvertently introduced during the procedure review process. Qualified personnel will perform and document the validation, which may be accomplished by detailed evaluation of the procedure as part of a walkthrough exercise or as part of a walkthrough drill. If the particular system or process is not available for a walkthrough validation, talk through may be performed in the particular shop or training environment.

11.3.4.2.7   Issuance and Distribution

The applicant will issue and control procedures in accordance with the applicant's records management and document control programs. The applicant states that line managers, or designees, will be responsible for ensuring personnel doing work that requires the use of procedures have access to controlled copies of the required procedures.

11.3.4.2.8   Procedures Process Change Control

The applicant described the process for changes to procedures. The preparer of the change will document the change and the reason for it. The applicant will perform an evaluation in accordance with 10 CFR 70.72, as appropriate, to determine whether the proposed changes will require an amendment to the license. If an amendment to the license is needed, the change will not be implemented until the applicant receives prior approval from the NRC. A designated reviewer will then review the procedure. The functional area manager will be responsible for approving procedure changes and for determining whether a cross-disciplinary review is necessary and by which departments. The applicant will perform interdisciplinary reviews, at a minimum, for changes involving chemical safety, radiation safety, facility safety, engineering, and QA or ESH. The applicant will maintain the records for cross-functional reviews for all changes to procedures involving licensed materials of QA Level 1 and QA Level 2 items and activities, in accordance with the LA, Section 11.7 (IIFP, 2012a), "Records Management."

The applicant described the process for temporary changes to procedures. The temporary changes will be issued for production activities that are of a nonrecurring nature or when revision of a production or other permanent procedure is not practical. Temporary changes to procedures will not involve a change to IROFS and will not alter the intent of the original procedure. Two members of the facility management staff, at least one of whom is a shift superintendent or designee, will approve the temporary changes to procedures. Temporary changes to procedures will have a designated expiration date and may be made permanent once the change is reviewed and approved through the normal procedure change and approval process.

11.3.4.2.9   Periodic Review

The procedure owner or subject matter expert will perform a complete administrative and technical review periodically to ensure that information is complete and accurate and that the procedure is usable as written. Specifically, operating procedures will be reviewed every three years while EPIPs will be reviewed annually. In addition to periodic reviews, procedures will be reviewed after unusual incidents to determine if changes are appropriate.

11.3.4.3   *Use and Control of Procedures*

The applicant described uses of the different designations of the procedures. IIFP staff will perform in-hand or continuous use procedures step by step without deviation, unless the procedure allows deviation. General Intent or Reference Use procedures will be followed as written, unless the procedure allows deviation. Information Use procedures will be followed to implement programmatic requirements.

The applicant will mark the controlled copies of procedures as "Controlled Copy" and information copies, which are not used to perform any work, as "Information Only." The applicant stated that working copies will be verified as the latest version by managing and limiting access to the most current revision of the document. Procedures will include provisions for operations to stop and place the process in safe condition, if a step of a procedure involving QA Level 1 and QA Level 2 SSCs cannot be performed as written.

11.3.4.4   Topics To Be Covered in Procedures

The LA, Section 11.4.4 (IIFP, 2012a), lists the topics covered by the different types of procedures including those used for: management controls, operations, maintenance, and emergencies. The list is not all inclusive but provides many of the activities that will be covered by operating, management control, maintenance, and emergency procedures.

11.3.4.5   Temporary Procedures

The applicant will have formal requirements governing the use of temporary procedures. The applicant will issue temporary procedures only when permanent procedures do not exist to (1) direct operations during testing, maintenance, and modifications, (2) provide guidance in unusual situations not within the scope of permanent procedures, and (3) ensure orderly and uniform production for short periods of time when the plant, a system, or component of a system is performing in a manner not covered by existing permanent procedures or has been modified or extended in such a manner that portions of existing procedures do not apply. The applicant stated that these temporary procedures may be used for a period of time which should not exceed 90 days or a period for which the temporary condition must exist, whichever is greater. If use of the temporary procedure exceeds the 90-day timeframe, the applicant will assess the procedure to ensure that it is appropriate to extend its use.

11.3.4.6   Records

The applicant will generate records that will be identified in the governing procedure and controlled according to the plant records management and document control program practices.

11.3.5 Audits and Assessments

The applicant stated that audits will be focused on verifying compliance with regulatory and procedural requirements and licensing commitments. Assessments will be focused on determining the effectiveness of activities and ensuring that IROFS are reliable and are available to perform their intended safety functions. The applicant will perform audits and assessments on critical work activities associated with facility safety, environmental protection, and other areas that it identifies.

11.3.5.1   Audits

The applicant stated that will conduct audits of the QA Level 1 and QA Level 2 work activities associated with IROFS, and any items that affect the function of the IROFS, in accordance with the QA Program, written plans, and checklists. Audits will be performed under the direction of a lead auditor. Lead auditors and staff auditors will be functionally and organizationally independent of the programs and activities audited or assessed. Audit teams may be supplemented with onsite or offsite technical specialists.

Plant procedures will specify the method in which audit results should be documented and reported. The applicant will have provisions in place for immediate reporting and implementation of corrective action, where warranted.

11.3.5.2   Assessments

As described by the applicant, management responsible for implementing the respective portions of the QA Program will perform and document assessments. They will assess the adequacy of the part of the program for which they are responsible and ensure its effective implementation. The application states that personnel may perform the assessment for their area, provided that they do not have direct responsibility for the specific area being evaluated. The assessment program contains management assessments and independent assessments.

Management assessments review occupational safety and health, radiological protection, environmental compliance, fire safety, emergency preparedness and security, safety requirements, conduct of operations, conduct of maintenance, training, QA, maintenance, and CM, among other programs. Qualified staff personnel that are not directly responsible for production activities will routinely perform independent assessments. Any deficiencies identified during the assessment will be forwarded to the corrective action program and addressed by the responsible manager of the applicable area.

11.3.5.3   Conduct of Audits and Assessments

The frequency of audits and assessments will be based on the safety importance of the activities, the status of the activity, and on the work history. IIFP independent qualified auditors from the QA and technical organizations within ESH will conduct audits in accordance with written plans, procedures, and checklists. Audit frequency will be determined by the status and safety importance of the activities being performed. As stated by the applicant, audit team members will be indoctrinated in audit techniques, in addition to being independent to the function and area being audited. The applicant will also require auditors to have technical expertise or experience in the area being audited.

Any deficiencies identified during the audit or assessment will be forwarded to the corrective action program and be addressed by the responsible manager of the applicable area or function in accordance with the Corrective Action Program (CAP) procedure. In addition, future audits and assessments must include a review to evaluate the effectiveness of corrective actions taken to address any deficiencies found.

The results of the audits will be provided in a written report to the Plant Manager and provided to the managers responsible for the activities audited or assessed. The applicant will maintain records of the instructions and procedures, persons conducting the audits or assessments, and corrective actions taken.

11.3.5.4   Activities Subject to Audit and Assessments

The applicant will conduct audits and assessments in the areas of radiation safety, chemical safety, industrial safety (including fire protection), environmental protection, emergency management, QA, CM, maintenance, training and qualification, procedures, CAP/incident investigation, and records management.

11.3.5.5   Scheduling of Audits and Assessments

The applicant will establish a schedule identifying the audits and assessments to be performed and the responsible organization assigned to conduct the activity. All major activities will be audited or assessed on a periodic basis. The applicant will review the schedule periodically and revise it as necessary to ensure coverage commensurate with current and planned activities.

11.3.5.6   Procedures for Audits and Assessments

The applicant stated that it will conduct internal and external audits and assessments using approved procedures that meet QA Program requirements. The procedures will provide requirements for audit and assessment activities, such as scheduling and planning, certification requirements of audit personnel, development of audit plans, any applicable checklists, performance of the audit and assessment, reporting and tracking of findings to closure, and closure of the audit and assessment. The applicable procedures will emphasize reporting and correction of findings to prevent recurrence.

The applicant will require the team leader to develop the audit or assessment report to document the findings, observations, and recommendations for program improvement. The applicant will require that responsible managers review the reports and provide any required responses based on the reported findings. The CAP will track the audit and assessment results data, which the applicant will periodically analyze for potential trends.

11.3.5.7   Qualification and Responsibilities for Audits and Assessments

The applicant stated that the QA Manager will initiate the audits and, in coordination with the responsible lead auditor, will determine the scope of each audit. The QA Manager may also initiate special audits or expand the scope of audits, if appropriate.

The lead auditor, as described by the applicant, will direct the audit team in developing checklists, instructions, or plans and performing the audit. Lead auditors will have to participate in a minimum of three QA audits or audit equivalent within a period of time not to exceed 3 years before the date of certification. Audit equivalents include assessments, pre-award evaluations, or comprehensive surveillances (provided the prospective lead auditor took part in the planning, checklist development, performance, and reporting of the audit equivalent activities). One of the audits must be a nuclear-related QA audit or audit equivalent within the year preceding certification.

Audits will be conducted in accordance with the checklists, but the audit team may expand the scope during the audit. The audit team will consist of one or more auditors who are responsible for performing audits in accordance with the applicable QA procedures. The auditors and lead auditors must hold certifications as required by the QA Program. Before certification under the IIFP QA Program, auditors will have to complete training on the IIFP QA Program; audit fundamentals, including audit scheduling, planning, performance, reporting, and follow-up action involved in conducting audits; objectives and techniques of performing audits; and on-the-job training. Auditors and lead auditors will be certified based on the QA Manager's evaluation of education, experience, professional qualifications, leadership, sound judgment, maturity, analytical ability, tenacity, and past performance and completion of QA training courses.

The applicant will not require personnel performing assessments to be certified; but they will be required to complete QA orientation training, as well as training on the assessment process.

11.3.6 Incident Investigations and Corrective Action Process

The applicant described the incident investigations and corrective action process in Section 11.6 of the application (IIFP, 2012a).

The applicant described the incident investigation process in the LA, Section 11.6.1 (IIFP, 2012a), as a simple mechanism available for reporting deficiencies, abnormal events, and potentially unsafe conditions or activities to assure that the upset condition is understood and appropriate corrective actions are implemented to prevent recurrence. The applicant will consider each event in terms of its requirements for reporting in accordance with regulations and will evaluate the event to determine the level of investigation required. Written incident investigation and CAP procedures will address the applicant's process of incident identification, investigation, root-cause analysis, recording, reporting, and followup. Incident investigation procedures will include guidance for classifying occurrences, including examples of threshold off-normal occurrences. The depth of the investigation will depend, as explained by the applicant, upon the severity of the classified incident in terms of the levels of uranium or chemicals released or the degree of potential for exposure of workers, the public, or the environment.

The QA Manager will maintain a record and track to completion corrective actions to be implemented as a result of off-normal occurrence investigations in accordance with CAP procedures.

The incident investigation process, as described by the applicant, will establish a process to investigate abnormal events that may occur during operation of the facility to determine their specific or generic root causes and generic implications, to recommend corrective actions, and to report to the NRC as required by 10 CFR 70.50, "Reporting Requirements," and 10 CFR 70.74, "Additional Reporting Requirements." The investigation process will include a prompt, risk-based evaluation; and depending on the complexity and severity of the event, one individual may suffice to conduct the evaluation. The investigators will be independent from the line functions involved with the incident under investigation. Investigations will begin within 48 hours of the abnormal event, or sooner, depending on the event's safety significance. The record of IROFS failures required by 10 CFR 70.62(a)(3) will be reviewed as part of the investigation. The applicant will make record revisions necessitated by post failure investigation conclusions, accordingly.

The applicant will appoint qualified internal or external investigators to serve on investigating teams when required. The teams will include at least one process expert and at least one team member trained in root-cause analysis.

The applicant will monitor and document corrective actions through completion. The applicant will maintain auditable records and documentation related to abnormal events, investigations, and root-cause analyses so that "lessons learned" may be applied to future operations of the facility. For each abnormal event, the incident report will include a description, contributing factors, a root-cause analysis, findings, and recommendations. Details of the event sequence will be compared with accident sequences already considered in the ISA. The applicant will modify the ISA Summary (IIFP, 2012b) to include evaluation of the risk associated with accidents of the type actually experienced.

The applicant will develop procedures for conducting incident investigations such as the following:

- a documented plan for investigating an abnormal event

- a description of the functions, qualifications, and responsibilities of the manager who would lead the investigative team and those of the other team members; the scope of the team's authority and responsibilities; and assurance of management cooperation

- assurance of the team's authority to obtain all of the information considered necessary and its independence from responsibility to the functional area involved in the incident under investigation

- retention of documentation relating to abnormal events for 2 years or for the life of the operation, whichever is longer

- guidance for personnel conducting the investigation on how to apply a reasonable systematic, structured approach to determine the specific or generic root causes and generic implications of the problem

- requirements to make available original investigation reports to the NRC upon request

- a system for monitoring the completion of appropriate corrective actions.

In the LA, Section 11.6.2 (IIFP, 2012a), the applicant described the corrective action program. IIFP identified the responsibilities and provided the authority for those individuals involved in quality activities to identify any condition adverse to quality. These individuals will identify and document conditions adverse to quality, analyze and determine how the conditions can be corrected or resolved, and take the necessary steps to implement corrective actions in accordance with documented procedures.

Employees will have the authority and responsibility to initiate the corrective action process if they discover deficiencies. The QA Program will contain procedures for identifying, reporting, resolving, documenting, and analyzing conditions adverse to quality. The applicant will analyze reports of conditions adverse to quality to identify trends in quality performance. Significant conditions adverse to quality and significant trends will be reported to senior management, in accordance with CAP procedures. The QA Manager will take follow-up actions to verify proper and timely implementation of the corrective actions.

11.3.7 Records Management and Document Control

The applicant will establish records management and document control programs to ensure that records and documents required by the IIFP QA Program are appropriately managed and controlled through standard methods and requirements for collecting, maintaining, and disposing of records. The programs will include administrative controls for the generation and revision of records and documents. The principal elements and a brief description of each are provided within the application and will be discussed below.

The baseline design requirements in 70.64 require applicants to provide for adequate records of management measures and IROFS to facilitate their remaining available and reliable. The records management program described below are consistent with 70.64(a)(1).

11.3.7.1 Records Management Program

The applicant will apply the Records Management Program to QA Level 1 and QA Level 2 SSCs and activities. Records management may also apply to records related to ESH, financial, quality, emergency response, or investigation records, as required by regulations or approved procedures. The Records Management Program direct the handling, transmittal, storage, and retrieval of records. Records will be categorized and handled in accordance with their relative importance to safety and storage needs. A records organization will be responsible for the administration of the Records Management Program. However, the managers and functional organizations that generate the records will be responsible for ensuring compliance with the Records Management Program.

The applicant described the program elements that will be implemented through procedures in Sections 11.7.1.1 through 11.7.1.14 of the application (IIFP, 2012a), including provisions for the following:

- legibility, accuracy, and completeness

- identification of items and activities

- authentication

- indexing and filing

- retention and disposition

- corrections

- protection of records

- storage requirements

- receipt of records

- access to records and accountability for removed records

- records requirements for procured goods or services

- types of records

- usage and control of computer codes and data

- assessment.

11.3.8 Quality Assurance Program Elements

The provisions provided by the applicant in the QAPD (IIFP, 2012c) apply during the design and construction and will remain in effect during operations. The QAPD will be mandatory for items identified as IROFS. The LA, Section 11.8, "Quality Assurance Program Elements" (IIFP, 2012a), includes a brief description of the QA Program elements that are discussed in detail in the LA, Appendix A, "Quality Assurance Program Description" (IIFP, 2012c). The introduction of Appendix A includes the IIFP commitment to ensure safety for employees and the public relative to its facility operations. The NRC staff conducted a review of the QAPD (IIFP, 2012c), and found it acceptable. The NRC staff's report of its review is included in Chapter 12 to this SER.

11.4   Evaluation Findings

11.4.1 Configuration Management

The NRC staff has reviewed the CM function for IIFP Facility according to Chapter 11 of NUREG-1520. The NRC staff found that the applicant's description of the overall CM program appropriately covered CM policy, design requirements, document control, change control, and assessments. Based on this evaluation, the NRC staff finds the applicant's CM Program acceptable.

The applicant has suitably and acceptably described its commitment to a proposed CM system, including the method for managing changes in procedures, facilities, activities, and equipment for IROFS. Management-level policies and procedures, including an analysis and independent safety review of any proposed activity involving IROFS, are described and will provide reasonable assurance that consistency among design requirements, physical configuration, and facility documentation is maintained as part of a new activity or change in an existing activity involving licensed material. The management measures will include the following elements of CM:

- CM Management: The applicant will develop the organizational structure, procedures, and responsibilities necessary to effectively implement CM.

- Design Requirements: The applicant will document the design requirements and bases, which will be supported by analyses. The applicant will maintain all design documentation current.

- Document Control: The applicant will appropriately store its documents, including drawings, so that they are accessible. Drawings and related documents captured by the system will include those necessary and sufficient to adequately describe IROFS.

- Change Control: Responsibilities and procedures will adequately describe how the applicant will achieve and maintain strict consistency among the design requirements, the physical configuration, and the facility documentation. Methods will be in place for suitable analysis, review, approval, and implementation of identified changes to IROFS, including appropriate CM controls.

- Assessments: The applicant has committed to an adequate function that includes both initial and periodic assessments. The assessments will verify and ensure the adequacy of the CM function.

11.4.2 Maintenance

The applicant committed to maintaining IROFS. The applicant's maintenance commitments contain the basic elements to ensure availability and reliability: surveillance/monitoring, corrective maintenance, PM, functional testing, equipment calibration, and work control for maintenance of IROFS. The applicant's maintenance function is proactive, using maintenance records, PM records, and surveillance tests to analyze equipment performance and to seek the root causes of repetitive failures.

The surveillance and monitoring, PM, and functional testing activities described in the LA provide reasonable assurance that the IROFS identified in the ISA summary will be available and reliable to prevent or mitigate accident consequences.

The maintenance function (1) will be based on approved procedures; (2) will employ work control methods that properly consider personnel safety; awareness of facility operating groups, QA, and the rules of CM; (3) will use the ISA summary to identify IROFS that require maintenance and determine the level of maintenance needed; (4) will justify the PM intervals in terms of the equipment reliability goals; (5) will provide for training that emphasizes the importance of IROFS identified in the ISA summary, regulations, codes, and personnel safety; and (6) will create documentation that includes records of all surveillance, inspections, equipment failures, repairs, and replacements of IROFS.

The NRC staff concludes that the applicant's maintenance functions meet the requirements of 10 CFR Part 70, provide reasonable assurance of public health and safety and the protection of the environment, and is acceptable.

11.4.3 Training and Qualification

Based on its review of the applicant's strategy for developing and implementing the training and qualification program and a comparison of this strategy to the acceptance criteria guidance in NUREG-1520, the NRC staff concludes that the applicant has adequately described and assessed its personnel training and qualification in a manner that (1) satisfies regulatory requirements and (2) is consistent with the guidance in NUREG-1520. The NRC staff's evaluation found that the applicant's description of its training program appropriately covered (1) organization and management of the training function, (2) analysis and identification of functional areas requiring training, (3) position training requirements, (4) training basis and objectives, (5) organization of instruction, (6) evaluation of trainee learning, (7) conduct of on-the-job training, (8) evaluation of training effectiveness, (9) personnel qualification, and (10) periodic personnel evaluation. Based on this evaluation, the NRC staff concludes that the applicant has adequately described and assessed its personnel training and qualification in a manner that satisfies the regulatory requirements and is consistent with the guidance in NUREG-1520.

There is reasonable assurance that implementation of the described training and qualification will result in personnel who are qualified and competent to design, construct, start up, operate, maintain, modify, and decommission the facility safely. The NRC staff concludes that the applicant's plan for personnel training and qualification meets the requirements of 10 CFR Part 70 and are acceptable.

11.4.4 Procedures

The application described a suitably detailed process for the development, approval, and implementation of procedures. The applicant addresses IROFS, as well as items important to the health of facility workers and the public and protection of the environment. The NRC staff concludes that the applicant's plan for developing and implementing procedures meets the requirements of 10 CFR Part 70 and are acceptable.

11.4.5 Audits and Assessments

Based on its review of the safety analysis report and a comparison of the applicant's plan and processes to the review acceptance criteria in NUREG-1520, the NRC staff concludes that the applicant has adequately described its audits and assessments. The NRC staff has reviewed the applicant's implementation process for audits and assessments and its description of IIFP's policy directives, plans, and procedural requirements, considering the following: (1) the general structure of the audits and assessments program, (2) the activities to be audited or assessed, (3) the scheduling of audits and assessments, (4) the procedures for audits and assessments, and (5) the qualifications and responsibilities for audits and assessments.

The NRC staff concludes that the applicant's plan for audits and assessments meets the requirements of 10 CFR Part 70 and provides reasonable assurance of protection of the health and safety of the public, workers, and the environment.

11.4.6 Incident Investigations

The applicant has committed to and established an organization responsible for (1) performing incident investigations of abnormal events that may occur during operation of the facility; (2) determining the root causes and generic implications of the event; and (3) recommending corrective actions for ensuring a safe facility and safe facility operations, in accordance with the acceptance criteria of Section 11.4 of NUREG-1520.

The applicant has committed to monitoring and documenting corrective actions through to completion. The applicant has also committed to taking follow-up actions to verify the proper and timely implementation of the corrective actions.

The applicant has committed to the maintenance of documentation so that "lessons learned" may be applied to future operations of the facility.

Accordingly, the NRC staff concludes that the applicant's description of the incident investigation process complies with applicable NRC regulations and is adequate.

11.4.7 Records Management

The NRC staff has reviewed the applicant's records management system against the acceptance criteria and concluded that the system is adequate because it (1) will be effective in collecting, verifying, protecting, and storing information about the facility and its design, operations, and maintenance; (2) will enable retrieval of the information in readable form for the designated lifetimes of the records; (3) will provide a records storage area with the capability to protect and preserve health and safety records that are stored there during the mandated periods, including protection of the stored records against loss, theft, tampering, or damage during and after emergencies; and (4) will provide reasonable assurance that any deficiencies in the records management system or its implementation will be detected and corrected promptly.

11.4.8 Other Quality Assurance Elements

In Appendix A, "Quality Assurance Program Description" to this SER, the NRC staff reviewed the QA elements that the applicant presented in the form of a QA Program. The elements included were organization; QA program; design control; procurement document control; instructions procedures and drawings; document control; control of purchased items and services; identification and control of material parts and component; control of special processes; inspection; test control; control of measuring and test equipment; inspection, test, and operating status; control of nonconforming materials; corrective action; QA records; and audits. Based on its review of the LA, the NRC staff concludes that the applicant has adequately described the application of the QA elements described in the QA Program (and the applicable QA elements of its principal contractors). Appendix A to this SER presents a detailed description of the NRC staff's findings. The NRC staff concludes that the applicant's use of other QA elements described in the QA Program meets the requirements of 10 CFR Part 70 and provides reasonable assurance of protection of public health and safety and of the environment. Specifically, the NRC staff concludes the following:

- The applicant has adequately established and documented a commitment to an organization responsible for developing, implementing, and assessing the management measures for providing reasonable assurance of safe facility operations, in accordance with the criteria in Section 11.4 of NUREG-1520.

- The applicant has adequately established and documented a commitment to QA elements, and the administrative measures for staffing, performance, assessment of findings, and implementation of corrective action are in place.

- The applicant has adequately developed a process for preparation and control of written administrative plant procedures, including procedures for evaluating changes to procedures, IROFS, and tests. The applicant will implement and maintain a process for review, approval, and documentation of procedures.

- The applicant has adequately established and documented surveillances, tests, and inspections to provide reasonable assurance of satisfactory in-service performance of IROFS.

- The applicant will conduct periodic, independent audits to determine the effectiveness of the management measures. Management measures will provide for documentation of audit findings and implementation of corrective actions.

- The applicant has adequately established and documented training requirements to provide employees with the skills to perform their jobs safely. Management measures will provide for the evaluation of the effectiveness of training against predetermined objectives and criteria.

- The organizations and persons performing QA element functions will have the required independence and authority to effectively carry out their QA element functions without undue influence from those directly responsible for process operations.

- QA elements cover the IROFS, as identified in the ISA summary, and the applicant adequately established measures to prevent hazards from becoming pathways to higher risks and accidents.

Accordingly, the NRC staff concludes that the applicant's use of other QA elements is adequate because it meets the requirements of 10 CFR Part 70, the baseline design criteria in 70.61(a)(1), and provides reasonable assurance of protection of public health and safety and the environment.

11.5   References

(IIFP, 2012a) International Isotopes Fluorine Products, Inc., "Fluorine Extraction Process and Depleted Uranium Deconversion Plant (FEP/DUP) License Application, Revision B," May 2012, Agencywide Documents Access and Management System (ADAMS) Accession No. ML12123A245.

(IIFP, 2012b) International Isotopes Fluorine Products, Inc., "ISA Summary Rev. B for IIFP," May 2012, Agencywide Documents Access and Management System (ADAMS) Accession No. ML12123A245.

International Isotopes Fluorine Products, Inc., "Appendix A QAPD Rev B of IIFP License Application," May 2012, Agencywide Documents Access and Management System (ADAMS) Accession No. ML12123A245.

(NRC, 2002) U.S. Nuclear Regulatory Commission, "Standard Review Plan for the Review of a License Application for a Fuel Cycle Facility," NUREG-1520, Rev. 0, March, 2002.

12.0 QUALITY ASSURANCE PROGRAM DESCRIPTION

12.1 Organization

The International Isotopes Fluorine Products, Inc.'s (IIFP), President is the responsible executive for quality assurance (QA), including policies and quality-related goals and objectives. However, the application clearly states that all IIFP employees will be responsible for ensuring safe facility design and operation and that its products and services meet all necessary requirements. During the licensing, design, and construction of the facility, the President of International Isotopes, Inc., (INIS) will be acting as the IIFP President.

The application states that a design and build (DB) contractor will be selected to be responsible for detailed design, engineering, procurement, and construction of the facility and will report to the Chief Operations Officer (COO) while the Engineering Manager position is being filled. Once the Engineering Manager position is filled, the DB Contractor will report directly to the Engineering Manager. The DB contractor will be responsible by contract for ensuring that the design meets all applicable regulations, codes, and standards. The work under the DB contractor will be in accordance with the IIFP's Quality Assurance Program Description (QAPD) (IIFP, 2012c) or an equivalent IIFP-approved program.

The IIFP President will report to the INIS Chief Executive Officer. The COO will report to the IIFP President and will be responsible for managing design, engineering, construction, QA, environmental, safety, and health (ESH), and training. The IIFP's QA Manager will report to and support the COO. The QA Manager's responsibilities will include ensuring compliance with the IIFP's QAPD and procedures; verifying compliance with the QAPD during design and construction of the IIFP,s facility; and reviewing the DB contractor's QA programs in accordance with the IIFP QAPD (IIFP, 2012c).

The applicant described the responsibilities during the project transitions in Section A.1.2 of the license application (LA) (IIFP, 2012a). The applicant also established that qualified contractors will perform the work in accordance with the IIFP's QAPD (IIFP, 2012c) or equivalent QA programs approved in accordance with written procedures. The IIFP's QA function will ensure that evaluation and preapproval of vendor qualification is performed in accordance with the IIFP's QAPD (IIFP, 2012c) and procedures or that increased inspection requirements are implemented where the procurement involves items relied on for safety (IROFS).

Section A.1.3 of the LA (IIFP, 2012a), also describes responsibilities during the operations phase. During operations, the COO will have the overall responsibility for implementation of the QA policies and programs by delegated authority from the IIFP President. The IIFP's Plant QA Manager will be responsible for the following:

- maintaining and complying with the corporate quality management system

- ensuring that identified project deficiencies are addressed by the appropriate management personnel and that containment actions, including interim corrective actions, are implemented until permanent corrective action are completed and verified

- ensuring that periodic audits are conducted to ensure compliance with the QA Program and to assess its effectiveness.

However, the applicant stated that while providing oversight for the corrective action, the QA Manager will not be directly involved in the affected project or production performance assignments.

The IIFP's President will be ultimately responsible for approval of company policies that could impact the quality system but may delegate approval authority to the IIFP's COO.

The applicant also described the responsibilities of the Plant Manager and QA lead positions. The QA lead position may be designated by the QA Manager to direct the QA effort within the production line organization and to exercise the necessary authority to fulfill all organization quality requirements.

12.1.1 Program

The applicant's QA Program will be applicable to all IIFP workers at all levels and to all IIFP products and services, in accordance with the applicable contract, and at the earliest time consistent with the project schedule. The IIFP's QA Program as described by the applicant sets forth the minimum requirements for those items, activities, and services within the scope of the IIFP QAPD (IIFP, 2012c) and will be established, maintained, and executed as described by the applicant in the QAPD(IIFP, 2012c). Supplementary manuals, procedures, and instructions may be implemented for project-specific quality standards and requirements.

12.1.1.1 Program Basis

The applicant committed, in Section A.2.1.2 of the LA (IIFP, 2012a), to incorporate into the QA Program a graded approach to QA (based on an item's importance to safety) that will ensure compliance with the regulatory requirements in Title 10 of the *Code of Federal Regulations* (10 CFR) Part 40. The corporate processes and procedures for the IIFP will be incorporated into the IIFP's QA program.

12.1.1.2 Quality Assurance Program Implementation

The applicant commits to implementing the QA Program through policies, procedures, instructions, specifications, drawings, procurement documents, contractual documents, among other appropriate documents, to ensure that activities within the scope of the QA Program are planned, accomplished, and monitored under conditions that ensure the achievement of project goals.

Quality-related activities will be controlled and conducted using documented procedures. The documents will provide details needed to accomplish quality-related activities under suitably controlled conditions. If work cannot be accomplished as specified in the implementing procedures, the applicant requires the work to be stopped until proper corrective action is taken.

12.1.1.3 Graded Application

The applicant will perform risk analysis reviews to determine the appropriate elements and principles to ensure that the necessary quality-related aspects of the facility are implemented. The applicant will take a risk-based, quality approach to determine to what extent the requirements of 10 CFR Part 70, Subpart H, apply.

QA levels will be assigned to the different components that are determined to be IROFS. The applicant described the three QA levels that will be used in its graded approach as discussed below.

QA Level 1 Requirements

The QA Level 1 requirements will conform to the criteria established in 10 CFR Part 70, Subpart H, and will apply to single IROFS preventing or mitigating a high-consequence event. All QA requirements will be applied to QA Level 1 IROFS.

QA Level 2 Requirements

The QA Level 2 requirements will be applied where two or more IROFS are credited to prevent or mitigate a high-consequence event, or where any single IROFS (sole IROFS) is credited to prevent or mitigate an intermediate-consequence event. The applicant will apply management measures and QA Program requirements to QA Level 2 IROFS, as appropriate, to ensure that IROFS remain reliable and available.

As described in Section A.2.1.3.2 of the LA (IIFP, 2012a), the following QA elements will be applied to QA Level 1 and QA Level 2 IROFS: design control; procurement control; document control; control of purchased items and services; identification and control of materials, parts, and components; control of measuring and test equipment; control of nonconforming items; corrective actions; and QA records. The remaining QA elements will be applied through a graded approach evaluating the factors contributing to the IROFS reliability and considering the following:

- risk significance

- relative importance to ESH, safeguards, and security

- applicable regulations, industry codes, and standards

- complexity or uniqueness of an item/activity and the environment in which it must function

- quality or safety history of the item in service or the activity

- degree to which functional compliance can be demonstrated or assessed by test, inspection, or maintenance methods

- anticipated life span

- degree of standardization

- importance of data generated

- reproducibility of results.

The applicant will incorporate results of the graded approach to quality into design requirement documents, specifications, procedures, instructions, drawings, inspection plans, test plans, procurement documents, and other documents that establish the requirements for items or activities.

For contractors, the applicant will describe QA Level 1 and QA Level 2 requirements in IIFP approved documents.

QA Level 3 Requirements

The applicant described the QA Level 3 requirements as commercial standard practice.

12.1.1.4   Indoctrination and Training

The applicant will provide and document training and indoctrination of personnel performing or managing activities affecting quality to ensure that they are knowledgeable about the applicability, purpose, scope, and implementation of the quality management system (QMS) and the appropriate implementing procedures. Methods for indoctrination may include formal classes, supervised on-the-job training and evaluation, or completion of required reading or self-study. Personnel will be trained as needed to achieve initial proficiency, maintain proficiency, and adapt to changes in the technology, methods, or job responsibility.

The QA Manager will be responsible for indoctrinating appropriate IIFP management and supervisory personnel in the basis for, objective of, and methods for ensuring the quality of IIFP's work. However, the IIFP's management team will be responsible for working together to determine the appropriate methods of indoctrinating and training company personnel.

12.1.1.5   Quality Improvement

The applicant stated that managers at all levels will be responsible for creating an atmosphere of continuous improvement as an integral part of the work activities. Managers will be expected to encourage the development and exploration of new ideas, increase all employees' awareness of the importance of quality, and emphasize the need for enhancing product and process safety and reliability.

The applicant will have measures in place for the detection and prevention of quality problems and the verification of the implementation of quality improvements. Sections A.11.1 and A.12.1 of the LA (IIFP, 2012a) establish the controls for the products, services, and processes that do not meet the specified requirements and the corrective actions to be implemented, as necessary.

The QA Manager will establish procedures to periodically perform trend analysis of nonconformances and corrective actions. Process trends and lessons learned, which are incorporated as a result of audit reports, surveillance reports, corrective action reports, management assessments, and other sources, will be reported regularly to management. Internal IIFP audits and management reviews will be used to identify opportunities for improvement. The Plant Manager will be responsible for managing process quality, identifying potential improvements, and emphasizing the responsibility of each project team member to understand how their processes contribute to the success of the overall project.

12.1.1.6 Review and Assessment

Managers of organizations that implement QMS elements will regularly assess the adequacy of that part of the program or any other sections assigned to them and will address any audit findings in their respective area by implementing appropriate corrective actions and ensuring their effective implementation in accordance with applicable procedures. Internal and external audits will be the responsibility of the QA Manager and will be performed in accordance with Section A.14.1 of the LA (IIFP, 2012a).

12.1.1.7 Qualification and Certification of Personnel

The principal objective of the applicant's training program system is to ensure job proficiency of all personnel through effective training and qualification. As described by the applicant, qualification will be achieved by successful completion of prescribed training, demonstration of the ability to perform assigned tasks, and maintenance of requirements established by the regulations and IIFP. The applicant will use a graded approach to systematic training when applied to the level of detail needed relative to safety.

The applicant stated that managers, with the support of the training organization, will have responsibility for, and authority to develop and conduct, training for their personnel. The QA Manager will be responsible for ensuring that personnel performing quality-related activities are adequately trained in activities associated with their work assignment and will periodically assess activities to ensure compliance with these QA Program requirements.

As discussed by the applicant, indoctrination will include the technical objective and requirements of the applicable codes and standards and the QA program elements that are to be employed. On-the-job training will be a systematic method of providing the required job-related skills and knowledge for a position in an environment as close to the work environment as feasible. Continuing training, defined as any training not provided as initial qualification or basic training and that maintains and improves job-related knowledge and skills, will be conducted as required.

QA/quality control inspections, examinations, surveillances, nondestructive examinations, and special processes will be performed by individuals qualified and certified in the discipline and/or method in which the activity is being performed. For special processes not covered by existing codes and standards or where quality requirements specified for an item exceed those of existing codes or standards, the necessary requirements for qualifications of personnel, procedures, and equipment will be specified or referenced in the procedures or instructions.

12.1.1.8 Work Control

The applicant stated that the QAPD will establish the requirements and define the procedures for controlling work activities for maintenance and future projects after the facility operations begin. This will ensure that the work activities will comply with the requirements of both the applicable contract and the QAPD itself (IIFP, 2012c). As described by the applicant, the degree of complexity and detail in instructions and procedures will be commensurate with the risk associated with the work being performed and the specific customer requirements.

Managers will have the responsibility to ensure that adequate controls are established over activities affecting quality and that personnel are properly trained and qualified and have the proper tools available before performing the work. Managers will also be involved in the work and the work processes to stay current and to create an environment that encourages employees to improve the quality of the work and work processes. The applicant stated that each individual must focus on his or her specific tasks and take responsibility for the quality of the work performed in order to meet work performance objectives and expectations. The Plant, QA, and ESH Managers are responsible for review and approval of the new procedures and procedural revisions before their implementation.

The applicant will conduct activities involving licensed materials or IROFS in accordance with approved procedures that are reviewed by the line managers, QA Manager, production staff, and the radiation protection staff to ensure that the activities are carried out safely and in accordance with regulatory requirements. Integrated safety analysis (ISA) results will also be considered in the identification of needed procedures. The applicant will use standardized methods for the identification, development, approval, implementation, and control of operating procedures. The applicant will also develop procedures for specific products in accordance with the requirements of the applicable portions of the QAPD (IIFP, 2012c).

The U.S. Nuclear Regulatory Commission (NRC) staff finds the applicant's QAPD (IIFP, 2012c) program acceptable because it complies with 10 CFR Pat 40, and it is implemented through procedures and policies. The NRC staff also determined that the personnel will be properly trained and qualified to engage in work activities.

12.1.2 Design Control

12.1.2.1 General

The design control provisions contained in the applicant's QAPD (IIFP, 2012c) will be applicable to activities during design, construction, and operation of the facility. The applicant stated that reconstitution of any prior conceptual design will not be required, but any deviation discovered from the design will be resolved by the Engineering Department.

The applicant will provide requirements and controls to ensure that (1) new design and design change activities will be carried out in a planned, controlled, and orderly manner; (2) design basis, regulatory requirements, and appropriate quality standards are correctly translated into design output, procurement requirements, and procedural documents; and (3) provisions will be established for verifying or checking the technical adequacy of design documents including computer codes.

12.1.2.2 Responsibilities

The Engineering Manager will be the design authority having responsibility for the implementation and execution of the design control system. The COO, with appropriate delegation of authorities to the DB Design Engineering Manager, will be the design authority with responsibility for implementation and execution of the design control system. Management will be responsible for ensuring that completed plant changes are tested and that personnel affected by the change are adequately trained as described in procedures.

12.1.2.3   Implementation

12.1.2.3.1   Design Inputs

Design inputs are those criteria, parameters, or other design requirements upon which the final design is established. Design inputs will be identified, documented, reviewed and approved by the responsible design organization. As described by the applicant, applicable design inputs will be appropriately specified, correctly translated into design documents, and approved on a timely basis and to the level of detail necessary to permit the design activity to be carried out correctly and to provide a consistent basis for making design decisions, accomplishing design verification measures, and evaluating design changes.

Changes to approved design inputs will include the reason for the changes, and they will be identified, approved, documented, and controlled before implementation.

12.1.2.3.2   Design Process

Design methods, materials, parts, equipment, and processes essential to the function of the IROFS will be selected and reviewed for suitability of the application. The applicant committed to having appropriate design documents to support facility design, construction, and operation. Appropriate quality standards and changes to design documents will be identified and documented and their selection reviewed, approved, and controlled before implementation.

The outputs of design and development will be provided in a format that enables verification against the design and development input requirements. Design and development outputs shall meet the input requirements for design and development; provide appropriate information for purchasing, production, and service provision; contain or reference product acceptance criteria; specify the characteristics of the product that are essential for its safe and proper use; and be reviewed and approved before design release.

12.1.2.3.3   Design Analysis

The applicant will perform design analyses in a planned, controlled, and documented manner. The design analyses documents will contain detailed description that will be sufficient to define the purpose, method, assumptions, design input, references, and units, such that a person technically qualified in the subject can review and understand the analyses and verify the adequacy of the results without recourse to the originator. Calculations within the design analysis documents will be identifiable by subject, originator, reviewer, and date; or by other data so that the calculations are retrievable.

12.1.2.3.4   Design Verification

The applicant stated that during the design/construction phase, the design requirements and associated design basis will be established and maintained by the design engineering organization (designated by the Configuration Management Lead and approved by the COO). After operations begin, the Plant Engineer will be responsible for designation and approval of design requirements and documents. The design requirements and basis of design documents will be controlled under the design control provisions of the configuration management program. Design documents associated with QA Level 1 or QA Level 2 IROFS will be subject to interdisciplinary reviews and design verification. The applicant will evaluate changes to the design to ensure consistency with the design basis. In addition, the applicant stated that the computer codes that will be used in the design of IROFS will also be subject to these design control measures, with additional requirements as appropriate for software control, verification, and validation.

Qualified individuals will prepare the design documents and will specify and include the appropriate codes, standards, and licensing commitments within the design documents. They will also note any deviations or changes from such standards within the design documentation package. Another individual qualified in the same discipline will review each design document for concept and conformity with the design inputs. The manager with overall responsibility for the design function will approve the document, and the Configuration Manager will ensure that the designated engineering organization documents the entire review process in accordance with approved procedures.

The applicant described the check and review process consistent with the description given in the application. The applicant discussed the independent design verification process and its requirements. The applicant requires independent design verification whenever appropriate; however, if the verification cannot be performed, the unverified portion of the document will be controlled and the design verification will be completed before the item is relied on to perform its functions. Any changes to the design and procurement documents will be reviewed, checked, and approved commensurate with original requirements.

Any deficiencies identified by the applicant affecting the design of IROFS will be documented and resolved in accordance with approved corrective action program procedures. Design interfaces will be maintained by communication among principals. During the operational phase, the applicant will provide measures to ensure that responsible facility personnel are made aware of design changes and modifications that may affect the performance of their duties.

12.1.2.3.5   Design Changes

The applicant will use procedures for controlling design, including the processes of preparation, review, design verification, approval, and release and distribution for use. Engineering documents will be assessed based on the QA level classification of the item being reviewed. In addition, changes to the approved design will be subject to a review to ensure consistency with the design basis of IROFS.

The procedures, as required by the QA Program, will specify that work will be performed in accordance with the requirements and guidelines imposed by applicable specifications, drawings, codes, standards, regulations, QA criteria, and site characteristics. These procedures, instructions, and drawings will also include any acceptance criteria established by the applicant.

The applicant will maintain documentation demonstrating that the work has been properly performed.

12.1.2.3.6   Design Interfaces

The applicant will identify, document, and control internal and external design interfaces. Any design change efforts will be coordinated among participating organizations. The responsibilities for the preparation, review, approval, release, distribution, and revision of documents involving design changes will require cross-functional team evaluation and must follow the standard document and configuration control requirements.

12.1.2.3.7   Design Documentation and Records

As stated by the applicant, design documentation and records will provide evidence that the design and design verification processes were performed in accordance with the program requirements. The documentation and records will be collected, stored, and maintained in accordance with documented retention policies and procedures. Before modification or extensive repair of equipment or systems, configuration management requirements must be followed to ensure that the planned, as-built condition is documented correctly and completely shown on the drawings, specifications, engineering change notices, and other descriptions of equipment and systems before implementation of the change.

The NRC staff finds that the design control of the proposed facility is adequate because implementation of design inputs, design process, design analysis and design verification will be reviewed, approved and documented The NRC staff determined that changes during the design will be consistent with approved procedures.

12.1.3 Procurement Document Control

12.1.3.1   General

The applicant stated that the procurement system requirements (1) ensure that applicable regulatory requirements, drawing and technical requirements, along with QA Program requirements, are included or referenced in procurement documents for the procurement of items and services for QA Level 1 and QA Level 2 control items; and (2) establish provisions for the preparation, review, approval, and control of procurement documents, including changes.

12.1.3.1.1   Responsibilities

The Engineering Department will be responsible for (1) the preparation and maintenance of design specifications and for identifying the technical and quality requirements necessary to ensure item acceptability; and (2) development of procedures that define these activities, including the criteria for developing the necessary technical and quality requirements for procurement.

12.1.3.1.2   Procurement Document Contents

The following are requirements for procurement documents as described by the applicant in Section A.4.1.3.1 of the LA (IIFP, 2012a):

• statement of work for procurement of services, or an engineering specification for the procurement of items for QA Level 1 and QA Level 2 items

- technical requirements by specific reference to drawings, specifications, codes, standards, regulations, procedures, or instructions, including revisions, that describe the items or services to be furnished for QA Level 1 and QA Level 2 items

- any special instructions and requirements for designing, fabricating, erecting, packaging, shipping, handling, storing, testing, inspecting, and accepting, if required

- requirements for suppliers to have a documented QA program consistent with the applicable requirements of the applicant's QA Program and other applicable codes and standards and commensurate with the item being procured and its importance to safety

- requirements for suppliers of noncommercial grade items and services to evaluate their lower tier suppliers that provide IROFS items or services within the scope of the statement of work or engineering specification

- documentation required to be submitted for information, review, or approval, as well as the time of submittal, where applicable

- requirements for reporting and obtaining disposition of nonconforming items and services, as appropriate.

12.1.3.1.3   Procurement Document Review

The applicant committed to performing document reviews to provide objective evidence of satisfactory accomplishment before contract award. If any changes are made as a result of the bid evaluation or precontract negotiations, those will be incorporated into procurement documents before contract award. Appropriate personnel who have access to pertinent information, and have an adequate understanding of the requirements and procurement documents, will perform the reviews and approvals.

The NRC staff finds the applicant's procurement and document control acceptable because a statement of work is required for procurement services and technical requirements will be referenced to drawings or regulations. Moreover, suppliers to IIFP will have a documented QA program that meets the applicant's QA Program.

12.1.4 Instructions, Procedures, and Drawings

12.1.4.1   General

The applicant described the requirements for instructions, procedures, and drawings as those applied to quality- and process-related activities and services. The applicant will have measures in place to ensure that activities affecting quality are prescribed by documented procedures, drawings, and instructions, appropriate to the circumstances, and are accomplished in accordance with these documents referencing quantitative and qualitative acceptance criteria.

12.1.4.2   Responsibilities

The IIFP's President will be responsible for approving all policy-level documents. The Training Manager or QA Manager will be responsible for reviewing the approval and use of procedures and instructions in accordance with the management oversight requirements defined in

Section A.5..2 of the LA (IIFP, 2012a). The Engineering Department will be responsible for the system of preparation, review, and approval of drawings.

Line and functional managers will be responsible for the development and approval of procedures controlling functions or activities within their area of responsibility. It will be the responsibility of all personnel to use and adhere to the requirements of applicable procedures, instructions, and drawings for activities.

12.1.4.3 Implementation

The NRC staff finds the applicant's implementation of instructions, and procedures acceptable because activities requiring skills normally possessed by qualified personnel are performed in accordance with work instructions, procedures, or drawings of a type appropriate to the circumstances for the control of maintenance and modification work. For activities known as "skill-of-the-craft," the applicant will not require detailed, step-by-step procedures. Written procedures for all of these activities will be prepared, reviewed, approved, implemented, and maintained in accordance with the IIFP's document control process.

12.1.5 Document Control

12.1.5.1 General

The applicant will establish a document control system to maintain policies, procedures, work instructions and any other documentation-related activities and services provided by the applicant to ensure that documents defining the performance of processes and quality-related activities will be controlled in order to maintain only current and correct information available at the location where the activity will be performed before the start of the work.

12.1.5.2 Responsibilities

The records/document lead will have the overall responsibility for the development and implementation of the records management and document control system. Managers will be responsible for (1) identifying documents to be included in the controlled document system; (2) ensuring that instructions, procedures, drawings, and other specified documents are reviewed for adequacy and approved for release; (3) complying with document distribution requirements; and (4) ensuring that personnel performing the prescribed activities maintain and use these documents.

12.1.5.3 Implementation for Document Control

The applicant will establish procedures for the control of document preparation, review, approval, and issuance to ensure that identification of documents will be controlled. In addition, the procedures will include provisions for (1) specific distribution; (2) identification of responsibilities for preparing, reviewing, approving, and issuing documents; and (3) review of documents for adequacy, completeness, and correctness before approval and issuance. The applicant will remove and/or replace obsolete or superseded documents in a timely manner.

The applicant will prepare drawings depicting as-built conditions, including changes, and related documentation in a timely manner reflecting the actual design. The applicant will identify controls to specify the current revision and any changes to instructions, procedures, specifications, drawings, and procurement documents. The applicant's document control system will also have provisions for update and distribution to predetermined personnel.

Changes to documents, other than minor changes, will be reviewed and approved by the same organization that performed the initial review and approval or will be delegated to other qualified organizations. Minor changes to documents will not require that the revised documents receive the same review and approval as the original documents. Procedures will specify the review and approval process for minor changes. Therefore, the NRC staff finds the document control program adequate.

12.1.6 Control of Purchased Items and Services

12.1.6.1   General

In Section A.7.1.1 of the application, the applicant committed to establishing a system for the control of purchased items and services.

12.1.6.2   Responsibilities

The QA Manager will be responsible for providing the necessary QA functions to support procurement, including (1) reviewing supplier quality documentation and evaluating the supplier's QA capability, supplier audits, and development and maintenance of the approved suppliers list; (2) providing support functions such as source verification or surveillance, receipt inspections, installation inspections, and review of procurement documents during receipt inspections; and (3) developing and implementing procedures.

The Engineering Department will assist the QA Manager by performing evaluations of supplier technical capabilities and determining the methods of acceptance to be applied to purchased items and services. The Engineering Department or QA will be responsible for the approval of dispositions and technical evaluations for supplier-generated nonconformances for items and services and for providing measures that ensure the proper selection, application, methods of acceptance, and use of items.

12.1.6.3   Implementation

12.1.6.3.1   Procurement Planning

The applicant will plan and document procurement activities to ensure a systematic approach to the procurement process that will result in the documented identification of procurement methods and organizational responsibilities.

12.1.6.3.2   Supplier Selection

As described by the applicant, suppliers will be selected in accordance with procedures, and selection will be based on evaluation of the supplier's capability to provide items or services in accordance with the requirements of the procurement documents before award of contract. The applicant described four methods to assess the supplier's technical and quality capabilities; upon acceptance, the suppliers are placed on the applicant's Approved Suppliers List. The methods described are the following:

- Evaluation of the supplier's history of providing an identical or similar product that performs satisfactorily in actual use. The supplier's history will reflect current capabilities.

- The supplier's technical and quality capability will be determined by an evaluation of its facility and personnel and the implementation of the supplier's QA program.

- The supplier implements a QA program accepted by the International Standardization Organization (ISO).

- The supplier maintains a valid ISO certification for the item or service being provided.

12.1.6.3.3 Source Verification

Source verification, when used as acceptance, will be performed at intervals consistent with the importance to safety and the complexity of the item or service, and it will be implemented to monitor, witness, or observe activities. This method provides plans to perform inspections, examinations, or tests at predetermined points. Upon purchaser acceptance of source verification, documented evidence of acceptance will be furnished to the receiving destination of the item, to the purchaser, and to the supplier.

12.1.6.3.4 Receiving Inspection

The applicant stated that receiving inspection will be performed for all purchased items to verify conformance to procurement documents. Receiving inspection will verify by objective evidence features as proper configuration; identification; dimensional, physical, or other characteristics; freedom of damage from shipping; cleanliness; and review of supplier documentation when procurement documents require the documentation to be furnished. Upon completion of receipt inspection, acceptable items will be released for storage or issued for installation or use. If any items are determined to be nonconforming after completion of the receipt inspection, those will be documented and processed.

12.1.6.3.5 Post Installation Testing

If post installation testing is used for acceptance of noncommercial-grade items, the purchaser and supplier will establish post installation test requirements and acceptance documentation.

12.1.6.3.6 Commercial-Grade Items

The applicant committed in Section A.7..3.3 of the LA (IIFP, 2012a) to establishing a method for determining whether an item can be purchased as commercial grade and dedicated for use in an IROFS application. The applicant stated that the criteria and methods selected will identify the critical characteristics that are essential to ensure that the item will perform its intended IROFS function. The applicant stated that Quality Level 1 and Quality Level 2 IROFS may be bought commercially, provided that they are subject to a dedication process. Moreover, the applicant stated that in cases where commercial-grade items are to be procured and then dedicated for use as IROFS or parts thereof, the procurement process procedures will include the requirements that IIFP will define to the supplier. Requirements for verifying acceptability of critical characteristics are also mentioned and could encompass inspection, tests, or analyses after delivery, supplemented as necessary by commercial-grade surveys; product inspections or witness at hold points at the manufacturer's facility; and analysis of historical records for acceptable performance.

As described by the applicant, as a minimum for acceptance of commercial-grade items, a receipt inspection will be performed to (1) determine that damage was not sustained during shipment, (2) ensure that the item received is the item ordered, (3) determine that inspection and testing were performed by the supplier as required by Engineering, (4) ensure conformance with the manufacturer's published requirements, and (5) ensure that required documentation is received and is acceptable.

Commercial-grade items will be identified as such in the contract or purchase order by the manufacturer's published product description. Alternative commercial-grade items will be allowed, provided that the Engineering Department verifies that the alternative commercial-grade item will perform its intended IROFS function.

The NRC staff finds the applicant's purchase of services and items acceptable because the supplier's history will be evaluated along with its technical and quality capability. In addition, the applicant will verify purchased items to ensure they meet all procurement documents.

12.1.7 Identification and Control of Materials, Parts, and Components

12.1.7.1   General

The applicant committed in Section A.8.1 of the LA (IIFP, 2012a) to establishing a system for the identification and control of IROFS within the scope of the QAPD (IIFP, 2012c), including associated materials, parts, spare parts, components, and subassemblies.

12.1.7.2   Responsibilities

The Engineering Department will be responsible for (1) specifying requirements for identification methods, traceability, shelf life, and operating life of items when required by codes, standards, or specifications; and (2) specifying these requirements during the generation of specifications, drawings, procurement documents, or other documents appropriate to the circumstances. Managers will be responsible for maintaining and implementing identification, traceability, and shelf life and operating life requirements for items under their jurisdiction. In addition, the Purchasing Manager will be responsible for receipt, delivery, storage, traceability, identification, and control of materials. The QA Manager will be responsible for verifying that items are correctly identified through receipt inspection.

12.1.7.3   Implementation

Procedures will be established depicting the requirements to be implemented for the identification, traceability, and control of materials, parts, and components, including partially fabricated assemblies or subassemblies. Some of the requirements that the applicant will establish in these procedures will include those described below.

- Where practical and required by codes, standards, or contractual documents, items or documents traceable to them will be identified in a manner that ensures that their identification is established and maintained.

- When physical identification is either impractical or insufficient to control the item, physical separation, procedural controls, or other means will be employed.

- Measures will be taken to prevent the use of defective, unapproved, incorrect, or incomplete materials or equipment. These measures will preclude the use of items whose shelf life or operating life has expired.

- Items will have unique identification and traceability by serial number, part number, batch, lot, or specified inspection, test, records, or other appropriate means.

- Productions of an item at any stage, from initial receipt through fabrication, installation, repair, modifications, and use, will be traceable to records such as applicable drawings, specifications, purchase orders, manufacturing and inspection documents, deviation reports, certified material test reports, or other pertinent applicable design-specifying documentation.

- Permanent physical identification will be placed on the item itself to the maximum extent possible, in a manner and location that will not impair or negate the item's intended use, quality, function, or service life.

- Physical separation, procedural control, or other appropriate means will be used where physical identification on the item is impractical or not sufficient.

- Correct identification of materials, parts, and components will be verified and documented on appropriate release documents, work packages, or controlling documents, and on materials before subdividing an item or material and before release for fabrication, assembly, shipping, and installation.

- Personnel performing quality activities will receive training, as required, to ensure understanding and proper implementation of the applicant's QA Program and use of approved procedures to ensure that improper, uncontrolled, damaged, incorrect, or nonconforming material or items are not used or installed.

- Audits, inspections and surveillances will ensure compliance with established procedures.

- Documents establishing and attesting to proper identification of IIFP-furnished items will be maintained as quality records in accordance with design, procurement, and process.

The NRC staff determined that the applicant's control of parts and materials is adequate because procedures to prevent the use of equipment are implemented. Items will have traceability that enables the applicant to determine whether the item is appropriate with its intended usage.

12.1.8 Control of Special Processes

12.1.8.1 General

The applicant stated in Section A.9 of the LA Appendix A (IIFP, 2012a) that a system for the control of special processes and its requirements will be established.

12.1.8.2 Responsibilities

The Engineering Department will be responsible for determining special processes, and the QA Manager will be responsible for the qualification of nondestructive examination personnel, including welder/brazing qualifications. Line managers have the responsibility of ensuring that identified special processes are performed by qualified personnel, using qualified and approved procedures or documents of a type appropriate to the circumstances.

12.1.8.3 Implementation

The applicant will control special processes affecting the quality of items and services such as welding, heat treating, and nondestructive examination through policies, plans, procedures, instructions, drawings, checklists, travelers, work orders, or other appropriate means. Procedures will ensure that special process parameters are controlled and specified environmental conditions are maintained.

The NRC finds the control of special processes adequate as quality processes, special processes will be performed by qualified personnel using qualified equipment when appropriate and approved written policies, plans, and/or procedures in accordance with specified requirements, codes, or standards.

Records of qualifications and equipment related to special processes will be maintained.

12.1.9 Inspection

12.1.9.1 General

The applicant will establish a system for inspection of IROFS that will provide measures to ensure that maintenance, repair, or modification work is completed satisfactorily. In addition, the applicant will identify the requirements for the certification of personnel who perform inspection, examination, surveillance, and testing.

12.1.9.2 Responsibilities

The QA Manager will be responsible for (1) inspection planning, (2) ensuring that inspections are performed, and (3) using qualified and certified inspection personnel, while the engineering or quality organization will be responsible for specifying "hold" and "witness" points for inclusion in applicable work control documents. Management will establish the appropriate measures to ensure that all requirements for inspection are met.

12.1.9.3 Implementation

As stated by the applicant, inspections will be planned and executed to verify conformance of an item or activity to specified requirements. The characteristics to be inspected and inspection methods to be employed will be specified. Inspection personnel will not report directly to the immediate supervisors who are responsible for performing the work being inspected. The depth and extent of inspections will be established by the Plant Engineering & Maintenance Manager and the quality history of the process and will be commensurate with the significance and complexity of the IROFS.

The applicant will document inspection results and records to identify the item inspected, date of inspection, inspector, type of observation, results or acceptability, and references to information on action taken in connection with nonconformances. The applicant will plan for inspection activities, including the documentation of the identification of characteristics, methods, and acceptance criteria. Documentation will also provide for recording objective evidence of inspection results.

The applicant will inspect items in process or under construction to verify quality for certain work activities where necessary. If mandatory inspection hold points are required beyond which work shall not proceed without the specific consent of the designated representative, appropriate documents will indicate the specific hold points; and any waiver will be recorded before the continuation of work beyond the designated hold point.

In-service inspection or surveillance of structures, systems, or components (SSCs) will be planned and executed by or for the organization responsible for construction or operation as specified in the QAPD (IIFP, 2012c) or the ISA. The inspection methods, including evaluation of the performance capability of essential emergency and safety systems and equipment, verification of calibration and integrity of instruments and instrument systems, and verification of maintenance, as appropriate, will be established and executed to verify that the characteristics of an item continue to remain within specified limits

The applicant will plan final inspections to arrive at a conclusion regarding conformance of the item to specified requirements. As described by the applicant, during these final inspections, (1) a records review will be performed to check the results and resolution of nonconformances identified by prior inspections; (2) completed items will be inspected for completeness, markings, calibrations, adjustments, protection from damage, or other characteristics as required to verify the quality and conformance of the item to specified requirements; (3) quality records will be examined for adequacy and completeness, if not previously examined; (4) the acceptance of the item will be documented and approved by authorized personnel; and (5) modifications, repairs, or replacement of items performed subsequent to final inspection will be required to undergo reinspection or retest, as appropriate, to verify acceptability.

The NRC staff finds the applicant's inspections of IROFS adequate because inspections will be carried out by approved procedures and the findings will be documented. The applicant will ensure actions are taken with respect to nonconformance to prevent any defective item from being used.

12.1.10   Test Control

12.1.10.1 General

The applicant will establish a system for design verification testing, acceptance testing, preoperational and operational testing, post maintenance testing, and special testing of IROFS to ensure that testing is completed satisfactorily. Personnel performing design verification testing, acceptance testing, preoperational and operational testing, post maintenance testing, and special testing of IROFS will be certified.

12.1.10.2 Responsibilities

The Engineering Department will be responsible for providing technical criteria for testing, evaluating test results, and resolving deficiencies identified from these tests, while line managers are responsible for the conduct of testing activities under their cognizance.

12.1.10.3 Implementation

The applicant will require tests for verification of conformance to specified requirements and to demonstrate satisfactory performance of services to be planned and executed. As described by the applicant, tests may include design verification tests, acceptance tests, preoperational and operational tests, post maintenance tests, and special tests.

The NRC staff finds the test control to be acceptable because the applicant will identify specific characteristics to be tested and test methods to be employed, and test results will be documented. The application specifies that planning for tests may include mandatory hold points, as required. In addition, the applicant listed the information that will be contained, as appropriate, within the test policies, plans, and procedures, including purpose or objective, references, any required provisions, and qualifications. In lieu of test policies, plans, and procedures, the applicant will use—when needed—appropriate methods taken from related documents, such as external manuals, maintenance instructions, or approved drawings— among others—provided that those include adequate instructions to ensure the required quality of work. The applicant described the contents of the test record information as item tested; test date; tester or data recorder; type of observation; test policy, plan, procedure, or reference; results and acceptability; actions taken in connection with any deviations noted; and the person evaluating the results.

12.1.11 Control of Measuring and Test Equipment

12.1.11.1 General

The applicant will establish a system for the control of measuring and test equipment (M&TE) used for measurement, test, and calibration items within the scope of the QA Program. The system will include requirements that will establish measures to (1) ensure that tools, gauges, instruments, reference and transfer standards, nondestructive test equipment, and other measuring and testing devices used in activities affecting quality will be properly controlled, calibrated, and adjusted at specified intervals to maintain equipment performance within required limits; and (2) ensure that devices and standards used for measurement, tests, and calibration activities will be of the proper type, range, accuracy, and tolerance to determine conformance to specified requirements.

12.1.11.2 Responsibilities

The Plant Engineering & Maintenance Manager will have the overall responsibility for the calibration control system for M&TE. In addition, individual area managers will be responsible for the implementation of the calibration control system for M&TE under their respective areas.

12.1.11.3 Implementation

The applicant will identify a list of devices within the calibration system including, as a minimum, the due date of the next calibration and any use limitations. The method and interval for calibration for each item will be defined, based on the type of equipment, stability characteristics, required accuracy, intended use, and other conditions affecting measurement control. The applicant clarified that calibration controls will not be required for rulers, tape measures, levels, and other such devices if the commercial equipment provides adequate accuracy.

M&TE will be properly handled and stored to maintain accuracy and will be calibrated at specified intervals or before use against certified equipment having known valid relationships to nationally recognized standards. If no nationally recognized standard exists, the bases for calibration will be documented. Also, calibrations may be performed when the accuracy of the equipment is deemed suspect by personnel performing measurements and tests. Records will be maintained, and equipment will be appropriately marked to indicate its calibration status.

If any M&TE is found to be out of calibration, the applicant will evaluate and document the validity of any previous inspection and test results and the acceptability of items previously inspected or tested. These devices will be tagged or segregated and will not be used until recalibrated. If consistently found to be out of calibration, the M&TE will be repaired or replaced.

The NRC staff finds M&TE acceptable because the NRC staff will have a scheduler with information on when the next calibration is due, and the equipment will be properly handled to ensure its purpose is not affected.

12.1.12   Inspection, Test, and Operating Status

12.1.12.1 General

The applicant requires the identification of the status of inspection and test activities. In addition, status indicators will provide for the operating status of IROFS systems and components to prevent inadvertent operation.

12.1.12.2 Responsibilities

The QA Manager is responsible for providing a status-indicating system for inspections that will be performed in accordance with these requirements, while line managers are responsible for the development and implementation of status-indicating systems.

12.1.12.3 Implementation

The NRC staff finds that inspections, tests and operating status are adequate because the applicant will establish policies, plans, and procedures to ensure that the status of inspection and test activities are either marked or labeled on the item or in documents traceable to the item. Status indicators are used, as appropriate, to prevent inadvertent operation. Only the persons specified by the applicant will authorize the application and removal of tags, markings, labels, and stamps.

12.1.13 Control of Nonconforming Items

12.1.13.1 General

The applicant will establish procedures for the control of items that do not conform to specified conditions and related activities and services. These will include requirements for identification, segregation, disposition, prevention of inadvertent installation or use, documentation, and notification to affected organizations.

12.1.13.2 Responsibilities

The applicant affirmed that all personnel participating in quality-affecting activities within the scope of the quality system will be responsible for reporting and documenting nonconforming items

The QA Manager will be responsible for implementing the nonconformance control system for items that do not meet the established specifications or technical requirements. The Engineering Department, ESH, and the quality organization will provide documented technical justification for the acceptability of nonconforming items (use-as-is or repair). These organizations will also be responsible for applying the design control measures of the QAPD (IIFP, 2012c) to nonconformances. The Engineering Department will also ensure that as-built records reflect the accepted change.

12.1.13.3 Implementation

Nonconforming items will be legibly identified in a manner that does not adversely affect the use of the item. If physical identification of the item is not practical, the container, package, or segregated storage area will be identified appropriately. Nonconforming items will be segregated, when practical, by placing them in a clearly identified and designated area until disposition. If segregation is impractical or impossible, other measures will be employed to prevent inadvertent use of the item.

Any nonconforming characteristics will be reviewed and disposition will be recommended based on the evaluation. The applicant will define in the appropriate documentation the applicable responsibilities and authorities. Specifically, personnel performing evaluations will have an adequate understanding of the requirements and have access to pertinent background information. Repaired or reworked items will be reexamined in accordance with applicable procedures and original acceptance criteria, unless the nonconforming item disposition has established alternate acceptance criteria. Further processing, delivery, installation, or use of the nonconforming item will be controlled pending an evaluation with an approved disposition and an approved notification to affected organizations.

Nonconforming items or services identified by suppliers will be reviewed to determine applicability. The NRC staff finds that identification of nonconforming items is acceptable because corrective actions will be initiated which will be evaluated to determine if reporting is required. When a nonconformance is identified after the product is shipped to a customer or recipient, IIFP management will evaluate whether the nonconformance is potentially reportable to the NRC under 10 CFR Part 21. In addition, the customer or recipient will be notified.

12.1.14　Corrective Action

12.1.14.1　General

The applicant described the corrective action system as a system established for those activities and services that are determined to have an adverse effect on the customer or have the potential for recurrence. The applicant stated that the corrective action system will (1) establish measures to ensure that conditions adverse to quality are identified and corrected as soon as practical, (2) ensure that a root cause analysis and corrective action report will be generated to document and report the analysis and action to the appropriate levels of management, and (3) ensure that follow-up actions are taken to verify implementation of the corrective action.

12.1.14.2　Responsibilities

The QA Manager will be responsible for maintenance and implementation of the corrective action system and for verifying that management reviews and assesses adverse conditions. Moreover, the QA Manager will be responsible for (1) reviewing deficiencies identified by a customer audit or assessment; (2) maintaining an orderly file of all audit findings and nonconformances related to products and services; (3) ensuring that the requirements of the customer and of the procedure are met, as a minimum; (4) evaluating any corrective action that results from suggestions by the customer; (5) ensuring that customer suggestions are appropriately addressed either by implemented change or resolution; and (6) reviewing and evaluating the effectiveness of corrective actions.

Line managers will be responsible for (1) evaluating and performing assigned corrective actions; (2) ensuring the identification and documentation of conditions adverse to quality promptly in accordance with applicable procedures; and (3) determining the cause of any deficiency, documenting it as a nonconformance, determining the necessary action, and reporting actions in a nonconforming report or in the corrective action report.

12.1.14.3　Implementation

The applicant committed, in Section A.16.1 of the LA Appendix A (IIFP, 2012a), to promptly identifying, documenting, reporting, and correcting conditions adverse to quality. The applicant defined "conditions adverse to quality" as failures, malfunctions, deficiencies, deviations, defective material and equipment, and nonconformances.

In addition, the applicant defined "significant conditions adverse to quality" as those for which the cause of the condition will be determined and appropriate corrective actions will be taken to preclude recurrence. The applicant will evaluate significant conditions adverse to quality to determine if stopping work is warranted. In addition, the identification, cause, and corrective action for significant conditions adverse to quality will be documented and reported to appropriate levels of management. Follow-up actions will verify the implementation of the corrective action.

The applicant defined conditions adverse to quality as follows:

- a deficiency that would seriously impact an item, activity, or service from meeting or performing its intended function or output of ensuring public health and safety

- a deficiency in design that has been approved for fabrication or construction where the design deviated from design criteria and bases

- a deficiency in the fabrication or construction of, or significant damage to, an SSC that requires extensive evaluation, redesign, or repair to establish the adequacy of the SSC to perform its intended function

- a deviation from performance specifications that will require extensive evaluation, redesign, or repair to establish the adequacy of the SSC to perform its intended function

- a significant error in a computer program used to support activities affecting quality after it has been released for use

- a deficiency, repetitive in nature, related to an activity or item subject to the IIFP's QA Program

- a condition that, if left uncorrected, has the potential to have a serious negative impact on activities or items subject to the IIFP's QA Program controls.

The NRC staff finds the corrective action program acceptable because it recognizes conditions adverse to quality and implements actions to prevent them from reoccurring.

The affected management, including the COO or Plant Manager, when appropriate, will be notified of any trends or additional remedial action.

12.1.15   Quality Assurance Records

12.1.15.1 General

The applicant will establish a records management system that will provide measures to control QA records.

12.1.15.2 Responsibilities

Records Management/Document Control will be responsible for the maintenance and implementation of the records control system. Managers will be responsible for (1) identifying QA records initiated by their organization, (2) controlling the records within their jurisdiction, and (3) transferring records to the Records Manager for retention. All activities related to records management and document control will be consistent with the requirements set forth in the QA Program.

12.1.15.3 Implementation

The applicant affirmed that through written procedures, QA records will be identified, prepared, indexed, stored, maintained, preserved, and kept safe in appropriate facilities. Records will be stored in authorized facilities that will provide appropriate measures to ensure continuous protection of the records. In addition, records will be legible, accurate, and complete and retrievable. The applicant will have appropriate measures in place to preclude entry of unauthorized personnel into record storage locations and to guard against larceny and vandalism. Records will be protected against damage, deterioration, and loss.

Documents will be considered valid records only if stamped, initialed, or signed and dated by authorized personnel or otherwise authenticated. The applicant will establish methods to ensure traceability of the record and the item(s) or activity(ies) to which it applies. The originating organization will approve and document any corrections to records.

The applicant will classify QA records as "lifetime" or "nonpermanent." Lifetime records must be maintained for the life of the particular item. Nonpermanent records will comply with the records retention requirements as defined in the IIFP records retention policy. As described by the applicant, lifetime records would be of significant value in (1) demonstrating that manufactured products meet requirements; (2) maintaining, reworking, repairing, replacing, or modifying critical items within the plant or manufactured products; and (3) determining the cause of an accident or malfunction of an item within the plant. Quality assurance Records have been determined by the NRC staff to be acceptable because the applicant has established procedures which keep them safe and ensure their quality is not compromised.

12.1.16   Audits

12.1.16.1 General

The applicant discussed, in Section A.181 of the LA Appendix A (IIFP, 2012a), an audit system for activities and services within the scope of the IIFP's QA Program. The audit system will establish planned and periodic audits to verify the compliance and the effectiveness of the QA Program in meeting system quality requirements.

Audits will be executed in accordance with established procedures and will be performed by personnel having no direct responsibilities in the areas being audited and with sufficient authority and organizational freedom to make the audit process effective.

As described by the applicant, internal audits will be performed in selected aspects of operational activities to ensure that audits of activities are completed within specified time periods. External audits to ensure compliance with applicable aspects of the IIFP's QA Program and procurement requirements will be performed for selected suppliers and service contractors to verify and evaluate their QA programs, procedures, and activities.

12.1.16.2 Responsibilities

The QA Manager will be responsible for the development, maintenance, scheduling, and performance of the internal audit and external supplier audit system. Audited organizations will be responsible for assisting during the planning and performance of audits (e.g., providing access to facilities, personnel, documents, and records) and ensuring that requests for corrective action are promptly answered and that actions taken to correct any discrepancy are adequate and timely.

12.1.16.3 Implementation

12.1.16.3.1 Training and Qualification

The applicant stated that audit personnel will be properly trained to perform the required audits. Any technical specialists participating as audit team members will have received the required indoctrination and guidance before the audit.

12.1.16.3.2 Scheduling

The applicant will schedule internal and external audits in a manner to provide coverage and coordination with ongoing QA Program activities at a frequency commensurate with the status and importance of the activities. The audit schedules will be reviewed periodically and revised as necessary to ensure that coverage is maintained current. To provide adequate coverage, additional audits or surveillance of specific subjects will supplement regularly scheduled audits.

Implementation audits may be scheduled and performed for initial evaluation of suppliers. External audits of approved noncommercial-grade suppliers will also be scheduled and performed based on the supplier certification program and established quality performance parameters. If the suppliers of services perform the work under the IIFP's QA Program and procedures, external audits will not be required. External audits will be required for suppliers of services based on the supplier quality certification program, established quality performance parameters, and quality level of the service provided.

12.1.16.3.3 Audit Plan

Each audit will require an audit plan, developed and documented by the auditing organization. Each audit plan will identify the audit scope, requirements, audit personnel, activities to be audited, organizations to be notified, applicable documents and written procedures, schedule, and approved checklists of questions covering the items to be audited.

12.1.16.3.4 Personnel and Selection of Audit Team

The applicant will establish measures for the selection of the audit team and audit team familiarization before the start of each audit to ensure consideration of special abilities, specialized technical training, prior experience, personal characteristics, and education. The applicant described audit teams as containing one or more auditors, with an appointed team lead. The team lead will organize and direct the audit, coordinate the preparation and issuance of the audit report, and evaluate the responses. The applicant will establish expected audit deliverables before the start of the audit.

12.1.16.3.5 Auditing

The applicant included requirements for the organizations being audited to provide access and assistance to the audit personnel. The audit team will examine objective evidence to determine if the QA Program elements are being implemented effectively. Any conditions requiring prompt corrective action will be reported immediately to the management of the audited organization.

12.1.16.3.6   Reporting

The audit team leader will sign the audit report, which will generally be issued within 30 days of the post audit conference. The audit report will be distributed to responsible management of both the auditing and the audited organizations. The applicant listed the following as information that will be contained in the report, as appropriate:

- description of the audit scope

- identification of the auditors

- identification of persons contacted during audit activities

- summary of audit results, including a statement of the implementation effectiveness of the QA Program elements that were audited

- description of each reported adverse audit finding in sufficient detail to enable the audited organization to take corrective actions.

12.1.16.3.7   Response and Follow-up Action

The management of the audited organization will investigate adverse audit findings. It will also identify and schedule corrective actions and measures to prevent recurrence. In addition, management will notify the appropriate organization in writing of the actions taken or planned. Follow-up action will be taken to verify that corrective action is completed as scheduled.

12.1.16.3.8   Records

Audit records will include audit plans, audit reports, written replies, and documented completion of corrective actions.

Audits of services within the QA program have been found acceptable by the NRC staff because audit plans will be made before each audit and audit personnel will have access to the required documentation to ensure the QA program is being implemented effectively.

12.2   Evaluation of Findings

The NRC staff reviewed the QA elements that were presented in the form of a QA Program. The elements included were organization; QA Program; design control; procurement document control; instructions, procedures, and drawings; document control; control of purchased items and services; identification and control of materials, parts, and components; control of special processes; inspection; test control; control of M&TE; inspection, test, and operating status; control of nonconforming items; corrective action; QA records; and audits. Based on its review of the LA, the NRC staff has concluded that the applicant has adequately described the application of other QA elements (and the applicable QA elements of its principal contractors). The NRC staff also concludes the following:

- The applicant has established and documented a commitment to an organization responsible for developing, implementing, and assessing the management measures for providing reasonable assurance of safe facility operations in accordance with the criteria in Section 11.4 of NUREG-1520, Revision 0, issued May 2002 (NRC, 2002).

- The applicant has established and documented a commitment to QA elements, and the administrative measures for staffing, performance, assessment of findings, and implementation of corrective actions are in place.

- The applicant has developed a process for preparation and control of written administrative plant procedures, including procedures for evaluating changes to procedures, IROFS, and tests. The applicant will implement and maintain a process for review, approval, and documentation of procedures.

- The applicant has established and documented surveillances, tests, and inspections to provide reasonable assurance of satisfactory in-service performance of IROFS.

- Periodic independent audits will be conducted to determine the effectiveness of the management measures. Management measures will provide for documentation of audit findings and implementation of corrective actions.

- Training requirements have been established and documented to provide employees with the skills to perform their jobs safely. Management measures have been provided for the evaluation of the effectiveness of training against predetermined objectives and criteria.

- The organizations and persons performing QA element functions have the required independence and authority to effectively carry out their QA element functions without undue influence from those directly responsible for process operations.

- QA elements cover the IROFS, as identified in the ISA Summary (IIFP, 2012b), and measures are established to prevent hazards from becoming pathways to higher risks and accidents.

Accordingly, the NRC staff concludes that the applicant's use of additional QA elements meets the requirements of 10 CFR Part 70 and provides reasonable assurance of the protection of public health and safety and of the environment.

12.3   References

(IIFP, 2012a) International Isotopes Fluorine Products, Inc., "Fluorine Extraction Process and Depleted Uranium Deconversion Plant (FEP/DUP) License Application, Revision B," May 2012, Agencywide Documents Access and Management System (ADAMS) Accession No. ML12123A245.

(IIFP, 2012b) International Isotopes Fluorine Products, Inc., "ISA Summary Rev. B for IIFP," May 2012, Agencywide Documents Access and Management System (ADAMS) Accession No. ML12123A245.

(IIFP, 2012c) International Isotopes Fluorine Products, Inc., "Appendix A QAPD Rev B of IIFP License Application," May 2012, Agencywide Documents Access and Management System (ADAMS) Accession No. ML12123A245.

(NRC, 2002) U.S. Nuclear Regulatory Commission, "Standard Review Plan for the Review of a License Application for a Fuel Cycle Facility," NUREG-1520, Rev. 0, March, 2002.

13.0 PHYSICAL PROTECTION AND MATERIAL CONTROL AND ACCOUNTING

13.1 Physical Protection

International Isotopes Fluorine Products, Inc. (IIFP), submitted, along with its license application (LA), Revision B of the Physical Security Plan (PSP), dated May 3, 2012 (IIFP, 2012b), for review and approval by the U.S. Nuclear Regulatory Commission (NRC) in accordance with Title 10 of the *Code of Federal Regulations* (10 CFR) 40.3, "License Requirements." This PSP provides for the protection of radiological material and chemicals of interest at IIFP.

Although 10 CFR Part 40, "Domestic Licensing of Source Material," does not include physical security requirements for licensees, the NRC has issued an order to another 10 CFR Part 40 licensee with similar materials. IIFP opted to implement physical security in accordance with 10 CFR 73.67, "Licensee Fixed Site and In-Transit Requirements for the Physical Protection of Special Nuclear Material of Moderate and Low Strategic Significance," for a Category III fuel cycle facility.

Details related to physical protection have been marked by the applicant as "Security-Related Information," pursuant to Title 10 of the Code of Federal Regulations (CFR) 2.390, and are therefore not included in the public version of this report.

13.1.1 Regulatory Guidance

The NRC staff's review and analysis was conducted in accordance with the following regulatory guide:

> Regulatory Guide 5.59, Revision 1, "Standard Format and Content for a Licensee Physical Security Plan for the Protection of Special Nuclear Material of Moderate or Low Strategic Significance," issued February 1983 (NRC, 1983).

13.1.2 Staff Review and Analysis

The IIFP submitted its PSP (IIFP, 2012b) as part of the LA for review and approval by the NRC in accordance with 10 CFR 40.3. After reviewing the initial PSP, the NRC submitted an RAI to IIFP regarding the security plan, which they responded to on January 10, 2011 (IIFP, 2011).

The NRC reviewed the PSP and RAI responses to ensure that the applicant committed that it will store chemicals of interest only in a controlled access area, which will be monitored by an intrusion detection system or other approved procedures. The applicant also committed to having a guard force and offsite response force available to respond to unauthorized penetrations or activities. Furthermore, the applicant committed to establishing and maintaining response procedures for dealing with threats of theft or theft of material and reporting safeguards events as required by regulation.

13.1.3 Evaluation of Findings for Physical Protection

The NRC staff's review of the applicant's PSP for the protection of special nuclear materials of low-strategic significance contains information that the applicant has marked as "Official Use Only—Security-Related Information," under 10 CFR 2.390, "Public Inspections, Exemptions,

Requests for Withholding." The NRC staff reviewed the applicant's PSP for fixed site physical protection of special nuclear materials of low-strategic significance and chemicals of concern. The NRC staff finds that the methods and procedures outlined in the PSP satisfy the performance objectives, systems capabilities, and reporting requirements specified in 10 CFR 73.67. The PSP for the facility is acceptable and provides reasonable assurance that the requirements for the physical protection of special nuclear materials of low-strategic significance and chemicals of concern will be met.

The NRC staff concludes that Revision B to the PSP (IIFP, 2012b), which incorporates the RAI responses (IIFP, 2011), is acceptable and recommends approval.

13.2 Material Control and Accounting

The U.S. Nuclear Regulatory Commission (NRC) staff's review of the applicant's nuclear Material Control and Accounting (MC&A) Program is documented in this appendix to the Safety Evaluation Report. Appendix H contains information that has been marked as "Proprietary Information" and "Export Control Information" by the applicant, pursuant to Title 10 of the *Code of Federal Regulations* (10 CFR) 2.390, "Public Inspections, Exemptions, Requests for Withholding," and 10 CFR 810, respectively.

The NRC staff concludes that the applicant provided an acceptable Fundamental Nuclear Material Control Plan (FNMCP) for the proposed facility that will meet the applicable requirements of 10 CFR Part 74, "Material Control and Accounting of Special Nuclear Material." The FNMCP describes acceptable methods for achieving the performance objectives in 10 CFR 74.33(a) and the system capabilities of 10 CFR 74.33(c). As a result, the NRC staff finds that the applicant will meet the requirements in the area of MC&A to operate the proposed facility under 10 CFR Part 74.

The purpose of this review was to verify that the applicant, IIFP, provided sufficient information to determine that the MC&A Program meets the applicable regulatory requirements in 10 CFR Part 40, "Domestic Licensing of Source Material."

13.2.1 Regulatory Requirements for Material Control and Accounting

The requirements for MC&A are specified in 10 CFR 40.64, "Reports."

13.2.2 Regulatory Acceptance Criteria

The application will be considered acceptable if it meets the regulatory requirements specified in the following:

- 10 CFR Part 40

- NUREG/BR-0006, Rev. 7, "Instructions for Completing Nuclear Material Transaction Reports (DOE/NRC Forms 741 and 740M)," effective January 2009 (NRC, 2009a)

- NUREG/BR-0007, Rev. 6, "Instructions for the Preparation and Distribution of Material Status Reports (DOE/NRC Forms 742 and 742C)," effective January 2009 (NRC, 2009b).

13.2.3 Staff Review and Analysis

The NRC staff reviewed and evaluated information provided by the applicant for the proposed MC&A Program in the LA (IIFP, 2012a). The following aspects of the MC&A Program were reviewed:

- MC&A organization

- measurements

- physical inventories

- recordkeeping and reports.

13.2.3.1 Material Control and Accounting Organization

The LA (IIFP, 2012a) describes a management structure and organization positions that have responsibilities related to the MC&A Program and that ensure clear responsibility for MC&A functions; the use of approved written MC&A procedures; and training for MC&A-related tasks. The NRC staff finds that the applicant is capable of implementing a management structure that permits effective functioning of the MC&A Program and ensures that the facility management structure will not adversely affect the MC&A Program performance.

13.2.3.2 Measurements

The LA (IIFP, 2012a) identified and described the measurement systems used for accounting purposes. The NRC staff reviewed the overall measurement program and finds the applicant's system of measurements appropriate and acceptable to ensure that all quantities of nuclear material in the accounting records are based on reliable measurements.

13.2.3.3 Physical Inventories

The applicant's physical inventory program contains the basic elements for scheduling, performing, and reporting annual physical inventories. The NRC staff finds that the physical inventory program demonstrates its ability to confirm the presence and quantities of nuclear materials.

13.2.3.4 Recordkeeping and Reports

The LA (IIFP, 2012a) describes the nuclear material accounting system for the facility and the control of this system. In addition, the applicant provided common reporting forms and formats to provide nuclear material data in accordance with current regulations. The NRC staff finds that the accounting system is secure and adequate. The NRC staff determined that the applicant's establishment, maintenance, and protection of a recordkeeping system are acceptable and meet the reporting requirements.

13.2.4 Evaluation Findings for Material Control and Accounting

The NRC staff concludes that the applicant provided an acceptable MC&A Program for the proposed facility that will meet the applicable 10 CFR Part 40 requirements. The application described acceptable methods for achieving the requirements of 10 CFR 40.64. As a result, the NRC staff finds that the applicant meets the requirements in the area of MC&A to operate the proposed facility under 10 CFR Part 40.

13.3 References

(IIFP, 2012a) International Isotopes Fluorine Products, Inc., "Fluorine Extraction Process and Depleted Uranium Deconversion Plant (FEP/DUP) License Application, Revision B," May 2012, Agencywide Documents Access and Management System (ADAMS) Accession No. ML12123A245.

(IIFP, 2012b) International Isotopes Fluorine Products, Inc., "Security Plan Rev B of IIFP License Application," Rev. B, 2012, Agencywide Documents Access and Management System (ADAMS) Accession No. ML12123A245.

(IIFP, 2011) International Isotopes Fluorine Products, Inc., "Submittal of Responses to Requests for Additional Information re Application to License a Depleted Uranium Hexafluoride De-Conversion and Fluorine Extraction Process Facility," January 10, 2011, Agencywide Documents Access and Management System (ADAMS) Accession No. ML110190058.

(NRC, 2009a) U.S. Nuclear Regulatory Commission, NUREG/BR-0006, Rev. 7, "Instructions for Completing Nuclear Material Transaction Reports (DOE/NRC Forms 741 and 740M)," effective January 1, 2009.

(NRC, 2009b) U.S. Nuclear Regulatory Commission, NUREG/BR-0007, Rev. 6, "Instructions for the Preparation and Distribution of Material Status Reports (DOE/NRC Forms 742 and 742C)," effective January 1, 2009.

(NRC, 1983) U.S. Nuclear Regulatory Commission, "Standard Format and Content for a Licensee Physical Security Plan for the Protection of Special Nuclear Material of Moderate or Low Strategic Significance," Regulatory Guide 5.59, 1983.

14.0 Human Factors

This appendix describes the human factors engineering (HFE) review of the International Isotopes Fluorine Products, Inc.'s (IIFP), license application (LA) (IIFP, 2012a) for a proposed depleted uranium deconversion facility. The purpose of this review as stated in NUREG-1520, Revision 1, "Standard Review Plan for the Review of a License Application for a Fuel Cycle Facility," Appendix E, page 3-E-1, issued May 2010 (NRC, 2010), is to determine whether the applicant has established a HFE Implementation Plan (HFEIP)that will be applied to personnel activities identified as safety significant, consistent with the findings of the integrated safety analysis (ISA), and the determination of whether an item relied on for safety (IROFS) has special or unique safety significance. Specifically, this review addresses the standards, guidance, and practices that specify the design and implementation of the human system interfaces (HSIs) (e.g., alarms, displays, and controls) for IROFS that require operator actions at the proposed facility.

14.1   Regulatory Requirements

Title 10 of the *Code of Federal Regulations* (10 CFR) 70.61(e) requires that IROFS remain available and reliable to perform their intended function when needed. NUREG-1520, Revision 1, Appendix E, page 3-E-1, states, "The application of HFE to personnel activities ensures that the potential for human error in the facility operations was addressed during the design of the facility by facilitating correct, and inhibiting wrong, decisions by personnel and by providing a means for detecting and correcting or compensating for error" (NRC, 2010). In addition, the baseline design criteria in 10 CFR 70.64(b)(2) require the facility to be designed to reduce challenges to IROFS. This requirement includes a significant human component for administrative IROFS. The applicant must demonstrate an adequate HFE program to ensure compliance with these requirements.

14.2   Regulatory Guidance and Acceptance Criteria

Since IIFP submitted their application prior to completion of NUREG-1520, Revision 1 (NRC, 2010), the bulk of the IIFP review was conducted consistent with NUREG-1520, Revision 0 (NRC, 2002). However, this version of the NUREG-does not contain guidance for the human factors review. Therefore, the NRC staff and IIFP agreed to conform to guidance in NUREG-1520, Revision 1 (NRC, 2010), for this portion of the review.

In addition, the applicant has committed in the LA to designing its HSIs in accordance with the guidelines in NUREG-0700, Revision 2, "Human-System Interface Design Review Guidelines," issued May 2002 (NRC 2002), and to performing its HFE program in accordance with the review guidance described in NUREG-0711, Revision 2, "Human Factors Engineering Program Review Model," issued February 2004 (NRC, 2004).

The key aspects of NUREG-0711 (NRC, 2004) applicable to fuel cycle facilities have been distilled into the nine acceptance criteria listed in Appendix E to NUREG-1520, Revision 1 (NRC, 2010). The NRC staff considers these nine HFE program elements to be the HFE activities needed to provide reasonable assurance that the control and maintenance systems will be designed to facilitate correct decisions and inhibit wrong decisions by personnel and provide a means for detecting, correcting, and compensating for error. HFE is an ongoing process that begins with design and continues in conjunction with the corrective action process after the startup of a facility.

During the licensing review, portions of the HFE program are often under development and the details of design are not available for review; therefore, a programmatic review of the design process is performed. At the programmatic level, the applicant is still developing detailed procedures. Thus, the applicant has provided the NRC staff with the commitments to be performed to achieve the objectives of the HFE program elements. The NRC staff reviewed the objectives of each HFE element, the sources on which the method will be developed, and the types of results the applicant expects the method to produce. The implementation of this program through detailed procedures will be verified by inspection during the readiness review.

The NRC staff used a graded approach, consistent with Footnote 2 in Appendix E to NUREG-1520, Revision 1 (NRC, 2010), to review the IIFP's HFE program. The proposed facility is a depleted uranium deconversion facility. While there are human actions associated with the IROFS, the LA, Section 3.1.4.4, indicates a design preference to minimize the need for human actions through use of active and passive-engineered controls for IROFS (IIFP, 2012a). The facility does not have criticality hazards, large quantities of high-activity beta and gamma emitters, or transuranic nuclides. Therefore, the proposed facility has significantly lower risk and a less active safety role for the human staff than are typical of a commercial nuclear power plant. The NRC staff reviewed the application with the awareness of the detrimental effects that automation may have on human performance, situation awareness, and workload, which can cause increases in the probability of human error and incorrect decisionmaking when automated systems are not appropriately designed for the tasks of the human operator (O'Hara, 2010).

The implementation of the applicant's proposed HFEIP will be reviewed in the inspections required prior to operations to ensure that the implementation of the HFEIP meets the guidance provided in NUREG-1520, Revision 1, Appendix E to Chapter 3 (NRC, 2010).

For this programmatic level review, the NRC staff used the criteria in NUREG-1520, Revision 1 (NRC, 2010), to review the applicant's HFEIP in terms of the overall objective of each HFE element, the technical basis (such as industry standards) that the applicant will use to develop the detailed HFE program, and the information that should be obtained from each HFE activity.

14.3   Staff Review and Analysis

The NRC staff reviewed the LA (IIFP, 2012a); the ISA Summary (IIFP, 2012b); and the Quality Assurance Program Description (QAPD) (IIFP, 2012c). IIFP provided formal responses to staff requests for additional information (RAIs) by letter dated August 22, 2011 (IIFP, 2011b); this letter contains the HFE IP, to be added as Section 3.1.4 to the IIFP LA, Chapter 3 (IIFP, 2012a).

This section presents the results of the review, which are organized by the acceptance criteria in Appendix E to Chapter 3 to NUREG-1520, Revision 1 (NRC, 2010), and the applicant's HFEIP (IIFP, 2012a).

14.3.1 Identification of Personnel Activities

14.3.1.1   Evaluation

The HFE program is described as an integral part of the ISA development, plant design, operation, and evaluation of operations. IROFS were determined, identified, and documented through the ISA process. The ISA process used a "What-if" process for the accident and consequence analysis process of the IIFP process hazard analysis (PHA) and relied on the considerable engineering and operations experience of the ISA team. In a What-if process, the subject matter experts consider "worst-case" scenarios in the "What-if" PHA method. The HSI was considered as part of the accident analysis review and discussion process.

14.3.1.2   Conclusion

The scope of the HFEIP is identified as the administrative IROFS for which human actions are relied on to ensure the performance of these administrative controls. Section 3.2.5.8 of the LA (IIFP, 2012a) states that safety controls used at the IIFP facility can be characterized as either administrative or engineered. Administrative controls may rely on human actions, and IIFP stated that they are generally not considered to be as reliable as engineered controls since human errors usually occur more frequently than equipment failures. Engineered controls may be categorized as being "passive" or "active" and are not reliant on human action. Therefore, the NRC staff finds that the focus of the HFEIP on administrative controls is appropriate.

Personnel actions will be identified and addressed during the operational experience review (OER); functional allocation analysis (FAA); task analysis (TA); HSI design, inventory, and characterization; procedures; training; and verification and validation (V&V) elements described in the HFE IP, Sections 3.1.4.3 through 3.1.4.9 of the LA (IIFP, 2012a). Section 3.1.4.1 of the LA indicates that this evaluation will also involve the identification of the structures, systems, and components (SSCs) and the boundary definitions of the active-engineered controls and passive-engineered controls IROFS. Table 6-3 of the IIFP ISA Summary (IIFP, 2012b) presents the IROFS that are subject to the IIFP's HFEIP and the HFE program. This table includes all IROFS that involve administrative components or human actions. In Section 3.1.4.2 of the LA, the applicant committed to updating the list through the ISA Summary revision and amendment process.

14.3.2 Human Factors Engineering Design Review Planning

14.3.2.1   Goals and Scope

Evaluation

The opening paragraph of the HFEIP in Section 3.1.4.2 of the HFEIP describes the applicant's goals for the plan (IIFP, 2012a). The HFEIP seeks to address the potential for human error in the design, procedures, and training for the facility. The goal is to provide the means for detecting and correcting or compensating for human errors so that IROFS will remain available and reliable. The HFEIP contains the following objectives:

- Critical personnel tasks are defined and accomplished within applicable time and performance criteria.

- The anthropomorphic standards for the relevant population are defined and applied.

- HSIs, procedures, staffing and qualifications, training, management, and organizational variables support a high degree of operating crew situational awareness.

- Allocation of functions accommodates human capabilities and limitations.

- Operator vigilance is maintained and distractions are minimized.

- The operator workload is acceptable.

- Operator interfaces contribute to an error-free environment.

- Error detection and recovery capabilities are provided.

- Control areas minimize stressors and fatigue while ensuring adequate communication.

This set of goals addresses HSI, procedures, ergonomics, time to respond/perform, vigilance, workload, error reduction and fatigue. Figure 3-1 of the LA provides an overview of the implementation of the HFEIP (IIFP, 2012a). Section 3.1.4.7 and other sections of the HFEIP state that the plan will be applied to procedures and training.

The LA, Section 3.1.4 (IIFP, 2012a), states that the HFE Program will be applied to personnel activities involving administrative IROFS to ensure that the potential for human error is addressed during the design of the facility, operating procedure development, and personnel training. Tracking and maintenance activities will transition to quality assurance (QA) and configuration management functions as operation nears and begins. Therefore, HFE is included in the entire life cycle of the facility.

Conclusions

The applicant has established appropriate goals and scope to ensure that HFE is incorporated into IROFS. These objectives are appropriate for an HFE program because they cover the full scope of human performance and human activities within the system. The transition of the HFE goals into the QA Program provides the commitment that HFE will be included in the complete design cycle. The focus of the HFE program on IROFS that include human actions and personnel activities that support safety is of sufficient scope to provide reasonable assurance that critical human performance in the plant will be reviewed and that human factors will be included in the design to support these activities. Therefore, reasonable assurance is provided that the critical aspects of human performance in the operation of the facility will be assessed by the HFE review, and the goals and scope of the HFEIP during design, construction and operation of the plant is satisfactory. Thus, the goals and scope of the HFE design review planning are acceptable and meet the acceptance criterion as described in Appendix E, Section (B)(i), to NUREG-1520, Revision 1 (NRC, 2010).

14.3.2.2   Composition and Experience of the Human Factors Engineering Team

Evaluation

Section 3.1.4.2, of the LA (IIFP, 2012a) states that an HFE expert is to be added to the team. Table 5-1 in the ISA Summary shows the composition of the HFE/ISA team and indicates that the team will be multidisciplinary, trained, knowledgeable, and experienced in a wide array of ISA methods including hazards identification, PHA, and safety analysis and risk (IIFP, 2012b).

The HFE expert will be relied on to achieve the stated HFE goals during design and construction of the facility. The HFE expert will provide oversight and develop advise on, the HFEIP for the IIFP facility before the design of the facility's IROFS SSCs begins. The HFE expert will be responsible for overseeing the incorporation and implementation of the HFEIP elements within the applicable HFE scope during the design, engineering, and startup of the facility. IIFP LA, Section 2.2.5.4, presents the minimum qualifications for the HFE professional position (IIFP, 2012a). The human factors engineer shall have as a minimum, a bachelor's degree in HFE, or a bachelor's degree in a scientific discipline or psychology with an emphasis on human factors, and at least 3 years of experience in human factors. As discussed in the ISA Summary the HFE/ISA team will have responsibility and authority for the following:

- Identify the HSIs issues and evaluate actions and consequences involved.

- Update the plan before the actual detailed design of IROFS.

- Perform needed interviews to collect information on facilities handling, processing, or manufacturing uranium hexafluoride (UF_6), uranium tetrafluoride (UF_4), and uranium oxide and related radiological or chemical hazards

- Coordinate and communicate with the design team, QA staff, and IIFP's facility project team to develop and implement criteria, requirements, design review (style) guidance, and findings of analyses derived from the plan's implementation processes.

Conclusions

The experience requirements for the HFE expert, the composition of the HFE/ISA team, and the authority and responsibility of the HFE expert and the HFE/ISA team are appropriate to provide reasonable assurance that HFE will be considered in the design of HSI for personnel activities. Given the breadth of responsibilities of the HFE/ISA team, the organizational authority of the team, and the interconnections to the QA and configuration management plans, the HFEIP provides reasonable assurance that there will be proper development, execution, oversight, and documentation of the HFE function.

14.3.2.3   *Provide a Structured Approach to Human Factors Engineering Using Established Processes and Procedures*

Evaluation

Figure 3-1 of the LA (IIFP, 2012a) illustrates an overview of the HFE process to be undertaken by IIFP. The figure indicates that the elements identified by NUREG-1520, Revision 1, Appendix E, will support and be supported by other design efforts.

The applicant stated that, for IROFS, an HFE review of the HSIs will be performed using NUREG-0700, Revision 2 (NRC, 2002), and NUREG-0711, Revision 2 (NRC, 2004), in accordance with the HFEIP (IIFP, 2012a). The results of the HFE review will be documented as part of the IROFS.

Other standards and established processes referenced by the IIFP's HFEIP to attain the goals and scope defined include the following:

- Institute of Electrical and Electronics Engineers (IEEE) 1023, "Recommended Practice for the Application of Human Factors Engineering to Systems, Equipment, and Facilities of Nuclear Power Generating Stations and Other Nuclear Facilities" (IEEE, 2004)

- Electric Power Research Institute NP-3659, "Human Factors Guide for Nuclear Power Plant Control Room Development," issued May 1991 (Kincaide, 1991)

- Military Standard MIL-STD-1478, "Task Performance Analysis," issued 1991 (DOD, 1991)

- MIL-HDBK-46855A, "Human Engineering Requirements for Military Systems, Equipment, and Facilities," issued 2011 (DOD, 2011)

- NUREG/CR-3331, "A Methodology for Allocation of Nuclear Power Plant Control Functions to Human and Automated Control," issued 1983 (NRC, 1983).

These standards provide a structured approach for incorporating HFE into the elements of design, including FAA and TA; HSI design, inventory, and characterization; staffing; procedure development; and training program development.

Conclusion

The HFEIP provides commitments to implementing the elements defined by NUREG-1520, Revision 1, Appendix E (NRC, 2010), in accordance with NRC standards. The LA, Section 3.1.4 (IIFP, 2012a), states that the HFE program will be developed during the initial design phase and implemented throughout the design and operation of the facility. As the detailed design progresses, the elements of the HFE analysis will be reviewed and revised when process changes are made through the design control process. This indicates that the HFE Program will be an iterative process that is updated throughout the design of the facility, which is a critical aspect of a good HFE program. The application provides adequate assurance that the HFEIP will follow established the processes and standards listed above. It also provides a structured approach to development, design, and evaluation of personnel activities and is, therefore, acceptable.

14.3.2.4   Conclusion to Human Factors Engineering Design Review Planning

The HFEIP describes the activities to be undertaken including implementing the HSI to prevent consequences for IROFS, conducting OERs to identify necessary actions, and designating the NRC staff groups responsible for implementing HFE for IROFS. The LA (IIFP, 2012a) provides appropriate goals and scope to ensure that HFE practices and guidelines are implemented during design as overseen by the HFE expert. Therefore, there is appropriate team composition, experience, and organizational authority to ensure that HFE is considered in the design of the HSI for IROFS.

14.3.3 Operating Experience

The main purpose of conducting an OER is to identify HFE related safety issues that have occurred previously in other similar facilities. The OER also seeks to identify relevant HSI and equipment that should be considered in the development and design of the facility. The OER should provide information on the past performance of similar designs. The issues and lessons learned from the operating experience provide a basis for improving the plant design in a timely way. The objective of this element is to verify that the applicant has a program to identify and analyze HFE-related problems and issues in previous designs that are similar to the design currently under review. In this way, negative features associated with predecessor designs may be avoided while positive features are retained. The applicant should do all of the following:

- Review the HFE-related events or potential events for relevance.

- Analyze the HSI technology employed for the relevant HFE events or potential events.

- Conduct (or review existing) operator interviews and surveys on the HSI technology for the relevant HFE events or potential events.

14.3.3.1 Evaluation

Section 3.1.4.3 of the LA describes the applicant's HFE OER (IIFP, 2012a). Figure 3-1 of the IIFP's LA illustrates that OER is the initial step in the design process once the HFE/ISA team has been established. The figure illustrates that OER provides direct input to the identification of personnel activities, the FAA, the TA, and staffing and qualifications. The OER will be used to determine staffing requirements and categories of personnel and to provide input into HSI design and procedure development.

The applicant committed to reviewing NRC and U.S. Department of Energy (DOE) event reports and the information available from the DOE's Office of Scientific and Technical Information Bridge Web site. The review of previous events will include searches for personnel actions, as well as HSI technology. In addition, operator interviews will be conducted with personnel who have experience in uranium processing. Interviews will be sought with personnel with relevant experience (e.g., operating UF_6 autoclaves, deconversion of UF_6, transporting UF_6 and radiological wastes, packaging fluorine products, refrigeration systems, and handling and transporting fluoride products). Potential sources identified include operating fuel cycle facilities, such as the DOE depleted uranium deconversion facilities at Portsmouth, OH; and Paducah, KY; the AREVA Eagle Rock Enrichment Facility and URENCO U.S.A. centrifuge enrichment facilities, and former plant personnel from the closed Sequoyah Fuels Corporation Depleted UF_4 Deconversion Facility. The applicant has hired several engineers from existing plants to import their experience into the existing design (see the LA, Sections 2.1.2 and 3.2.7 [IIFP, 2012a]).

14.3.3.2 Conclusion

The applicant's OER will include searches for events related to human error and human action in previous events at relevant facilities via accepted databases. It also includes interviews of individuals with industry experience and searches for information on HSI and technology. Information derived from the OER serves as the basis for the HFE Program and will be incorporated into the development of procedures, training, and HSI. Therefore, the operating experience meets the acceptance criteria in Appendix E, Section C, of NUREG-1520, Revision 1 (NRC, 2010). The NRC staff concludes that the HFE operating experience is satisfactory and meets the requirements in 10 CFR 70.61(e) and 10 CFR 70.64(b)(2).

14.3.4 Functional Allocation Analysis and Task Analysis

Functional requirements analysis is the identification of functions that must be performed to satisfy safety objectives. FAA is the review of the requirements for plant control and the assignment of control functions to (1) personnel (e.g., manual control), (2) system elements (e.g., automatic control and passive, self-controlling phenomena), and (3) combinations of personnel and system elements. The purpose of the functional requirements analysis and function allocation review is to verify that the plant's safety functions have been defined and that the allocation of those functions to human and system resources takes advantage of human strengths and avoids human limitations. The FAA should be based on the OER. Function allocations should take advantage of human strengths and avoid allocating functions that would be negatively affected by human limitations.

The functions allocated to personnel define their roles and responsibilities. Human actions (HAs) are performed to accomplish these functions. HAs can be further divided into tasks. A task is a group of related activities that have a common objective or goal. TA is the identification of requirements for accomplishing these tasks (i.e., for specifying the requirements for the displays, data processing, controls, and job support aids needed to accomplish tasks). TA should be iterative and include the scope, identification of critical tasks, detailed description of personnel demands, and the incorporation of job design issues. It should address each operating mode and be used to support function allocation.

14.3.4.1   Functional Analysis

Evaluation

In Section 3.1.4.4 of the HFEIP in the LA (IIFP, 2012a), the applicant discusses its FAA. The safety philosophy of the ISA team was to minimize human involvement in the activation and performance of IROFS as much as available technology would allow. The concept of preferential use of active- and passive-engineered controls for IROFS is consistent with NRC guidance. Where engineered-controlled IROFS and administrative action IROFS are used in the same accident sequence, the engineered-control IROFS take precedence over human actions.

IIFP proposes to follow the guidance and practices for FAA contained in NUREG/CR-3331 (NRC, 1983). NUREG/CR-3331 is one of the technical bases for the guidance criteria on function allocation and functional analysis provided in NUREG-0711, Revision 2 (NRC, 2004), and NUREG-1520, Revision 1, Appendix E (NRC, 2010). It provides best practices for team composition, methods for performing the function allocation and functional analysis, assessment of the quality of the analysis, and guidance on appropriate allocation of functions between human and system so that the strengths of the human are leveraged and human weaknesses are avoided.

Conclusion

The IIFP's HFEIP indicates that the OER, ISA, and actual operating experience form the basis for determination of allocation of functions. The use of the processes, procedures, and guidance defined in NUREG/CR-3331 (NRC, 1983) provide reasonable assurance that personnel activities will be allocated to take advantage of human strengths and avoid demands that are not compatible with human capabilities. Therefore, the NRC staff concludes the functional analysis meets the acceptance criteria in Appendix E, Section d(i), of NUREG-1520, Revision 1 (NRC, 2010).

14.3.4.2   Task Analysis

In the LA, Section 3.1.4.4, of the HFEIP (IIFP, 2012a), the applicant discussed TAs and stated that the TA involves determining the requirements of personnel to successfully and safely perform complex real-time control actions as part of the job assignments resulting from function allocation.

The scope of the TA is defined as those IROFS that involve human actions. The IROFS that involve administrative components are defined in ISA Summary Table 6.3 (IIFP, 2012b). Section 3.2.5.8 of the LA (IIFP, 2012a) states that safety controls used at the IIFP facility can be characterized as either administrative or engineered. Administrative controls may rely on human actions, and IIFP stated that they are generally not considered to be as reliable as engineered controls, since human errors usually occur more frequently than equipment failures. Engineered controls may be categorized as being "passive" or "active" and are not reliant on human action. Therefore, the scope of the TA is appropriately defined as IROFS that involve administrative components. TA is defined as an ongoing, and thus iterative, activity that extends through the training process. The TA scope will include startup, shutdown, and emergency operations. TA is an input to function allocation and HSI design. Section 3.1.4.4 of the HFEIP indicates the applicant's commitment to considering job design issues (IIFP, 2012a).

IIFP proposes to follow the methods described in military standards MIL-STD-1478 (DOD, 1991) and MIL-HDBK-46855A (DOD, 2011). MIL-STD-1478 provides a process for incorporating FAA and TA into the design, which allows for flexibility in the performance of TA. Users are directed to select the established method of TA that is most cost effective if other guidance does not specify a method. The format and content of a task performance analysis product are described, inputs from TA to other design aspects such as HSI design and function allocation are indicated, and inputs from TA to procedure development are described. MIL-STD-1478 indicates that for each task in the TA, the minimum data collected and analyzed should be equipment acted on, consequence of the action, and feedback information resulting from the action, which would support HSI design and functional allocation. MIL-HDBK-46855A contains recommended HFE program tasks and procedures and specifies the need for TA, the inputs from other processes to TA, and the outputs from TA to other elements. Section 8.3 of MIL-HDBK-48655A summarizes a number of widely accepted TA methods and the stage of design to which they apply. The TA meets the acceptance criteria in Appendix E, Section D(ii), to NUREG-1520, Revision 1 (NRC, 2010).

14.3.4.3   Functional Allocation Analysis and Task Analysis Conclusion

Based on the NRC staff's review of the applicant's FAA and TA, using the criteria described in Appendix E, Section D, to NUREG-1520, Revision 1, the NRC staff concludes that the FAA and TA are adequate and meet the requirements in 10 CFR 70.61(e) and 10 CFR 70.64(b)(2).

14.3.5 Human System Interface Design, Inventory, and Characterization

HSI design is the process by which HSI design requirements are developed and HSI designs are identified and refined. HSI should incorporate FAA and TA into the safety-significant HSI (alarms, displays, controls, and operator aids) via systematic application of HFE. The work environment, workspace layout, control panel and console design, control and display device layout, and information and control interface design details should be included. HFE should be applied to HSIs required to perform personnel activities. The HSI design process should exclude the development of extraneous controls and displays. HSI documentation should include a complete HSI inventory and the basis for the HSI characterization.

14.3.5.1   Evaluation

The LA, Section 3.1.4 (IIFP, 2012a), contains the applicant's proposal to follow the guidance provided in NUREG-0700, Revision 2 (NRC, 2002), which provides the guidelines for the development of HSI. The guidelines address the physical and functional characteristics of HSIs. Part I contains guidelines for the basic HSI elements: displays, user interface interaction and management, and controls. Part II contains the guidelines for the alarm system, group view display system, soft control system, computer-based procedure system, computerized operator support system, and communication system. Part III provides guidelines for workstations and workplaces. Part IV provides guidelines for HSI support (i.e., maintainability of digital systems). In the LA, Section 3.1.4.5, the applicant proposed to follow the design process methods in IEEE 1023, (IEEE, 2004). IEEE 1023 provides industry-accepted guidance and methods for the development of HSI, including considerations of the work environment and testing and validation of the designs.

In the LA, Section 3.1.4.5, of the HFEIP (IIFP, 2012a), the applicant discussed HSI design. In this section, the applicant stated that the minimum inventory of HSIs, displays, alarms, and control instruments will be developed from the IROFS control function, the IROFS functional information, and input derived from the OER, FAA, and TA results, which would provide a basis for the characterization of the HSI. This minimum inventory will be updated during the design stage and during facility maintenance. It will be documented in accordance with the QA Program. The applicant committed to considering the work environment. The applicant also made a commitment that the implementation process for each HFE element will be applied accordingly to ensure consideration of the HFE aspects for the IROFS SSCs once those SSCs are defined and before actual design and design review.

The IIFP's QAPD will document the methodology for creation of the inventory of HSIs that are implemented to support IROFS and its results (IIFP, 2012c). This formal methodology is part of the facility's Configuration Management Program and will ensure that the inventory list of HSIs is kept up to date. Extraneous displays and controls will be identified during task support verification.

IIFP's LA, Section 11.1.5.3 (IIFP, 2012a), states that the facility modification procedure will include HFE as a review and evaluation activity for any modifications that may impact HSIs. The applicant stated that if the assessment reveals that the modification affects HSI, the HFE process will be applied. The IIFP's HFEIP states that if any HSIs are identified that have no associated tasks because the FAA or TA was incomplete, then the FAA or TA will be reviewed and shortcomings will be resolved.

14.3.5.2   Conclusion

The applicant committed to using the OER, FAA, and TA to establish the types of HSI to incorporate into the IROFS design. The HSI will be applied, as part of the HFE Program, to each control process involving the administrative components identified in the IROFS and their work environments. The inventory of HSIs will be tracked through the QA Program. The HSIs will be inventoried as part of the QA process. Based on the review of the applicant's HSI design, inventory, and characterization using the acceptance criteria in Appendix E, Section E, to NUREG-1520, Revision 1 (NRC, 2010);, and to follow the structured approach provided in

IEEE 1023 (IEEE, 2004) and the development guidance provided in NUREG-0700, Revision 2 (NRC, 2002), the NRC staff concludes that the HSI design, inventory, and characterization are adequate and meet the requirements in 10 CFR 70.61(e) and 10 CFR 70.64(b)(2).

14.3.6 Staffing

Staffing and qualification analyses identify the requirements for the number and qualifications of personnel. These analyses should reflect a thorough understanding of task and regulatory requirements. The applicant should conduct this review in a systematic manner that incorporates FAA and TA results.

14.3.6.1   Evaluation

Multiple sections of the LA and HFEIP address staffing. Sections 4.1.1 and 4.3 of the LA (IIFP, 2012a) state that staffing will be consistent with guidance in Regulatory Guide 8.2, "Guide for Administrative Practice in Radiation Monitoring," issued February 1973 (NRC, 1973); and Regulatory Guide 8.10, Revision 1-R, "Operating Philosophy for Maintaining Occupation Radiation Exposures as Low as Is Reasonably Achievable," issued May 1977 (NRC, 1977). Chapter 2 of the IIFP's LA describes the qualifications for staff positions (IIFP, 2012a). The LA, Section 2.3.3 (IIFP, 2012a), provides information on the training and qualifications of staff.

In the LA, Section 3.1.4.6 (IIFP, 2012a), the HFEIP discusses staffing and qualification. The HFEIP states that the applicant will ensure adequate staffing to operate the facility safely; furthermore, the OER, FAA, and TA will provide the initial basis for determining staffing requirements. These requirements will be considered throughout the design process. The initial estimates of staffing requirements and acceptability of staffing goals will be reassessed as the detailed design progresses. Therefore, assessment of staffing and use of the TA will be iterative. Categories of personnel activities will be derived from the TA, FAA, and OER. HSI design will be considered in the determination of staffing requirements and goals. During the design and build phase of operations, contractors will have the primary responsibility for ensuring adequate staffing, with oversight from an HFE expert or staff. As part of the HFE team, the HFE expert will review and provide oversight and advice on staffing needs. Staffing assignments, personnel tasks, and ergonomics will be validated during the V&V portion of the HFE IP.

14.3.6.2   Conclusion

The staffing and qualification levels are based on issues derived from the OER, the FAA, and the TA. Procedure development, HSI design, and V&V will be used to inform staffing levels, along with appropriate regulatory requirements. The number and qualifications of staff will be reviewed and updated throughout the design and build phase of operations, with oversight and review from the HFE team. Categories of tasks will be derived from the FAA and TA. Therefore, the NRC staff concludes the staffing plan meets the acceptance criteria in Appendix E, Section F, to NUREG-1520, Revision 1 (NRC, 2010). Thus, based on the review of the applicant's staffing using the criteria described in Appendix E, Section F, of NUREG-1520, Revision 1, the staffing commitments are adequate and meet the requirements in 10 CFR 70.61(e) and 10 CFR 70.64(b)(2).

14.3.7 Procedure Development

Procedure development addresses the verification that human engineering principles and guidance are applied, along with all other design requirements, to develop procedures that are technically accurate, comprehensive, explicit, easy to use, and validated. Procedures should be derived from the same design process and analyses as the HSI and be subject to the same evaluation processes.

14.3.7.1   Evaluation

Chapter 11 of the LA covers nine elements for procedure development: identification, development, verification, review and comment resolution, approval, validation, issuance, change control, and periodic review (IIFP, 2012a). Section 11.1.5 of the LA provides a formal change control process for modifying procedures. The LA, Section 3.1.4.7, of the HFEIP (IIFP, 2012a), states that procedures will govern the production work aside from routine custodial work and office duties. ISA results, including IROFS, will be used to identify needed procedures. The adequacy and fidelity of procedures, and the allocation of functions in procedures, will be validated during the HFE integrated system validation as described in the LA, Section 3.1.4.9 (IIFP, 2012a). One objective of the HSI task support verification is to confirm that all procedures required for personnel tasks are provided.

In a letter dated August 22, 2011 (IIFP, 2011b), IIFP stated that procedure development and refinement are ongoing activities. As construction nears completion, technical subject matter experts will perform walkthroughs to validate and refine procedures before use. Flowcharts of procedure steps will be created as part of V&V, as described in the LA, Section 3.1.4.9 (IIFP, 2012a). Procedures will include required actions for normal and off-normal conditions of production. The QA elements and Configuration Management Program at the facility provide reasonable assurance that current procedures are available and used at all work locations.

The applicant committed to considering the HFE program elements identified in NUREG-1520, Revision 1, Appendix E (NRC, 2010), in the development of procedures. IROFS and other safety-related items will be highlighted in work procedures, typically as "cautions" and "warnings." The responsible organizations will develop and approve procedures. Employees will be trained on all procedures they follow as part of their work assignments. Work procedures and supplemental safety-related procedures are expected to be located in the general work areas. Facility and process changes will require procedure updates in the form of revisions, and such revisions shall be in place before the operation can recommence. The relevant procedures will incorporate changes to safety systems and safety basis documentation. Refinements and changes to procedures that are within the scope of the HFE Program will receive review for consideration of human factors. Employees will be retrained on the revised procedures before the restart of work.

Procedures throughout the facility will identify IROFS and associated management measures. Procedures go through a validation process as part of the QA and Configuration Management Programs. The functional managers must also approve procedures. The LA, Section 3.1.4.2 (IIFP, 2012a), states that the IIFP staff will implement procedure development and the HFE team will provide review, advice, and oversight.

14.3.7.2 Procedure Development Conclusion

Each technical review area described in the application contains commitments to developing detailed procedures to ensure safe operations; the HFEIP contains commitments to developing procedures with consideration of the elements defined in NUREG-1520, Revision 1, Appendix E (NRC, 2010). The procedures will be developed and maintained consistent with the change control process and configuration management defined in Chapter 11 of the LA (IIFP, 2012a). In addition the HFE team will oversee and review the procedures to ensure that they incorporate HFE principles. The procedures are also reviewed by the V&V Program and evaluated during development via walkthrough. Therefore, based on the review of the applicant's procedure development using the criteria described in Appendix E, Section G, to NUREG-1520, Revision 1, the NRC staff concludes that the procedure development commitments are adequate and meet the requirements in 10 CFR 70.61(e) and 10 CFR 70.64(b)(2).

14.3.8 Training Program Development

The applicant's training program is designed to help ensure that personnel have the knowledge and skills necessary to perform their tasks. The development of the training program indicates how knowledge and skill will be evaluated, how training development will be coordinated with the HFE design process, and how the training program will be implemented in a manner consistent with human factors principles and practices. The training program development addresses the areas of review in Chapter 11 of NUREG-1520, Revision 1 (NRC, 2010), and results in a training program that provides personnel with qualifications commensurate with their activities.

14.3.8.1 Evaluation

In the LA, Section 3.1.4.8, of the HFEIP (IIFP, 2012a), the applicant discussed training program development. The objective of the training program is to ensure the job proficiency of facility personnel through effective training and qualification. The training development will address all personnel activities in the performance of job tasks involving IROFS. The training program will be coordinated with other activities of the HFE design process.

The training combines both classroom and on-the-job training and involves formal evaluation to ensure that individuals demonstrate a minimum level of proficiency. The training uses a graded approach to emphasize safety-significant procedures, such as those that involve IROFS. Chapter 11 of the LA and each technical area throughout the application (IIFP, 2012a) describe additional commitments to the training program. These descriptions demonstrate that the training program is comprehensive and ensures that staff members have the proper skills to meet their responsibilities.

The HFE is incorporated into the training program by regular job TAs to ensure that the training imparts the knowledge required to do the tasks and to iteratively improve the courses. In an RAI response (IIFP, 2011a), IIFP indicated that the methods used to perform job TA are contained in the same response as the description of the TA. LA Figure 3-1 (IIFP, 2012a) indicates that operation TA includes HFE TA as a subpart and that TA, functional allocation, staffing and qualifications, and OER form the basis for HSI development and procedure development and thus for HFE V&V and training program development.

The HFEIP makes a commitment that the HFE team will review and provide oversight and advice to the training program as the facility nears operating status. Thus, the HFE team will review the training program to ensure that training on IROFS incorporates human factors principles.

The training also ensures that the specific tasks and details associated with each IROFS are addressed. The training program covers all appropriate staff for the facility and must be approved by the cognizant management. The training is reviewed regularly and incorporates input from the job task analyses. The development of training and regular review receives input from the HFE expert.

14.3.8.2   Conclusion

Each technical review area described in the application contains commitments to appropriate training. In addition, Chapter 11 of the LA states commitments to a comprehensive training program, which emphasizes safety-significant procedures. The training incorporates HFE practices and principles and is verified by the HFE expert or staff. Specific training is provided for actions related to IROFS. Therefore, based on the review of the applicant's training program development and using the criteria described in Appendix E, Section H, to NUREG-1520, Revision 1 (NRC, 2010), the NRC staff concludes that the training program development commitments are adequate and meet the requirements in 10 CFR 70.61(e) and 10 CFR 70.64(b)(2).

14.3.9 Human Factors Verification and Validation

V&V confirms that the design incorporates HFE into HSI in a manner that enables the successful completion of personnel activities. The V&V should be applied to personnel activities and HSI design as described in the V&V section of Chapter 3, Appendix E to NUREG-1520, Revision 1 (NRC, 2010).

14.3.9.1   Evaluation

Section 3.1.4.9 of the HFEIP (IIFP, 2012a) indicates that the purpose of the V&V Program will be to verify that the HSIs have the appropriate controls, procedures, and alarms so that the IROFS functions and controls remain available and reliable. The V&V Program will be used during the design phase of operations and to review individual actions needed to support IROFS.

HSI Task Support Verification. The HSI task support verification (TSV) portion of the LA, Section 3.1.4.9 (IIFP, 2012a), provides a high-level overview of the process to be followed to perform TSV. This portion states that the criteria for TSV come from TAs of HSI requirements. HSIs and their characteristics will be compared to the personnel task requirements derived from the TA. Human engineering discrepancies (HEDs) are identified when needed HSIs are not available, characteristics do not match requirements, or HSIs are unnecessary or have no associated task. On this basis, the TSV demonstrates regulatory compliance with the acceptance criteria in Appendix E, Section (I)(i), to NUREG-1520, Revision 1 (NRC, 2010).

HFE Design Verification. HFE design verification will be performed to determine that each HSI identified for personnel activity in the IROFS has HFE incorporated into its design and that the HSI has been verified to conform with HFE criteria and style guidance. Section 3.1.4.5 states that design review criteria and style guidance, with consideration of human factors for affected layouts, locations, and configurations of alarms, displays, and control instrumentation will be developed for use by the design engineers before they begin the actual design. The HFE issue resolution verification portion of the HFEIP states that the HFEIP will be subject to the requirements of the QAPD. All HFE requirements and design documents will be controlled under the design control provisions of the Configuration Management Program, subject to the same change control as analysis, specifications, and drawings. The LA, Section A.3.3.4, of the QAPD (IIFP, 2012c), states that testing used for design verification shall demonstrate adequacy of performance under conditions that simulate the most adverse design conditions. The tests used for design verification must meet all the design requirements.

Integrated System Validation (ISV). The applicant stated in LA, Section 3.1.4.9, that the purpose of the ISV is as follows:

- validate the role of plant personnel;

- validate staffing assignments;

- validate each human ergonomic function;

- validate specific personnel tasks;

- validate that the integrated system performance is tolerant of failures; and

- validate procedure adequacy, allocation, and fidelity.

IIFP committed to having ISV done by knowledgeable personnel performing task walkthroughs and talk throughs in a part-task simulator, which provides a representation of the process control rooms and local control panels. The part-task simulator will have sufficient fidelity to be used as a training tool and to upgrade software or design new software. Walkthroughs and talk throughs are widely accepted HFE techniques for validating processes. When combined with simulation, they provide an effective method of assessing the integration of all aspects of design in a realistic task setting, which provides reasonable assurance that HFE issues will be identified. Reviews of TA and functional analysis findings, flowcharts of procedure steps, and operational task charts will also be used in combination with the part-task simulation walkthroughs and talk throughs to provide reasonable assurance of a link to the HSI and task requirements developed earlier.

HFE Issue Resolution Verification. The QAPD governs HFE issue resolution verification. As stated in the QAPD and in the HFE IP, qualified reviewers who were not involved in the design process will perform HFE issue resolution. All HFE requirements and design documents will be controlled under the design control provisions of the Configuration Management Program and will be subject to the same change control as analysis, specifications, and drawings.

Results of V&V activities will be summarized in a report, and discrepancies will be identified as HED. An HED report will include the results of resolution. The QAPD contains the facility's existing Corrective Action and QA Programs described in the LA, Chapter 11, Section 11.6 (IIFP, 2012a), and in the QAPD (IIFP, 2012c), respectively. The V&V involves a check and review of the HFE program and HSIs. The report will identify the changes needed and describe how they are resolved. Identification of HFE issues will be incorporated into the Root-Cause Analysis Program, and issues will be resolved via the corrective action process. The commitment to include a root-cause analysis of HFE into the Corrective Action Program provides reasonable assurance of regulatory compliance with the acceptance criteria in Appendix E, Section (i)(iv), to NUREG-1520, Revision 1 (NRC, 2010).

14.3.9.2 Conclusion

Based on the review of the applicant's human factors V&V and using the criteria described in Appendix E, Section I, to NUREG-1520, Revision 1 (NRC, 2010), the NRC staff concludes that the human factors V&V commitments are adequate and meet the requirements in 10 CFR 70.61(e) and 10 CFR 70.64(b)(2).

14.4 Evaluation Findings

The NRC staff reviewed the applicant's LA (IIFP, 2012a); ISA Summary (IIFP, 2012b); the QAPD (IIFP, 2012c); and the Configuration Management Program description in the LA for the application of HFE to personnel activities for the proposed facility. The NRC staff conducted the review at a programmatic level, as described in Appendix E, Section E.2, above.

The documents named above and the HFEIP (IIFP, 2012a), with reliance on the HFE standards identified in the HFEIP elements, provide acceptable objectives, expected results, and plans for documentation. The NRC staff concludes that the applicant has committed to incorporating accepted HFE guidance and practices to design and implement those HSIs that support IROFS requiring operator actions. The guidance and practices, when implemented as described in the LA (IIFP, 2012a), demonstrate compliance with the regulatory criteria in Appendix E to Chapter 3 of NUREG-1520, Revision 1 (NRC, 2010), and should result in IROFS that will perform their intended functions and meet the requirements of 10 CFR Part 70, "Domestic Licensing of Special Nuclear Material."

The detailed procedures for implementation of the HFE program will be reviewed and verified during the inspection program.

14.5 References

(DOD, 2011) U.S. Department of Defense, MIL-HDBK-46855A, "Department of Defense Standard Practice: Human Engineering Requirements for Military Systems, Equipment, and Facilities," Washington, DC, May 2011.

(DOD, 1991) U.S. Department of Defense, Military Standard MIL-STD-1478, "Task Performance Analysis," Washington, DC, May 1991.

(IEEE, 2004) Institute of Electrical and Electronics Engineers, IEEE 1023, "Recommended Practice for the Application of Human Factors Engineering to Systems, Equipment, and Facilities of Nuclear Power Generating Stations and Other Nuclear Facilities," Piscataway, NJ, December 8, 2004.

(IIFP, 2012a) International Isotopes Fluorine Products, Inc., "Fluorine Extraction Process and Depleted Uranium Deconversion Plant (FEP/DUP) License Application, Revision B," May 2012, Agencywide Documents Access and Management System (ADAMS) Accession No. ML12123A245.

(IIFP, 2012b) International Isotopes Fluorine Products, Inc., "ISA Summary Rev. B for IIFP," May 2012, Agencywide Documents Access and Management System (ADAMS) Accession No. ML12123A245.

(IIFP, 2012c) International Isotopes Fluorine Products, Inc., "Appendix A QAPD Rev B of IIFP License Application," May 2012, Agencywide Documents Access and Management System (ADAMS) Accession No. ML12123A245.

(IIFP, 2011a) International Isotopes Fluorine Products, Inc., "Official Responses to Human Factors RAIs," Rev. A, May 2011. Agencywide Documents Access and Management System (ADAMS) Accession No. ML11130A128.

(IIFP, 2011b) International Isotopes Fluorine Products, Inc., "Official Responses to Human Factors RAIs," Rev. B, August 2011. Agencywide Documents Access and Management System (ADAMS) Accession No. ML11250A055.

(Kincaide, 1991) Kincaide, R.G., and J. Anderson, Electric Power Research Institute, "Human Factors Guide for Nuclear Power Plant Control Room Development," Electric Power Research Institute EPRI/NP-3659, MIL-STD-1478, "Military Standard: Task Performance Analysis," May 1991.

(NRC, 2010) U.S. Nuclear Regulatory Commission, NUREG-1520, "Standard Review Plan for the Review of a License Application for a Fuel Cycle Facility," Rev. 1, Washington, DC, May 2010.

(NRC, 2004) U.S. Nuclear Regulatory Commission, NUREG-0711, Rev. 2, "Human Factors Engineering Program Review Model," Washington, DC, February 2004.

(NRC, 2002) U.S. Nuclear Regulatory Commission, NUREG-0700, Rev. 2, "Human-System Interface Design Review Guidelines," Washington, DC, May 2002.

(NRC, 1983) U.S. Nuclear Regulatory Commission, NUREG/CR-3331, "A Methodology for Allocation of Nuclear Power Plant Control Functions to Human and Automated Control," Washington, DC, 1983.

(NRC, 1977) U.S. Nuclear Regulatory Commission, Regulatory Guide 8.10, Rev. 1-R, "Operating Philosophy for Maintaining Occupational Radiation Exposures as Low as Is Reasonably Achievable," Washington, DC, May 1977.

(NRC, 1973) U.S. Nuclear Regulatory Commission, Regulatory Guide 8.2, "Guide for Administrative Practice in Radiation Monitoring," Washington, DC, February 1973.

(O'Hara, 2010) O'Hara, J., and J. Higgins, "Human-System Interfaces to Automatic Systems: Review Guidance and Technical Basis," Brookhaven National Laboratory Technical Report BNL-91017-2010, Upton, NY, 2010.

15.0 SAFETY EVALUATION REPORT PREPARERS

The individuals and organizations listed below are the principal contributors to the preparation of this Safety Evaluation Report. The U.S. Nuclear Regulatory Commission's (NRC's) staff directed the effort and contributed to the technical evaluations. The NRC staff also used contractor support from the Center for Nuclear Waste Regulatory Analyses in the preparation of this document.

Table 15-1 U.S. Nuclear Regulatory Commission Contributors

Matthew Bartlett, Office of Nuclear Material Safety and Safeguards (NMSS)	NRC Project Manager General Information and Organization and Administration
Maria Guardiola, NMSS	NRC Project Manager
Patti Silva, NMSS	Branch Chief, Conversion, Deconversion and Enrichment Branch (CDEB)
Mary Adams, NMSS	Decommissioning
Damaris Arroyo, NMSS	Management Measures and Quality Assurance
Julie Marble, Office of Nuclear Regulatory Research (Research)	Human Factors Engineering
Gregory Chapman, NMSS	Health Physics and Accident Analysis
Jonathan De Jesus-Segarra, NMSS	Chemical Safety
James Downs, NMSS	Fire Safety
Stan Echols, NMSS	Environmental Protection
Madhumita Sircar, Research	Seismic and Structural
Roman Przygodzki, Office of Federal and State Materials and Environmental Management Programs (FSME)	Financial Qualification and Decommissioning Financial Assurance
Mike Norris, Office of Nuclear Security Incidence Response (NSIR)	Emergency Management
James Anderson, NSIR	Emergency Management
Tom Pham, NMSS	Material Control and Accountability
Steve Ward, NMSS	Material Control and Accountability
Yawar Faraz, NMSS	Integrated Safety Analysis
Mary Adams, NMSS	Integrated Safety Analysis
Barry Wray, NSIR	Physical Protection

Table 15-1 Contributors to the Safety Evaluation Report (continued)

Center For Nuclear Waste Regulatory Analyses Contributors	
Asadul Chowdhury	Structural Design
Sui-Min (Simon) Hsiung	Natural Phenomena
John Stamatakos	Seismic Engineering
George Adams	Electrical Engineering and Instrumentation and Controls
Stanley Consultants Contributors	
Chris Storms	Stanley Consultants, Muscatine, Iowa

Appendix A   ACCIDENT ANALYSIS

The U.S. Nuclear Regulatory Commission's (NRC) staff independently evaluated the possible consequences of a subset of potential accident sequences identified in the applicant's Integrated Safety Analysis (ISA) Summary. The primary purpose of this analysis is to ensure that the modeling used by the applicant for assessing accidents is acceptable. This evaluation is documented in Appendix A to this Safety Evaluation Report. Appendix A contains information that has been marked by the applicant as "Security Related Information" and is designated by the NRC as "Official Use Only—Security Related Information," pursuant to Title of the *Code of Federal Regulations* (10 CFR) 2.390.

The five accident sequences analyzed by the NRC staff are presented in Appendix A. The selected accident consequences vary in severity from high- to low-consequence events, and include accidents initiated by natural phenomena, operator error, and equipment failure.

The NRC staff concluded that the proposed International Isotopes Fluorine Products, Inc.'s (IIFP), design would reduce the risk (likelihood) of these accidents by using items relied on for safety (IROFS) and defense-in-depth measures. In addition, the facility's Emergency Plan (EP) addresses these types of events. The NRC staff independently verified the applicant's accident analysis by performing confirmatory calculations and modeling. The NRC staff concluded that through the combination of plant design, passive- and active-engineered IROFS, administrative IROFS, and defense-in-depth features, the consequences of potential accidents at the proposed IIFP will pose an acceptably low safety risk to workers, the public, and the environment. As a result, the NRC staff determined that the applicant meets the requirements to operate the proposed facility under 10 CFR Part 40, "Domestic Licensing of Source Material."

Details related to the accident analysis have been marked by the applicant as "Security-Related Information," pursuant to Title 10 of the Code of Federal Regulations (CFR) 2.390, and are therefore not included in the public version of this report.

Appendix B   STRUCTURAL AND GEOTECHNICAL DESIGN

The U.S. Nuclear Regulatory Commission (NRC) staff independently evaluated the protection of building structures against natural phenomena described in the applicant's License Application (LA) (IIFP, 2012a); Integrated Safety Analysis (ISA) Summary (IIFP, 2012b); and IIFP responses to NRC requests for additional information (IIFP, 2011a, 2011b, and 2011c). This evaluation is documented in this appendix of the Safety Evaluation Report (SER). Appendix B contains information that has been marked as "Security Related Information" by the applicant and is designated by the NRC as "Official Use Only—Security Related Information."

The NRC staff reviewed the structural and geotechnical design of the proposed facility. This included reviewing the building designs, general design criteria, loading conditions, and applicable American Society of Civil Engineers (ASCE) standards. The proposed International Isotopes Fluorine Products, Inc.'s (IIFP), design reduces the risk (likelihood) of the accident by identifying items relied on for safety (IROFS), and defense-in-depth features.

The NRC staff concluded that, through the combination of plant design, passive- and active-engineered IROFS, administrative IROFS, and defense-in-depth features, building structures will have adequate protection against natural phenomena so that the safety risk to workers, public, and the environment will be acceptably low. As a result, the NRC staff finds that the applicant meets the requirements to operate the proposed facility under Title 10 of the *Code of Federal Regulations* (10 CFR) Part 70, "Domestic Licensing of Special Nuclear Material."

Details related to the structural and geiotechnical design have been marked by the applicant as "Security-Related Information," pursuant to Title 10 of the Code of Federal Regulations (CFR) 2.390, and are therefore not included in the public version of this report.

Appendix C   INSTRUMENTATION AND CONTROL

The U.S. Nuclear Regulatory Commission (NRC) staff's review of the instrumentation and control (I&C) is documented in Appendix C to this Safety Evaluation Report (SER). Appendix C contains information that has been marked by the applicant as "Security-Related Information" and is designated by the NRC as "Official Use Only—Security Related Information," pursuant to Title 10 of the *Code of Federal Regulations* (10 CFR) 2.390.

The NRC staff reviewed accident sequences related to I&C. This included reviewing the baseline design criteria (BDC), general design criteria, applicable Institute of Electrical and Electronics Engineers (IEEE) standards, and applicable American National Standards Institute standards. The proposed International Isotopes Fluorine Products, Inc.'s (IIFP), design reduces the risk (likelihood) of the accident by identifying items relied on for safety (IROFS), and defense-in-depth features. The NRC staff concludes that through the combination of plant design, passive and active engineered IROFS, administrative IROFS, and defense-in-depth features, accidents related to the electrical system and I&C at the proposed IIFP will pose an acceptably low safety risk to workers, public, and the environment. As a result, the NRC staff determined that the applicant meets the requirements to operate the proposed facility under Title 10 of the *Code of Federal Regulations* (10 CFR) Part 70, "Domestic Licensing of Special Nuclear Material."

Details related to instrumentation and control have been marked by the applicant as "Security-Related Information," pursuant to Title 10 of the Code of Federal Regulations (CFR) 2.390, and are therefore not included in the public version of this report.

Appendix D   ELECTRICAL SYSTEMS

The U.S. Nuclear Regulatory Commission (NRC) staff's review of the electrical system is documented in Appendix D to this Safety Evaluation Report. Appendix D contains information that has been marked by the applicant as "Security-Related Information," pursuant to Title 10 of the *Code of Federal Regulations* (10 CFR) 2.390 and is designated by the NRC as "Official Use Only-Security Related Information.".

The NRC staff reviewed the electrical power system for the facility. The review verified that the design of the electrical system will meet the regulatory requirements specified in 10 CFR Part 70, "Domestic Licensing of Special Nuclear Material," Subpart H. The NRC staff evaluated the adequacy of the proposed conceptual design and intended operations of these systems as reflected in the applicant's commitments and goals with respect to that design. The NRC staff evaluated the applicant's commitments for completing the design of the electrical system in a manner that addresses specific regulatory acceptance criteria, identified in Section 3.2 of this report. These commitments and goals are described within the Integrated Safety Analysis (ISA) Summary (IIFP, 2012b) and the applicant's responses to the NRC staff's request for additional information (RAI) (IIFP, 2011a).

Details related to electrical systems have been marked by the applicant as "Security-Related Information," pursuant to Title 10 of the Code of Federal Regulations (CFR) 2.390, and are therefore not included in the public version of this report.

NRC FORM 335
(12-2010)
NRCMD 3.7

U.S. NUCLEAR REGULATORY COMMISSION

BIBLIOGRAPHIC DATA SHEET

(See instructions on the reverse)

1. REPORT NUMBER
(Assigned by NRC, Add Vol., Supp., Rev., and Addendum Numbers, if any.)

NUREG-2116

2. TITLE AND SUBTITLE

Safety Evaluation Report for the International Isotopes Fluorine Products, Inc. Fluorine Extraction Process and Depleted Uranium Deconversion Plant in Lea County, New Mexico

3. DATE REPORT PUBLISHED

MONTH	YEAR
May	2012

4. FIN OR GRANT NUMBER

5. AUTHOR(S)

Maria Guardiola, Patti Silva, Mary Adams, Damaris Arroyo, Julie Marble, Gregory Chapman, Jonathan De Jesus Segarra, James Downs, Stan Echols, Madhumita Sircar, Roman Przygodzki, Mike Norris, James Anderson, Tom Pham, Steve Ward, Yawar Faraz, Mary Adams, Barry Wray, Asadul Chowdhury, Sui Min (Simon) Hsiung, John Stamatakos, George Adams, Chris Storms, and Matt Bartlett

6. TYPE OF REPORT

Final

7. PERIOD COVERED (Inclusive Dates)

8. PERFORMING ORGANIZATION - NAME AND ADDRESS (If NRC, provide Division, Office or Region, U. S. Nuclear Regulatory Commission, and mailing address; if contractor, provide name and mailing address.)

Division of Fuel Cycle Safety and Safeguards
Office of Nuclear Material Safety and Safeguards
U.S. Nuclear Regulatory Commission
Washington, DC 20555-0001

9. SPONSORING ORGANIZATION - NAME AND ADDRESS (If NRC, type "Same as above", if contractor, provide NRC Division, Office or Region, U. S. Nuclear Regulatory Commission, and mailing address.)

Same as above

10. SUPPLEMENTARY NOTES

11. ABSTRACT (200 words or less)

This report documents the U.S. Nuclear Regulatory Commission's (NRC's) staff review and safety and safeguards evaluation of International Isotopes Fluorine Products, Inc.=s (IIFP), application for a license to construct a Fluorine Extraction Process & Depleted Uranium Deconversion Plant (FEP/DUP) and possess and use source materials. INIS proposes that the FEP/DUP be located in Lea County, New Mexico, approximately 24 kilometers (15 miles) west of the city of Hobbs. The facility will possess natural and depleted uranium, and significant quantities of hazardous chemicals derived from licensed material, e.g., hydrogen fluoride.

The objective of this review is to evaluate the facility=s potential adverse impacts on worker and public health and safety, under both normal operations and accident conditions. The review also considers the management organization, administrative programs, and financial qualifications provided to ensure safe design and operation of the facility.

The NRC staff concludes, in this safety evaluation report, that the applicant's descriptions, specifications, and analyses provide an adequate basis for safety of facility operations and that operation of the facility does not pose an undue risk to the worker and public health and safety.

12. KEY WORDS/DESCRIPTORS (List words or phrases that will assist researchers in locating the report.)

nuclear fuel cycle
depleted uranium deconversion
uranium hexafluoride
International Isotopes
Part 40
ISA

13. AVAILABILITY STATEMENT

unlimited

14. SECURITY CLASSIFICATION

(This Page)

unclassified

(This Report)

unclassified

15. NUMBER OF PAGES

16. PRICE

NRC FORM 335 (12-2010)

Federal Recycling Program

UNITED STATES
NUCLEAR REGULATORY COMMISSION
WASHINGTON, DC 20555-0001

OFFICIAL BUSINESS

NUREG-2116

Safety Evaluation Report Related to the International Isotopes Fluorine Products, Inc. Fluorine Extraction Process and Depleted Uranium Deconversion Plant in Lea County, New Mexico

May 2012